STABLE ISOTOPE TECHNIQUES IN THE STUDY OF
BIOLOGICAL PROCESSES AND FUNCTIONING OF ECOSYSTEMS

Current Plant Science and Biotechnology in Agriculture

VOLUME 40

Scientific Editor
R.J. Summerfield, *The University of Reading, Department of Agriculture, P.O. Box 236, Reading RG6 2AT, Berkshire, UK*

Scientific Advisory Board
J. Hamblin, *Research Director, Export Grains Centre Ltd., WA, Australia*
H.-J. Jacobsen, *Universität Hannover, Hannover, Germany*

Aims and Scope
The book series is intended for readers ranging from advanced students to senior research scientists and corporate directors interested in acquiring in-depth, state-of-the-art knowledge about research findings and techniques related to all aspects of agricultural biotechnology. Although the previous volumes in the series dealt with plant science and biotechnology, the aim is now to also include volumes dealing with animals science, food science and microbiology. While the subject matter will relate more particularly to agricultural applications, timely topics in basic science and biotechnology will also be explored. Some volumes will report progress in rapidly advancing disciplines through proceedings of symposia and workshops while others will detail fundamental information of an enduring nature that will be referenced repeatedly.

The titles published in this series are listed at the end of this volume.

Stable Isotope Techniques in the Study of Biological Processes and Functioning of Ecosystems

Edited by

MURRAY UNKOVICH

Department of Botany, The University of Western Australia,
Centre for Legumes in Mediterranean Agriculture, The University of Western Australia,
Victorian Institute for Dryland, Agriculture, Mallee Research Station, Walpeup

JOHN PATE

Department of Botany, The University of Western Australia,
Centre for Legumes in Mediterranean Agriculture, The University of Western Australia

ANN MCNEILL

Centre for Legumes in Mediterranean Agriculture, The University of Western Australia,
The University of Adelaide, Roseworthy Campus – Agronomy and Farming Systems

and

D. JANE GIBBS

Centre for Legumes in Mediterranean Agriculture, The University of Western Australia

KLUWER ACADEMIC PUBLISHERS
DORDRECHT / BOSTON / LONDON

Library of Congress Cataloging-in-Publication Data

Stable isotope techniques in the study of biological processes and functioning of ecosystems / Murray Unkovich ... [et al.].
    p. cm. -- (Current plant science and biotechnology in agriculture ; v. 40)
    Based on papers from a workshop held at the University of Australia and the CSIRO Floreat Laboratories, Perth, Western Australia, in Feb. 1999.
    ISBN 0-7923-7078-3 (alk. paper)
    1. Crops--Ecophysiology--Congresses. 2. Stable isotopes in plant physiology research--Congresses. 3. Stable isotopes in ecological research--Congresses. I. Unkovich, Murray. II. Current plant science and biotechnology in agriculture ; 40.

SB106.E25 S73 2001
571.2--dc21

2001038175

ISBN 0-7923-7078-3

Published by Kluwer Academic Publishers,
P.O. Box 17, 3300 AA Dordrecht, The Netherlands.

Sold and distributed in North, Central and South America
by Kluwer Academic Publishers,
101 Philip Drive, Norwell, MA 02061, U.S.A.

In all other countries, sold and distributed
by Kluwer Academic Publishers,
P.O. Box 322, 3300 AH Dordrecht, The Netherlands.

Printed on acid-free paper

Printed in the Netherlands.

Table of Contents

Preface

In the last two decades technological advances in isotope ratio mass spectrometry have been very rapid, opening up new possibilities for analysis of biological and environmental materials. The new instrumentation has facilitated faster analysis of samples via automated sample preparation and multi-isotope analysis of single samples, resulting in considerable cost savings, and enabling access to isotope analysis for many more researchers. These changes are reflected in the rapidly growing international literature on stable isotopes. While there have been some excellent books and review papers aimed at interpreting isotope signals in biology and environmental science, there have been fewer attempts to provide practical tools for researchers making forays into this exciting new arena. This book aims to address this inadequacy by providing a set of practical guidelines for the application of a range of novel and well proven stable isotope techniques to the fields of plant physiological ecology, agriculture, marine ecology and palaeoecology.

The book is the outcome of a weeklong workshop held under the auspices of the Cooperative Research Centre for Legumes in Mediterranean Agriculture (CLIMA 1992 - 2000) at The University of Western Australia and the CSIRO Floreat Laboratories, Perth, Western Australia, in February 1999. The workshop was designed to provide practical tools and experiences for researchers and students concerned with how one goes about using stable isotopes in field investigations. The success of the week was very much due to the support that we received from CLIMA, the Grains Research and Development Corporation, and The Crawford Fund for International Agricultural Research. The support of these organisations is gratefully acknowledged and we hope that this volume will add to the lasting legacy of their investments.

The topics covered in the book principally reflect expertise carried within Australia, a country which we believe remains a major player in the application of stable isotope techniques to field-based studies, whether in relation to natural variations in isotopes or experiments involving application of enriched isotopic tracers. Indeed, the authors of the chapters contained in this book collectively express well over 200 years of practical experience using stable isotopes, and it is this knowledge that they were asked to share with you. The guiding principles that they were asked to consider were (i) what are the processes that one can measure, either directly or indirectly, with the isotope in question, (ii) why is it preferable to use stable isotopes rather than other techniques, (iii) how does one go about the experimental design and sampling of experiments, (iv) what are some of the practical considerations, calculations and special equipment that one should be aware of before starting out (v) what are the pitfalls of the methodology, and importantly, how can these be successfully circumvented, and (vi) using examples, illustrate the potential usefulness of the data for specific experimental procedures. Although almost all of the examples cited are in an Australian setting, the emphasis is on methodology rather than the specific details regarding the particular ecosystems or regions in which the studies were conducted. It should thus be easy to translate the message of each chapter to other locations and systems in space and time.

Our book commences with a guest contribution by Todd Dawson and Paul Brooks (Chapter 1) which will be particularly useful for those venturing into stable isotope territory for the first time, as it introduces basic issues of stable isotope chemistry, measurement of isotope enrichment, and the notation commonly used when referring to measurements involving stable isotopes. The chapter concentrates on isotopes of nitrogen, oxygen, hydrogen and carbon, since these are the only ones referred to in subsequent chapters of the book.

This introductory chapter is followed by three chapters that relate to water use of plants. The first by John Pate (Chapter 2) examines specifically the relationship between natural carbon isotope discrimination of C_3 plants and plant water use efficiency. It presents a number of novel sampling procedures for evaluating plant water stress in which carbon isotope discrimination is partnered with conventional assessments of water relations of plants. The following two chapters illustrate how stable isotopes of oxygen, hydrogen and to a lesser extent carbon, might be used to infer the differing sources of water taken up by plant roots using both natural variations in isotopic abundances and enriched tracer techniques. Chapter 3 by Jeff Turner and colleagues, focuses on some new analytical opportunities and problems and how they might best be exploited, while Glen Walker and colleagues (Chapter 4) provide broad ranging case studies to highlight ways in which the techniques can be applied in the field to examine plant water sources.

Our attention then turns towards plant nitrogen acquisition and metabolism with a chapter by George Stewart (Chapter 5) exploring the possible relationships between $\delta^{15}N$ signatures of supposed sources of nitrogen in natural ecosystems and consumer biota. The reader is left with the impression of guarded optimism concerning the value of the approach in evaluating nitrogen dynamics of ecosystems. Chapter 6 by Murray Unkovich and John Pate demonstrates how ^{15}N natural abundances of annual legumes and associated non legume reference plants can be used effectively to assess proportional dependency of the legume on symbiotically fixed N. This discussion is continued in Chapter 7 by Mark Peoples and colleagues with a foray into measurement of N_2 fixation of woody tree legumes in agroforestry. Chapter 7 also highlights how both ^{15}N enriched and natural abundance techniques can be used to study N_2 fixation.

The use of enriched isotopic tracers are then explained in the next three chapters, which deal with the fate of applied isotope(s) within and between components of plant and soil. The first of these chapters (Chapter 8) is by Jairo Palta and outlines ^{13}C and ^{15}N feeding techniques for studying the effectiveness with which carbon and nitrogen assimilated at various times in the life of crop plants, are transferred to filling seeds. The following chapter by Ian Fillery (Chapter 9) provides an excellent guide to the practice of using ^{15}N enrichment to follow soil nitrogen transformations in relation to current N uptake by crop plants. Ann McNeill (Chapter 10) then provides some new techniques and opportunities for addressing the somewhat neglected issue of turnover of root biomass and subsequent incorporation of N into soil organic matter using steady state *in situ* ^{15}N enrichment of root biomass of pasture and crop plants.

The final two chapters of the book switch to aquatic rather than terrestrial ecosystems. Albertus Smit (Chapter 11) reviews the extensive literature on use of natural ^{13}C and ^{15}N signals to trace food webs in pristine marine ecosystems and in systems experiencing pollution. Kliti Grice (Chapter 12) illustrates the potential of a relatively new analytical technique (compound specific isotope analysis) and provides intriguing accounts of the use of carbon isotope signatures of specific biochemical components isolated from fossil deposits as indicators of biotic interrelationships in palaeoenvironments.

The wide range of practical applications of stable isotope techniques presented in this volume bears testimony not only to the versatility and reliability of the high precision mass spectrometric instruments now regularly available, but also to how practical issues of experimental design, sampling, laboratory analysis and data interpretation have been progressively resolved through experience in the particular area and system being examined. The book exposes many situations where the success or otherwise of an experiment will clearly depend on what specific entities or components of a system are selected for analysis, how samples are harvested, and at what frequencies and scales data need to be obtained before a convincing picture emerges. In virtually all of the case studies, authors point to the fragility of a data base relying exclusively on isotope enrichments, and urge experimenters to interpret their data alongside other more traditional measurements of functioning and fluxes of commodities within a system. It is our hope that this multivalent approach to stable isotope investigations will be applied increasingly in the future.

Finally we stress again that this book is designed principally as a practical guide to application of stable isotope techniques rather than as yet another academic treatment summarising and reviewing the results of many studies from the international literature. Accordingly, we hope that our volume will be used more often in the laboratory and field than as a library of theoretical information. We encourage you to question and refine the applications and approaches detailed in the book, so as to make them more useful when defining and answering biological and ecological issues at field and whole ecosystem level.

Murray Unkovich
John Pate
Ann McNeill
Jane Gibbs

March 2001

Chapter 1

Fundamentals of Stable Isotope Chemistry and Measurement

Todd E. Dawson[1] and Paul D. Brooks[1,2]
[1]Center for Stable Isotope Biogeochemisty, Department of Integrative Biology, University of California - Berkeley, Berkeley, CA 94720 USA. Email: tdawson@socrates.berkeley.edu
[2]Department of Environmental Science, Policy and Management, University of California - Berkeley, Berkeley, CA 94720 USA. Email: isotopes@uclink.berkeley.edu

Key words: stable isotope ratio, fractionation, nuclide, isotope ratio mass spectrometer, isotope chemistry

1.   INTRODUCTION

Understanding the power of stable isotopes in research requires general knowledge of the fundamental principles of stable isotope chemistry as well as, more specifically, how the ratios of biologically relevant lighter stable isotopes of H, C, N, O, and S in the biosphere are measured using modern isotope ratio mass spectrometers (IRMS). Here, we provide a *primer* to these topics. The chapter serves as a springboard to the following chapters in this book. We make no attempt to be comprehensive. For more detailed reading on stable isotope chemistry or geochemistry we encourage readers to see treatments by Criss (1999), Hoefs (1997), Faure (1986) or Gat and Gonfiantini (1981). For in-depth discussions on particular topics we refer you to recent works by Kendall and McDonnell (1998; hydrology), Griffiths (1998; biology, ecology, atmospheric science, biogeochemistry), Clark and Fritz (1997; hydrogeology), Boutton and Yamasaki (1996; soils), Lajtha and Michener (1994; environmental sciences), Ehleringer *et al.* (1993; plant eco-physiology) and Rundel *et al.* (1989; ecology).

M. Unkovich et al. (eds.),
Stable Isotope Techniques in the Study of Biological Processes and Functioning of Ecosystems, 1–18.
© 2001 *Kluwer Academic Publishers. Printed in the Netherlands.*

2.   ISOTOPE DEFINITIONS, TERMINOLOGY AND UNITS

2.1   Background Definitions

The word 'isotope' comes from the Greek *isos* = equal and *topos* = place which is presumed to refer to their common place at a specific 'elemental address' within the Periodic Table of nuclides or elements (see Fig. 1). Recall that an atom is composed of a nucleus surrounded by electrons. The nucleus is composed of protons (Z) and neutrons (N) that constitute most of the mass of an atom. Protons are positively charged (Z^+), electrons negatively charged (e^-) and neutrons have no charge (N). The *atomic number* (A) of an atom is determined by its unique Z number, plus the number of electrons that balance its charge. The *atomic mass*, however, is the sum $Z + N$ for each nuclide. A nuclide, or isotope-specific atom, is a 'species' of an element, either stable or radioactive, defined by its unique number of protons (Z) and neutrons (N). Isotopes are atoms of the same element, and differ from isotones or isobars in the following manner (Fig. 1);
— *isotopes* are atoms which have the same Z and e but different N,
— *isotones* are atoms which have the same N but different Z and e, and
— *isobars* are atoms which have the same A but different combinations of Z and N

For example, all isotopes of carbon have 6 protons, but the radioactive isotope, ^{14}C has two more neutrons $(N = 8)$ than the stable and more common isotope ^{12}C $(N = 6)$. An isotope tends to be 'stable' when (a) it has an N/Z ratio below approximately $1.0 - 1.5$, particularly for nuclides with N or Z below ~ 25, or (b) it has an 'even' Z-number, for nuclides with N or Z greater than ~ 25 (Criss 1999). Stable isotopes do not appear to decay to any other nuclide once they are formed. In contrast, radioactive isotopes decay spontaneously, emitting alpha or beta particles, to become stable isotopes such as when ^{14}C beta decays to its stable form ^{14}N (see pathway 'a' in Fig. 1 as well as discussions by Kendall and Caldwell 1998).

2.2   Notation, Units and Terminology

The stable isotopes of most compounds are composed of one overwhelmingly abundant isotope and one or two isotopes of relatively minor abundance (see Table 1). The low abundance of these isotopes provides opportunities to use enriched sources of the isotopes as 'tracers' in biochemical, biological and environmental studies. In such cases the enrichment of the isotope is usually expressed as atom% excess, that is the % of atoms as the minor isotope in excess of their background abundance.

For example, for ^{15}N, which has a natural abundance in atmospheric N_2 of 0.3663 atom% ^{15}N (Table 1), an enriched compound containing 1 atom% ^{15}N has 0.6337 atom% ^{15}N excess.

Table 1. The (**A**) isotope abundance ratios measured and their internationally accepted reference standards, and (**B**) percent abundance in terrestrial environments of the principle stable isotopes discussed in this review.

A.

Isotope	Ratio measured	Standard	Abundance ratio of reference standard
$^2H^1$	$^2H/^1H$ (D/H)	V-SMOW[2]	1.5575×10^{-4}
^{13}C	$^{13}C/^{12}C$	V-PDB[3]	1.1237×10^{-2}
^{15}N	$^{15}N/^{14}N$	N_2-atm.[4]	3.677×10^{-3}
^{18}O	$^{18}O/^{16}O$	V-SMOW	2.0052×10^{-3}
		V-PDB	2.0672×10^{-3}
^{34}S	$^{34}S/^{32}S$	CDT[5]	4.5005×10^{-2}

B.

Element	Isotope	Percent abundance	Terrestrial range	Technical precision
Hydrogen	^1H	99.985	‰; 700	0.8 – 0.2
	^2H (D)	0.015	ppm; 109	0.16
Carbon	^{12}C	98.98	‰; 100	0.05
	^{13}C	1.11	ppm; 1123	0.56
Nitrogen	^{14}N	99.63	‰; 50	0.10
	^{15}N	0.37	ppm; 181	0.72
Oxygen	^{16}O	99.759	‰; 100	0.10
	^{17}O	0.037	ppm; 200	0.20
	^{18}O	0.204		
Sulfur	^{32}S	95.00	‰; 100	0.20
	^{33}S	0.76	ppm; 4580	9.16
	^{34}S	4.22		
	^{36}S	0.014		

[1] The hydrogen stable isotope with mass two (^2H) is also called deuterium, D; [2] the original standard was 'Standard Mean Ocean Water' (SMOW) which is no longer available; however, 'Vienna'-SMOW is available from the IAEA; [3] the original standard was a Belemnite from the PeeDee formation (PDB) which is no longer available; however, 'Vienna'-PDB is available from the IAEA; [4] atm. = atmospheric gas; [5] the standard still used is troilite (FeS) from the 'Canyon Diablo' meteorite (CDT)

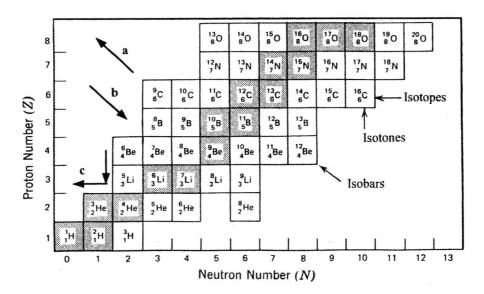

Figure 1. Partial chart of nuclides (atom specific isotopes). Shaded squares are the stable nuclides and unshaded squares are the radioactive or unstable nuclides. The arrows labelled 'a', 'b' and 'c' indicate the changes in proton and neutron number associated with beta decay (a), positron decay and beta capture (b) and alpha decay (c) respectively. From Kendall and Caldwell (1998), modified from Faure (1986).

There are very small variations in the background levels of most isotopes in the different components of the biosphere, and for these variations stable isotope composition is usually reported as the parts per thousand (‰) deviation from an internationally accepted standard. This notation is used since the absolute abundance of these minor isotopes is usually less than 1 percent of the total for a given element (e.g. ^2H typically accounts for only 0.015576% of hydrogen). These relative values of deviation from an arbitrary (but internationally accepted) standard are known as delta (δ) values with units of ‰ (this is also denoted 'permil', 'permille', 'per mil', or 'per mill'). The δ value of any material is calculated as:

$$\delta = (\frac{R_{sample}}{R_{standard}} - 1) \times 1000 \qquad (1)$$

in ‰ units, where R is the ratio of the heavy to light (or rare to abundant) isotope (e.g., $^{13}C/^{12}C$) and R_{sample} and $R_{standard}$ are the heavy to light isotope ratios of the sample and the standard respectively. A positive δ value indicates that the sample is higher than the standard, that is has more of the heavier isotopic species compared to the standard, while a negative δ

value indicates that the sample has a lower isotope ratio than the standard. For example, we might find that an unknown water sample has a δ^2H value of -50‰, which means that it has a $^2H/^1H$ (or D/H) isotope ratio 50 part-per-thousand or 5% lower than the $^2H/H$ of standard water (ocean water; see below). For clarity, a higher isotope ratio or value generally means the material is 'heavier' and therefore more positive in its δ value relative to the material to which it is being compared. To avoid misunderstanding when writing or speaking about isotope values we recommend always defining what you mean and keep your language simple. For accuracy without ambiguity, use 'high' or 'low' or 'more/less' positive or negative when referring to δ values. This language leaves no room for confusion. If you must use 'enriched' or 'depleted', state which isotope you are referring to and refer to the heavier, in most cases also the rarer, of the two isotopes in any particular ratio (e.g. "The material was depleted in ^{18}O . . .").

2.3   Standards

A list of internationally accepted reference standards for the lighter isotopes in the biosphere is shown in Table 1A. This table also includes information about the abundance ratio measured for each standard. Table 1B shows the percent abundance of each stable isotope of these same elements as well as the range of natural variation one can expect to find in terrestrial ecosystems (in ‰ and ppm). It also lists the current technical precision one can expect to obtain when analyzing the listed isotopic species using modern isotope ratio mass spectrometry (IRMS) instrumentation. The δ value of all internationally accepted reference standards is by definition 0‰. These standards can be obtained from either the International Atomic Energy Agency (IAEA) in Vienna, Austria (website: http://www.iaea.org.at) or the National Institute of Standards and Technology (NIST) in the United States (website: http://www.nist.gov).

Values are expressed against these standards when isotope ratios are determined using an IRMS. However, since IAEA or NIST standards are expensive to use on a daily basis, most stable isotope facilities routinely use their own 'internal working standard' (IWS). An IWS is used on a daily basis and is most often unique to that particular analytical laboratory and usually matched to the particular types of analyses being performed. An IWS must always be calibrated against an international standard. Moreover, it is a common and wise practice to be sure that each IWS has been run 'blind' at several other laboratories conducting similar types of analyses. Inter-lab comparison of any IWS is an essential part of running a stable isotope facility because it ensures that the IWS value is verified and provides confidence in the analysis. Values obtained relative to a specific IWS are converted to their final value (relative to the internationally accepted reference standard) following the calculation provided by Craig (1957), as:

$$\delta_{final} = \frac{(\delta_{sample} - \delta_{IWS}) \times (\delta_{IWS} - \delta_{standard})}{1000} \qquad (2)$$

Occasionally, it may be necessary or desirable to *normalize* the δ value of any reference material to the per mil scale and the procedure for this can be found in several papers by Coplen (1994, 1995, 1997).

3.   PROPERTIES OF ISOTOPES

3.1   Fundamental Isotope Chemistry

Because the *chemical properties* of any element are largely determined by the number of electrons (e^-) and their configuration in the e-shell, which do not vary between isotopes of an element, each isotope has common chemical properties. However, because neutrons (N) which influence the weight of an isotope are relatively massive, isotopes will differ in atomic mass (A_m) and thus react in a chemical reaction in different ways because of their *physical properties*; this occurs despite the fact that each isotope has the same atomic number (A). Hence, any change that comes about in the relative abundances of nuclides in an isotopic mixture results mostly from mass-dependent isotope effects. This is particularly true for the elements of low mass (A_m) where the mass differences among the different isotopes and the molecules that contain these isotopes can be quite large (2 – 100%; Table 2). Such differences can have a marked effect on the physicochemical properties of the molecules and thus their relative proportions in various reactants, products, phases and mixtures.

3.1.1   Isotope Behaviour

Physicochemical differences which are brought about because of differences in N usually lead to differences in the *chemical behaviour* of an isotope involved in a chemical reaction and therefore its representation in the product of a reaction. Thus differential chemical behaviour of the isotopes involved in a reaction, which results in their differential representation in reactant(s) and product(s), is the primary reason isotopes exist in different proportions and with different δ values in each pool. The process that leads to differential representation, and therefore, different isotope ratios (δ values), is called *isotopic fractionation*. Differences in the N-number of an isotope can influence the rate of a chemical reaction in which it is involved (higher N usually depresses the rate). This is especially true for the lightest elements. For example, Table 3 shows how slight differences in the stable isotope (^1H or ^2H) present in water can influence its physicochemical properties.

Table 2. The atomic mass, relative difference among the stable isotopes of the lighter elements, and relative difference among the molecules composed of these isotopes that are analyzed with an isotope ratio mass spectrometer. The masses of the molecules are shown in parentheses.

Element	Isotope	Atomic mass	Elemental relative mass difference	Molecular relative mass difference
Hydrogen	^1H	1.0078	$^2H/H$	$^1H^2H/^1H^1H$
	^2H	2.0141	100%	(3/2); 50.0%
Carbon	^{12}C	12.0000	$^{13}C/^{12}C$	$^{13}C^{16}O^{16}O/^{12}C^{16}O^{16}O$
	^{13}C	13.0034	8.3%	(45/44); 2.3%
Nitrogen	^{14}N	14.0031	$^{15}N/^{14}N$	$^{15}N^{14}N/^{14}N^{14}N$
	^{15}N	15.0001	7.1%	(29/28); 3.6%
Oxygen	^{16}O	15.9949	$^{18}O/^{16}O$	$^{12}C^{16}O^{18}O/^{12}C^{16}O^{16}O$
	^{17}O	16.9991	12.5%	(46/44); 4.5%
	^{18}O	17.9992		
Sulfur	^{32}S	31.9721	$^{34}S/^{32}S$	$^{34}S^{16}O^{16}O/^{32}S^{16}O^{16}O$
	^{33}S	32.9714	6.3%	(66/64); 3.1%
	^{34}S	33.9679		
	^{36}S	35.9671		

Table 3. Characteristic physical properties of $H_2^{16}O$, $^2H_2^{16}O$, and $H_2^{18}O$ (after Hoefs 1997). $H_2^{16}O$ is the most common form of water on earth.

Property	$H_2^{16}O$	$^2H_2^{16}O$	$H_2^{18}O$
Density (20°C, in g cm^{-2})	0.997	1.1051	1.1106
Temperature of greatest density (°C)	3.98	11.24	4.30
Melting point (@760 Torr, in °C)	0.00	3.81	0.28
Boiling point (@760 Torr, in °C)	100.00	101.42	100.14
Vapour pressure (@ 100°C, in Torr)	760.00	721.60	758.07
Viscosity (@ 20°C, in centipoise)	1.002	1.247	1.056
Molar volume (@ 20°C, in cm^3/mole)	18.049	18.124	18.079

3.1.2   Energy States of Isotopes

Different physicochemical properties of isotopes are the result of quantum mechanical effects and how the strength of the chemical bonds changes as the mass of the molecule that contains different isotopes changes. For example, molecules which contain heavier isotopes have chemical bonds that are more stable, and so take a greater amount of energy to break. For this reason, the isotopic fractionation which results between two pools that contain different relative proportions of isotopes can be explained by differences in each isotope's individual energy state, or their so-called *zero-point energies* (ZPE). At a temperature of absolute zero, a vibrating element or molecule, composed of a unique mixture of protons, neutrons and/or isotopes will have a certain zero-point energy content (Fig. 2) which lies above the minimum potential energy level of that molecule (bottom of the curve in Fig. 2). The vibrational frequency of the molecule depends upon the mass of the isotopes. Thus a molecule with the same chemical formula but composed of different isotopic species (nuclides) will have different vibrational motions and a different zero-point energy content. This means there is a greater difference between the zero-point energy state and the minimum (continuous) energy state of a molecule when it is composed of 'lighter' nuclides than when it is composed of 'heavier' nuclides (Fig. 2). For example, a water molecule composed of hydrogen with ^2H-^2H bonds is more stable, and thus the bonds more difficult to break, than water composed of H-H bonds. This is illustrated in Figure 2 which shows the ZPE of a ^2H-^2H - bond to be lower (105.3 kcal/mole) compared to the ZPE of an H-H bond (103.2 kcal/mole). If bonds are broken differentially, the chemical reactions which involve molecules with different isotopes will show an isotope effect. Such isotope effects are more apparent at low than at high temperatures.

3.2   Isotopic Fractionation

The partitioning of isotopes between two substances (e.g. reactant vs. product) or two phases of the same substance (e.g. liquid vs. vapour) with different isotopic ratios (δ values), is called *isotopic fractionation*. There are two main phenomena that lead to isotopic fractionations:
1. Isotopic exchange reactions (also referred to as equilibrium or thermodynamic fractionation)
2. Kinetic processes which are determined by reaction rates of molecules with unique isotopic composition

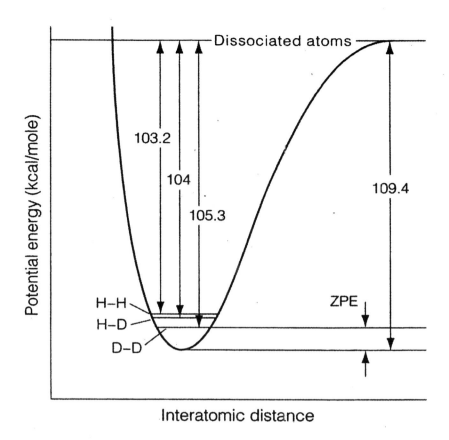

Figure 2. An example of the relationship between interatomic distance and the potential energy associated with stable hydrogen isotopes of a molecule. Higher zero-point energies (ZPE), or lower numerical values, result in a particular molecule being less stable and therefore its chemical bonds being more easily broken. From Kendall and Caldwell (1998).

Isotopic exchange/equilibrium fractionation reactions are those in which the distribution (representation) of isotopes differs between chemical substances, phases, or individual molecules, but the reaction which produces these distributions comes to a chemical equilibrium such that:

$$aA_1 - bB_2 = aA_2 + bB_1 \qquad\qquad (3)$$

where the subscripts indicate that species A and B contain either the light (1) or heavy (2) isotope. In theory this only occurs in well-mixed, closed systems where the forward and back reaction rates of any isotope are the same *but* the ratios of the different isotopes are different in each compound or phase of a substance (e.g. liquid water *vs.* water vapour) at any particular moment and temperature. During isotopic exchange reactions the heavier isotope generally accumulates in the compound or phase which is the most dense (and/or with the highest oxidation state), becoming *enriched* and heavier, while the less dense compound or phase becomes *depleted* and therefore lighter. For example, it is known that at equilibrium, of the phases of water, ice is more enriched in ^{18}O than is liquid water which is more enriched than water vapour.

Kinetic fractionation depends primarily on differences in the reaction *rates* of isotopes or molecules with unique isotopic composition. Kinetic fractionation events are predominantly unidirectional (meaning that the forward and back reaction rates are not identical) and occur because of (a) differences in the *diffusional flux* of molecules impinging on a surface where they are adsorbed (determined by Graham's Law), and (b) by differences in the rates of *diffusion* of molecules through a substance (e.g. air, water) by mass transfer. The diffusion rate is determined by an isotope's diffusional coefficient (determined by Fick's Law). Diffusional flux and diffusion are *not* the same, being determined by different laws, a full mathematical treatment of which can be found in Hoefs (1997).

For our purposes, it is most important to note that kinetic fractionation is mass dependent and that the kinetic isotope effect, β, defined as the difference in the chemical or physical properties arising from differences in the atomic mass of an element, is written as:

$$\beta = \frac{1}{2}mv^2 \qquad\qquad (4)$$

where 1 and 2 are the heavy and light isotopes respectively and m = mass and v = velocity. For CO_2 with $^{12}C^{16}O_2$ that has a mass = 44 and $^{12}C^{18}O_2$ that has a mass = 48, we then say that:

$$\beta = \frac{V_{heavy}}{V_{light}} = (\frac{48}{44})^{1/2} = 1.0445 \qquad\qquad (5)$$

so that the average velocity of $^{12}C^{16}O_2$ is ~4.4% greater than $^{12}C^{18}O_2$. If the kinetic fractionation which results in different isotopic ratios in the reactants and in the products of the reaction is enzyme-mediated, the enzyme can 'discriminate' against one isotopic species over another which then results in isotope fractionation. Stable isotope discrimination is therefore defined as enzyme-mediated fractionation.

3.2.1 Fractionation Factors

A fractionation factor (α) is the ratio of two isotopes in one chemical compound or phase, X, divided by the ratio of the same two isotopes in the other chemical compound, Y:

$$\alpha_{X-Y} = \frac{R_X}{R_Y} \tag{6}$$

where R is the heavy/light isotope ratio (e.g., $^{13}C/^{12}C$) in X and Y. Within compounds X and Y if the isotopes are randomly distributed, then α can be related to the equilibrium constant of a reaction, K, by:

$$\alpha = K^{1/n} \tag{7}$$

where n is the number of atoms exchanged. The quantity α expresses the magnitude of isotopic fractionation between two substances, either, in equilibrium (called an 'equilibrium' fractionation factor) or the substrate and product of a unidirectional reaction (called a 'kinetic' fractionation factor). Fractionation factors are obtained in three ways (Criss 1999, Hoefs 1997):
1. Spectroscopic (vibrational frequency) data and partition functions derivable from statistical mechanics,
2. Laboratory calibration studies,
3. Measurements of natural samples whose formation, as well as the conditions under which the formation occurs, are well known or highly constrained.

3.2.2 Temperature and Fractionation

Fractionation factors (α) are temperature dependent. For example, at a given temperature, when water evaporates from a wet surface and undergoes the liquid-to-vapor phase change, the vapor phase is enriched in the lighter isotopes H and ^{16}O because $H_2{}^{16}O$ has a higher vapor pressure than either $H^2H\,{}^{16}O$ or $H_2{}^{18}O$ (Table 3; see Gat and Gonfiantini 1981). Consequently, the liquid phase becomes enriched in ^2H and ^{18}O, but depleted in H and ^{16}O and the δ^2H and $\delta^{18}O$ of water vapor in the atmosphere takes on a negative value. At standard temperature and pressure, the fractionation factors (α) for evaporation of water (at equilibrium) are:

$$\alpha_{18} = (^{18}O/^{16}O)_{liquid}/(^{18}O/^{16}O)_{vapor} = 1.0092 \text{ or } 9.2\text{‰} \qquad (8)$$

$$\alpha^2H = (^2H/H)_{liquid}/(^2H/H)_{vapor} = 1.074 \text{ or } 74\text{‰} \qquad (9)$$

The temperature dependence of these fractionation factors are shown in Figure 3 and as stated above, the higher the temperature, the less differences between the isotope species because the difference in the ZPE becomes smaller.

4.   MEASUREMENT OF ISOTOPE RATIOS BY MASS SPECTROMETRY

Stable isotope abundances are most commonly measured using mass spectrometry. The past 20 years have seen rapid growth in the development of better gas-phase stable isotope instruments, interfaces and measurements. Automated systems for preparation of H, C, N, O and S have seen extensive development and coupled with the newest continuous-flow isotope ratio mass spectrometers, provide for a growing array of applications in the fields of agriculture, natural and atmospheric sciences, forensic science and biomedical research.

4.1   The instrumentation

A mass spectrometer is an instrument which separates charged atoms or molecules based on their mass-to-charge ratios and motions in the presence of a strong magnetic field. A stable Isotope Ratio Mass Spectrometer (IRMS) is, in principle, fairly simple, with four main components: (1) inlet system, (2) ion source, (3) mass analyzer [with a magnet and flight-tube assembly], and (4) ion detector system (Fig. 4).

Samples are introduced into the IRMS via the inlet system as a gas. Under ambient conditions gases would naturally move into the IRMS by molecular flow and if they were of different masses, fractionation would occur. To avoid this problem the gas is either (a) temporarily stored in a variable volume bellows for dual inlet-IRMS (DI-IRMS), or (b) placed within a continuous flow of an inert carrier gas (He) for continuous flow-IRMS (CF-IRMS). In both systems gas enters the analyzer chamber through a set of capillaries ensuring viscous flow in which no mass separation (fractionation) occurs (Fig. 4). In the IRMS 'source' chamber an ionizing filament causes the ejection of an electron from the outer shell of the gas molecule, leaving the resulting particle positively charged. The beam of ionized gas is then repelled towards the flight tube, focused and then accelerated with the aid of a powerful magnet down the tube toward the ion detector/collector assembly. Because the ions have different mass-to-charge ratios, they are deflected differently as they pass the magnet. The ions separate into beams based on their isotope ratios and at the exit of the flight tube are captured by different, and specific collectors (the Faraday cups). The radii of curvature within the flight tube are proportional to the square root of the mass:charge ratio ($\sqrt{m/e}$) Thus nuclides with higher m/e will be deflected less as they move towards the Faraday cups, which are carefully spaced to collect each ion beam separately and with minimal interference from the other beams. The detector system converts the ion impacts into a voltage which is in turn converted by a multiplexing system into frequency of impacts. The ratio of the frequency of impacts for any particular sample is termed the R value for that sample and from this value the final δ value is determined using Equation 1. In dual-inlet mode, a changeover valve is used to switch between the sample and a standard gas, each of which is held in variable volume bellows. The difference in the signals between the sample and the standard is used to calculate the isotope ratio for the sample. Although the absolute ratios (e.g. 45/44 and 46/44 for CO_2) are measured for both the sample and the standard, the values are not of direct importance and in fact depend in part on the amount of gas in the analyser. Thus, it is the difference between the reference and sample gas that is used to calculate the δ value relative to the appropriate internationally accepted standard (Table 1A).

The conventional IRMS geometry requires only two cups (often on a tighter and shorter spur of the flight tube/magnet assembly) for measuring hydrogen isotope ratios since there are only two ionized gases ($^1H^1H$, mass 2 and $^2H^1H$, mass 3). For CO_2 and N_2, however, there are three isotopic forms (for CO_2, masses 44, 45, and 46; and for N_2, $^{14}N^{14}N$, mass 28, $^{14}N^{15}N$, mass 29 and $^{15}N^{15}N$, mass 30) which requires one cup to detect each form. The newest IRMS instruments (such as the Finnigan, model Delta^{plus}XL^{TM}) do not use a short H-spur but instead use precise electromagnetic control to focus the very light HH and H^{2}H isotopes onto one of the widely dispersed Faraday cups that can detect masses 2, 3, 28, 29, 30, 40, 44, 45, 46, 64 and 66. In all instruments the entire system is kept under very high vacuum ($\sim 10^{-8}$ Torr = 10^{-5} Pa = -5 atmospheres) to ensure that there are extremely low levels of contaminating molecules, and because the mean free path length of a gas molecule moving in space without striking another is inversely proportional to the pressure; a low pressure (vacuum) of 10^{-8} Torr ensures a mean free path length of 500 m. Because the flight tubes on a typical IRMS are 0.3 to 2.3 m long, extremely few molecular collisions will occur.

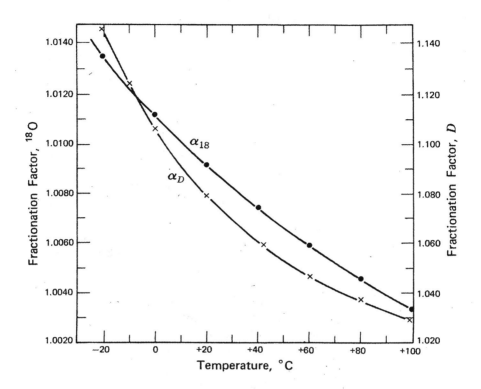

Figure 3. The relationship between a change in temperature and the associated change in the isotope fractionation factors for ^{18}O (α_{18}; left-hand y-axis) and ^2H (D) (α_D; right-hand y-axis) during the evaporation of water. For both isotopic species (^{18}O and ^2H) the fractionation factor is defined as the ratio of the heavy to light isotope ratio in the liquid phase to the heavy to light isotope ratio in the vapour phase of water. As stated in the text, at standard temperature and pressure, $\alpha_{18} = (^{18}O/^{16}O)_{liquid} / (^{18}O/^{16}O)_{vapour} = 1.0092$ while $\alpha_D = (^2H/H)_{liquid} / (^2H/H)_{vapour} = 1.074$. As temperature increases, the differences between the isotope species become less because the differences in the ZPE each possesses become smaller. From Faure (1986).

4.1.1   Available Analysis Methods

Recent advances in automated analyses using continuous-flow IRMS (CF-IRMS) have now made it possible to analyze, with high accuracy, a large number of samples as well as many different sample types. Some of the greatest advances using CF-IRMS have been made in interface designs that couple the IRMS with automated preparation and introduction of samples to the ion source. Most of these interfaces use well tested methods (e.g. gas chromatography) and have the advantage of being automated for ease of operation. One of the original CF interface systems was for processing samples for C and N analysis using an elemental (C/N) analyzer interfaced with an IRMS and includes the software to automatically change the accelerating voltage. This enables the instrument to 'jump' between analyzing nitrogen for a $\delta^{15}N$ isotope analysis (masses 28, 29, 30) to analyzing carbon for $\delta^{13}C$ analysis (masses 44, 45, 46) on a single sample. This procedure has now been extended to measurements of S isotopes (of solid samples) using an C/N/S analyzer, the pyrolysis of samples in a modified C/N analyzer to measure either $\delta^{2}H$ or $\delta^{18}O$ in solid samples, to direct analysis of gases, such as CO_2 in breath and to trace gases such as N_2, N_2O and CH_4 using a cold-finger or GC trapping interface. In addition, Gas Chromatography Combustion IRMS (GC-C-IRMS) for $\delta^{13}C$ and $\delta^{15}N$ analysis of organic compounds and GC-Pyrolysis-IRMS (GC-P-IRMS) for hydrogen isotope analysis of organic compounds is possible using modern CF-IRMS technology (see chapter in this volume by Grice).

4.1.2 Comparison of Continuous Flow and Dual Inlet Systems for N and C analysis

For continuous flow C and N analysis of solid samples, a carbon nitrogen analyzer connected to an IRMS offers the advantage of a fully automated 5 – 10 minute analysis time per sample for %N, $\delta^{15}N$, %C and $\delta^{13}C$. However, depending on the design of the system the sample size is limited to about 10 mg of plant material or 50 mg of soil. These samples must be weighed out to high accuracy into small tin capsules and then wrapped up using forceps. Because of the small sample size it is essential that samples be well homogenized, which usually requires ball milling. Older mass spectrometers fitted with a C/N analyzer can have a non-linear response of δ compared to the amount of C or N in the sample, which may require confining samples to a narrow elemental weight. Usually 30 – 300 μg N and 0.2 – 2 mg C can be analyzed in a tin capsule. There are numerous laboratories that, as a service, will analyse samples provided in pre-weighed tin capsules. Precision is usually better than 0.2‰ for N and 0.1‰ for C.

For analyses using a DI-IRMS, samples are weighed into a sealed quartz or Vicor glass tube, mixed with other compounds to facilitate combustion/reduction, and then heated at a known and high temperature for a set time period in a muffle furnace. The tubes must then be placed in a manifold, evacuated and processed by hand to purify either N_2 or CO_2. This purified gas is collected in a separate vial which is placed on the dual inlet IRMS and either analyzed manually or via a programmable inlet system.

This more 'classical' method requires a skilled operator, takes more time than CF-IRMS and is thus generally more expensive. A few laboratories provide such a service with the precision of this type of analysis usually better than 0.1‰ for N and 0.05‰ for C.

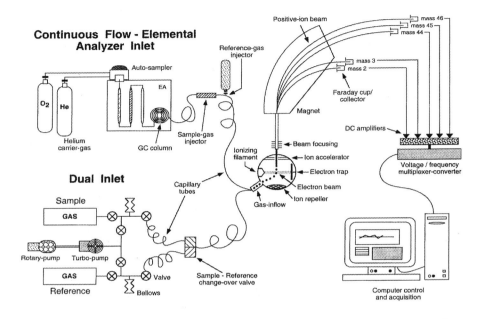

Figure 4. Diagrammatic representation of the continuous flow-elemental analyzer inlet (top left) and dual inlet (bottom left) interfaces to the ion source of an isotope ratio mass spectrometer (center). In some cases when using the CF-IRMS, a reference-gas injector is used. In other cases, reference sample material is introduced at the auto-sampler along with the samples. Ionized gases leaving the source are accelerated and focused with the aid of the magnet towards the collector array.

4.2   Terminology with IRMS performance

There are a few terms that are worth familiarizing yourself before using a mass spectrometer to make stable isotope measurements. You may also want to request these data from the stable isotope facility that processes your samples; the information will help you evaluate your data more fully.

Accuracy is the measure of how true the measurement made is to its known value. It is always problematic to make accurate measurements using mass spectrometry because of mass-dependent isotope fractionations that can occur during sample introduction into the IRMS. These effects are largely overcome, however, by comparing the sample to a matched standard (the basis of the 'delta' notation) as stated above. Therefore, when making actual measurements it is more important to have accurate (and precise) *differences* compared to the standard, than absolute accuracy since this has a greater influence on your δ values.

Precision is the reproducibility of an instrument. *Internal precision* is the measure of how the mass spectrometer performs a reproducible analysis in isolation (e.g. using repeated analysis of the same sample of gas from a single aliquot). The best way to determine the internal precision of an IRMS is in the dual-inlet configuration. The units used to evaluate internal precision are twice the standard deviation (2 stdv) of the mean of 6 – 12 repeated comparisons of the sample to the known standard. A highly precise instrument can have a 2 stdv value of 0.005‰ for 13-carbon. *External precision* is a measure of the entire systems performance, including all of the analytical processes leading up to the introduction of a sample gas into the mass spectrometer (see Unkovich *et al.* 1993). Systematic errors can give results that are very precise but not necessarily accurate.

Sensitivity of an isotope ratio mass spectrometer is defined as its ability to resolve and measure a single ion within the lowest number of sample molecules introduced into the detector system. The precision improves as the sensitivity of an instrument to detect ions in the sample increases. Thus more sensitive instruments can analyze smaller samples with greater accuracy. The sensitivity of an instrument is often rated by the smallest raw sample needed to produce one ion. Of course this number will be determined by both the mass spectrometer and the inlet system, and it is important that any inlet system interfaced with the IRMS is designed to optimize sensitivity.

ACKNOWLEDGMENTS

We thank the editors for inviting us to contribute to this volume and Murray Unkovich for his comments on an earlier draft of this chapter. The U.S. National Science Foundation and the University of California provided financial support.

REFERENCES

Boutton, T.W. and Yamasaki, S-i (1996). Mass Spectrometry of Soils. (Marcel Dekker Inc: New York.)

Criss, R.E. (1999). Principles of Stable Isotope Distribution. (Oxford University Press: New York.)

Clark, I. and Fritz, P. (1997). Environmental Isotopes in Hydrogeology. (Lewis Publishers: Boca Raton, New York.)

Coplen, T.B. (1994). Reporting of stable hydrogen, carbon and oxygen isotopic abundances. *Pure and Applied Chemistry* 66, 273-276.

Coplen, T.B. (1995). Discontinuance of SMOW and PDB. *Nature* 375, 285.

Coplen, T.B. (1996). New guidelines for reporting stable hydrogen, carbon and oxygen isotope-ratio data. *Geochimica et Cosmochimica Acta* 60, 3359-3360.

Craig, H. (1957). Isotopic standards for carbon and oxygen and coorection factors for mass-spectrometric analysis of carbon dioxide. *Geochimica et Cosmochimica Acta* 12, 133-149.

Ehleringer, J.R., Hall, A.E., and Farquhar, G.D. (1993). Stable Isotopes and Plant Carbon-Water Relations. (Academic Press, Inc.: San Diego.)

Faure, G. (1986). Principles of Isotope Geology. Second Edition, (John Wiley & Sons: New York.)

Gat, J.R. and Gonfiantini, R. (1981). Stable Isotope Hydrology: Deuterium and Oxygen-18 in the Water Cycle. IAEA Technical Report Series, No. 210, Vienna.

Griffiths, H. (1998). Stable Isotopes: Integration of Biological, Ecological and Geochemical Processes. (BIOS Scientific Publishers Limited: Oxford.)

Hoefs, J. (1997). Stable Isotope Geochemistry. Fourth Edition (Springer-Verlag: New York.)

Kendall, C. and Caldwell, E. A. (1998). Fundamentals of isotope geochemistry. In 'Isotope tracers in Catchment Hydrology.' (Eds C. Kendall and J.J McDonnell.) pp 51-86. (Elsevier: Amsterdam.)

Kendall, C. and McDonnell, J.J. (1998). Isotope Tracers in Catchment Hydrology. (Elsevier Science B.V: Amsterdam.)

Lajtha, K. and Michener, R.H. (1994). Stable Isotopes in Ecology and Environmental Science. (Blackwell Scientific Publications: Oxford.)

Rundel, P.W., Ehleringer, J.R. and Nagy, K.A. (1989). Stable Isotopes in Ecological Research. (Springer-Verlag: New York.)

Unkovich, M.J., Pate, J.S., and Sanford, P. (1993). Preparation of plant samples for high precision nitrogen isotope ratio analysis. *Communications in Soil Science and Plant Analysis* 24, 2093 - 2106.

Chapter 2

Carbon Isotope Discrimination and Plant Water-Use Efficiency
Case Scenarios for C₃ Plants

Case Scenarios for C_3 Plants

John S. Pate

The Department of Botany, and Centre for Legumes in Mediterranean Agriculture (CLIMA), The University of Western Australia, Crawley, WA 6009 Australia. Email: johnpate@cyllene.uwa.edu.au

Key words: $\delta^{13}C$ signals, water-use efficiency, transpiration ratio, phloem sap, growth rings, photosynthate partitioning, sampling for $\delta^{13}C$ analysis

1.   INTRODUCTION

It has been known for a long time that the isotope fractionation of carbon formed by photosynthesizing plants varies radically depending on the pattern of photosynthesis in which the species is engaging (see O'Leary 1988, Farquhar *et al.* 1989, and chapter by Dawson and Brooks in this volume). The principal reason for this is that carbon fixed by the photosynthetic CO_2 assimilating enzymes RUBP carboxylase [Rubisco] (C_3 plants) and PEP carboxylase (C_4 and CAM plants) show different isotope fractionation ($\delta^{13}C$) values, since the former enzyme discriminates 27 to 30‰, the latter only 0 to 2‰. As a result $\delta^{13}C$ values for C_3 plants utilizing Rubisco range from -20 to -35‰ while C_4 species utilizing PEP carboxylase exhibit an equivalent range of from -7 to -17‰ (see Deines 1980, Ehleringer 1989). Those 'obligate' CAM plants which operate by fixing CO_2 only at night using PEP carboxylase have $\delta^{13}C$ values similar to C_4 plants, whereas 'facultative' C_3:CAM species which are able to shift back and forth between C_3 and CAM-type photosynthesis depending on conditions, exhibit $\delta^{13}C$ values resembling those of C_3 plants when under well watered conditions but values close to C_4 plants when in dry or saline environments (see Teeri 1982, Ehleringer 1989). Table 1 summarizes some of the principal differences

19

M. Unkovich et al. (eds.),
Stable Isotope Techniques in the Study of Biological Processes and Functioning of Ecosystems, 19–36.
© 2001 *Kluwer Academic Publishers. Printed in the Netherlands.*

between C_3, C_4 and CAM plants in respect of water use, carbon isotope discrimination and other physiological features.

For plants with a C_3 pattern of photosynthesis, the topic of this chapter, the carbon isotope ratios are traditionally related to leaf gas exchange in the following way:-

$$\delta^{13}C_{plant} = \delta^{13}C_{air} - a - (b - a) \times c_i/c_a \qquad (1)$$

where a = 4.4‰ and represents the isotope fractionation associated with differential diffusivities of $^{13}CO_2$ versus $^{12}CO_2$ in air and through plant stomata, c_i and c_a refer respectively to ambient CO_2 concentrations in the internal gas space of the leaf and the surrounding atmosphere, b=27‰ and denotes the net fractionation by Rubisco during the photosynthetic CO_2 fixation reaction, and $\delta^{13}C$ of air is taken to be -8‰.

In essence the ratio c_i/c_a is determined by interaction of CO_2 assimilation rate and the extent of stomatal opening i.e. by the supply and utilization of CO_2 at sites of carboxylation in the mesophyll of the leaf. Thus, a photosynthetically active plant with relatively closed stomata will tend to exhibit low c_i/c_a ratios and accordingly less negative $\delta^{13}C$ values for the photosynthate which it is producing than in a corresponding plant with relatively open stomata, higher c_i/c_a , and hence a more negative $\delta^{13}C$ value. To date, c_i/c_a values for typical C_3 species have been shown to range from 0.4 to 0.8, or in $\delta^{13}C$ terminology corresponding to $\delta^{13}C$ values approximately in a range from -21 to -30‰ (see Ehleringer 1989, Pate and Dawson 1999).

In a series of far ranging studies by Farquhar, his colleagues and other groups of researchers it has been found that a broad relationship exists between carbon isotope discrimination of plant dry matter and the efficiency with which this dry matter is laid down relative to amounts of water which the plant transpires (see Farquhar *et al.* 1988). Expressed in terms of leaf functioning, the instantaneous water-use efficiency of the foliage canopy of a plant is defined simply as A/E, that is the molar ratio of photosynthetic capture of CO_2 (A) to transpirational loss of water (E). Then because both $\delta^{13}C$ and A/E are partly determined by the function c_i/c_a, measurement of $\delta^{13}C$ should provide a relative but, it must be stressed, not absolute index of water-use efficiency of foliage at the time of measurement.

Because of the relative nature of these relationships between $\delta^{13}C$ and E, $\delta^{13}C$ values for the photosynthetic products of leaves of the same or different species of C_3 plants are indicative of differences in instantaneous water-use efficiency only if the plants concerned are experiencing identical light, temperature and aerial humidity conditions over the time of measurement (see Condon *et al.* 1992). In general then, the more negative the $\delta^{13}C$ value of a plant, the poorer is water-use efficiency (see Farquhar and Richards 1984).

Table 1. Principal differences in carbon isotope discrimination, water-use efficiency in dry matter production, and ecophysiological characteristics of C_3, C_4, and CAM plants.

Plant type	$\delta^{13}C$ (‰)	WUE (mL H_2O transpired per g dry matter gain)	Ecophysiological characteristics
C_3 (e.g. wheat, lupins)	-20 to -34	400 – 600	• Use Calvin Cycle with Rubisco to fix CO_2 in leaf • Stomata open in day • Perform best at moderate temperatures and light intensities

C_4 (e.g. maize, sugar cane	-9 to -17	200 – 300	• Use PEP carboxylase to fix CO_2 in leaf
			• Stomata open in day
			• More efficient water users than C_3 plants especially at high temperature because:
			  1. PEP carboxylase is better scrubber of CO_2
			  2. C_4 plants lack photorespiration
			  3. C_4 plants less prone to photoinhibition
			  4. Better N use efficiency through lesser investment in photosynthetic enzyme

CAM/C3 (e.g. pigface, pineapple)	As for C_4 plants when operating in CAM mode, but close to C_3 species when in C_3 mode.	50 (when in CAM mode)	• Fix CO_2 at night with stomates open, using PEP carboxylase to form malate • Close stomates in day and convert malate to sugar • Succulent and very slow growing in CAM mode • Can convert to C_3 mode when water is available and salinity is reduced

1.1 $\delta^{13}C$ values and whole plant water-use efficiency

Additional complications arise whenever one attempts to use $\delta^{13}C$ values of the total carbon of plant dry matter to assess whole plant water-use efficiency (WUE). For example, let us consider two hypothetical plant species showing virtually identical $\delta^{13}C$ values for their dry matter, but very different water usage in terms of water transpired per unit dry matter gain. The first of the pair of species, of fast growth rate and living under high nutrient conditions, might for sake of argument show almost twice the WUE of the second, but the differences would be caused principally by the first species having lost only a small fraction of its net daily gain of carbon in maintenance respiration and activation of nutrient uptake. By contrast the second species, say a slow growing deep rooted one of low shoot:root ratio and living in a nutrient impoverished environment, might well be expending over half of its photosynthetic carbon gain in respiration, particularly for example in support of microbially-assisted uptake of nutrients by below ground biomass. Not surprisingly, this second species would then display a poorer WUE than the first.

Extending the above interpretation of whole plant water-use efficiency (WUE) further, one would expect annual species to show better WUE than perennials of similar stature and growth form and younger plants of woody tree and shrub species to be more efficient than older counterparts of the same species. Similarly, fire sensitive (seeder) species with high ratios of shoot to root biomass would be more efficient than fire resistant (resprouter) species with low shoot : root ratios.

Despite the above complications, it is still useful to examine quantitative relationships between $\delta^{13}C$ of whole plant carbon and total water transpired per unit dry matter gain (so called transpiration ratio). Examples of such data based on gravimetric estimates of water usage, dry matter gain and $\delta^{13}C$ analyses of pot grown plants under well watered conditions show considerable inter- and intra-specific differences in transpiration ratios and corresponding $\delta^{13}C$ values when cultured under a mediterranean-type climate such as experienced in Perth, Western Australia (see Pate and Dawson 1999). For example in the data shown in Figure 1, field peas compared under identical conditions in spring showed lower transpiration ratios and marginally less negative $\delta^{13}C$ values (i.e. lesser c_i/c_a ratios) in conventionally-leaved than semi-leafless genotypes (Armstrong *et al.* 1994). On the other hand, between season comparisons for Tasmanian blue gum (*Eucalyptus globulus*) and tagasaste (*Chamaecistus proliferus*) demonstrated considerably more efficient use of water by both species in winter than summer, but good evidence from their respective $\delta^{13}C$ values of tighter stomatal control (lesser values for c_i/c_a) under summer than winter conditions (Pate and Arthur 1998). Thus, whereas the $\delta^{13}C$ values obtained would suggest that stomatal mechanisms of both species had been highly effective in maximising leaf carbon gain per unit water loss during summer conditions (i.e. low c_i/c_a ratios), the parallel assessments of water-use efficiency clearly showed that such adjustments had failed by a considerable margin to offset the excessively high water loss expected under hot and dry conditions of summer (Pate and Dawson 1999).

2. TYPES OF PLANT MATERIALS WHICH MAY BE SAMPLED FOR CARBON ISOTOPE ANALYSIS

Types of plant material typically used for $\delta^{13}C$ analysis have ranged from (a) simple bulk assays on total plant dry matter, (b) dry matter of defined plant parts such as roots, stems, specific age groups of leaves and growth rings in wood, and (c) the recently developed approach of using isotope discrimination of the translocated carbon recovered from phloem exudates. As shown in the sections which follow, the rationale behind which component is selected for assay depends very much on the type of information which one would hope to gain and the relevance of such information to the objectives of the study which is being undertaken.

2.1 $\delta^{13}C$ of total carbon of a plant

Firstly, assaying the bulk carbon of a whole plant would be fully justified where one is simply wishing to determine the photosynthetic pattern to which a plant species is conforming over the time frame in which it has been growing. Bulk measurements of this kind would thus enable one to readily distinguish C_3 from C_4 species or C_3 species from those operating in CAM mode (see Ehleringer 1989, Boutton 1991). Similarly, for facultative C_3:CAM species, the closeness or otherwise of the C_4-like $\delta^{13}C$ signal to that expected of the species in nocturnally-mediated CAM mode would indicate the relative extent over the life of the plant to which this photosynthetic attribute had been operating relative to a conventional diurnally operating C_3 pattern of photosynthesis.

Secondly, bearing in mind that $\delta^{13}C$ signals of C_3 species should provide integrated measures of c_i/c_a and are hence related to mean instantaneous water-use efficiency across a broad time scale, differences in signals between genotypes of the one species grown under identical conditions can be justifiably used to assess relative effectiveness in usage of water. For example, this approach has been extended to broad scale evaluations of water use characteristics of parasitic plants and the hosts which they parasitize. Generally more negative $\delta^{13}C$ values for parasites denote more profligate use of water, as shown for example in a number of mistletoes relative to their hosts (see review of Pate 1995).

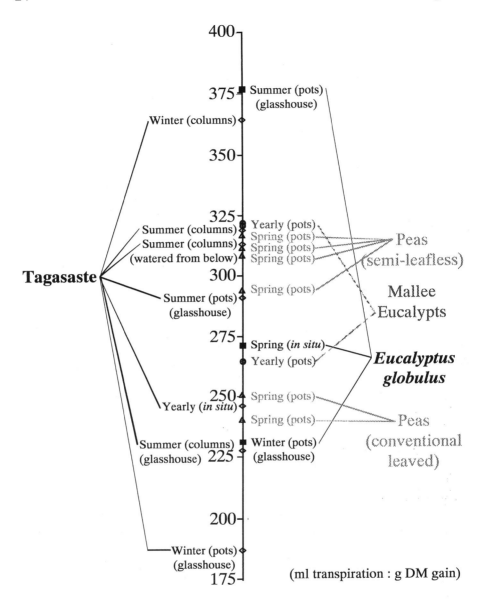

Figure 1. Experimental data comparing transpiration loss per unit whole plant dry matter gain in a range of C₃ species growing in the mediterranean-type environment of south west Western Australia (collated from Armstrong *et al.* 1994, Pate and Dawson 1999 and unpublished data of EC Lefroy, DJ Arthur and JS Pate).

There are, however, a number of problems with a simplistic approach based on total plant carbon. Firstly, if one is dealing with C₃ plants growing naturally outdoors or in a glasshouse, carbon isotope discrimination is likely to change significantly and progressively with season, say achieving a highly negative $\delta^{13}C$ value for dry matter laid down under cool moist conditions to a much less negative one when hot dry conditions are promoting more conservative stomatal behavior in terms of c_i/c_a characteristics. Secondly, following from the above, comparisons between or within species in terms of stomatal responses and presumed water-use efficiency will be valid only if time courses and relative amounts of carbon accumulated by all members of the study group of species are closely matched. Thirdly, even when making cross species comparisons under situations in which phenologies of carbon gain are closely similar, differences in $\delta^{13}C$ signals may still not reflect differences in stomatal behavior and water use unless the species concerned are of similar growth and life form (see comments in respect of $\delta^{13}C$ values for different life forms by Brooks *et al.* 1997). Finally, the situation may be further complicated due to isotope discrimination as secondary products are synthesized from photosynthetic products such as sugars (see Brugnoli *et al.* 1988). Thus, lignin and cellulose are typically of different $\delta^{13}C$ value from one another (see Sheu and Chiu 1995) and from the photoassimilated sugars from which they are derived, and $\delta^{13}C$ signatures of other major constituents such as lipids and nitrogenous solutes and protein may differ appreciably from that of the primary photosynthetic product of the species (e.g. see Gleixner *et al.* 1993). While such differences in $\delta^{13}C$ have been generally found to be relatively small, normally less than 1 – 2‰, they may still cause complications, for instance when comparing a woody plant rich in lignin with one rich in cellulose.

2.2 $\delta^{13}C$ assays of plant foliage

Regardless of whether all foliage or targeted leaves of a specific age group are harvested and assayed, interpretations of $\delta^{13}C$ values for leaf dry matter require detailed consideration of a number of issues. As a general rule for herbaceous annual species, each newly formed leaf is first fed with phloem translocate formed elsewhere in the plant after which it quickly graduates to a fully autotrophic condition and incorporates sufficient of its own photosynthate to satisfy its growth. Some of the early imported carbon might come from older adjacent leaves or it might be mobilized from carbohydrate reserves laid down in stems and roots from earlier episodes of photosynthesis. The latter situation would apply for instance, to nourishment of leaves forming on deciduous trees following bud break in spring, or say to the new coppice shoots sprouting from a woody species after aboveground parts had been destroyed by fire or cutting.

With the possibility that imported and autotrophically-generated carbon of a young leaf might carry different isotope signatures, use of $\delta^{13}C$ values of bulk leaf dry matter for assessing stomatal responses and instantaneous water-use efficiencies will obviously pose problems in interpretation. Nevertheless, for those cases in which it can be proven that the bulk of the structural fabric of the leaf has been derived from current photosynthesis of the leaf and adjacent older nurse leaves experiencing similar environmental conditions, errors due to differently discriminated sources of carbon should be relatively minor.

Once a leaf is mature and actively translocating carbon and also importing carbon through its xylem connection with the parent plant, further complications arise if these current two pools of carbon are discriminated from each other, or more importantly, appreciably different from the structural carbon accumulated earlier during the young life of the leaf. However, the soluble fraction currently produced in photosynthesis or cycling through the leaf would typically comprise only relatively small proportions of the total carbon resident in the leaf, so the leaf should still carry a $\delta^{13}C$ signal more closely related to photosynthetic conditions experienced during its early growth than are obtaining currently in the leaf.

Complications of potentially greater significance have been shown to relate to the site and micro-environment on a plant at which a particular leaf is developing when accumulating most of its structural carbon. For example, shaded regions of a tree canopy generally carry leaves whose bulk carbon may be several δ units more negative than in comparably aged leaves on well-illuminated regions of the same tree. This presumably reflects less stress in terms of temperature, light intensity and relative humidity in shaded than in well-illuminated sites in a canopy (see Garten and Taylor 1992, Waring and Silvester 1994, Hanba *et al.* 1997, Pate and Arthur 1998). Additional more subtle effects can relate to the height above ground at which leaves are developing, since closely spaced trees with poorly stirred lower air layers are likely to capture appreciable amounts of carbon dioxide coming from root and rhizosphere respiration. $\delta^{13}C$ signals will then become appreciably perturbed if the latter CO_2 sources carry a different isotope signal from the CO_2 of well stirred upper ambient atmosphere (see Medina and Minchin 1980, Buchmann *et al.* 1997, Berry *et al.* 1997, Schleser and Jayasiura 1985).

A final source of potential variation, again applicable especially to large woody species, stems from effects on leaf functioning caused by variations in hydraulic architecture between xylem pathways serving different regions of a leafy canopy. Thus, it has been shown that as a general rule slow-growing short shoots of conifers or side branches of other tree species show relatively poor hydraulic conductivity and, in therefore being more stressed, are likely to produce foliage of less negative signal than leaves sited on rapidly-extending main branches of better conductivity (Panek 1996, Walcroft *et al.* 1996 and 1997, Waring and Silvester 1994).

With all of the above in mind it is suggested that the following principles should be followed when sampling plant foliage between or within species.

1. Leaf samples should be of strictly comparable age within and between study plants. This is best validated by marking specific groups of leaves early in their development and restricting harvests to these tagged specimens after a preselected time interval judged to coincide with completion of their growth.

2. Leaves to be compared should be of strictly comparable position in relation to aspect and illumination received and preferably well above poorly stirred layers of air at lower elevations in the canopy.

3. Branching patterns of study plants should be examined carefully and where possible, foliage should be selected from each plant by restricting collections of leaves to branches of similar diameter and therefore presumably of similar hydraulic characteristics.

4. When foliage is being sampled during a period of seasonal extension growth followed by onset of water stress conditions, there is the likelihood that earlier formed leaves will carry significantly more negative (less stress) $\delta^{13}C$ signals than those formed later. One should then either obtain comparable composite samples of the foliage produced on extension growth of study plants over the interval of study or, more definitively, sample age sequences of leaves and plot changes in $\delta^{13}C$ signals against time of production. Figure 2 provides an example illustrating the latter approach for *Banksia prionotes*, a species in which leaves are retained on shoots for up to three years and each season's extension of the shoot and cohort of new leaves can be readily distinguished from one another. It can be seen that for the three seasons studied, $\delta^{13}C$ values for leaves successively laid down in the shoot growth increment for the year became progressively less negative. This indicated increasing stress as soil dried out with the approach of summer. Note also the generally lesser stress of new shoot growth following a wet winter season of 1995 and 1996–7 than in the dry 1997–8 season.

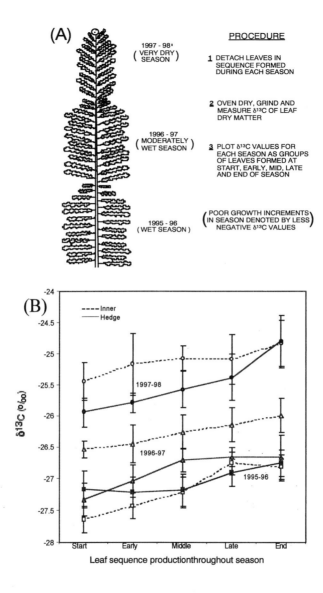

Figure 2. Assessment of δ^{13}C signals of sequences of leaves (A) formed on trees of *Banksia prionotes* over the successive growing seasons at Moora, Western Australia. Data in B marked as 'hedge' refer to trees sited at the interface with agriculture and in receipt of nutrient-laden runoff. Data denoted 'inner' refer to comparable trees sampled 150 m into the bush and out of range of influence of agriculture. Note greater stress (less negative δ^{13}C) for leaves produced at end than beginning of spring/summer growing season (October-February), consistently greater stress of inner than hedge trees in all seasons, and generally much greater stress of all trees in the very dry 1997–8 season than the two other wetter seasons (see Grigg, *et al.* (2000) for general characteristics of hedge and inner bush zones, δ^{13}C data from JS Pate and AM Grigg unpubl.).

2.3 $\delta^{13}C$ **signatures of growth rings of wood**

The rationale of this approach is that wood in xylem of trunks of woody species should be composed of carbon whose isotopic signatures reflect that of the photosynthate produced in the foliar canopy at the time when the wood in question was being laid down. Thus, any changes in c_i/c_a ratios during functioning of the leaves donating carbon to wood should be reflected in fluctuations in $\delta^{13}C$ value of the wood laid down within and between successive seasons. When analyses are conducted on trunks of old trees across several hundred years of annual rings, the technique offers promise as a means of studying response to climatic change and increasing CO_2 levels in the atmosphere, while also being used to identify effects of years or sequences of seasons of unusual climatic qualities on tree performance (eg. see studies of Bert *et al.* 1997, Leavitt 1992, Panek and Waring 1995, Sauer *et al.* 1995). Correlations between $\delta^{13}C$ of the wood of different years and the width of the rings produced over the same time frame can also be useful as a means of testing whether 'good' or highly 'stressful' years in terms of photoassimilate production affect only that year's growth or whether the benefits or penalties concerned spill over into subsequent seasons (see McNulty and Swank 1995).

As shown in a number of studies, difficulties in interpretation of data sets of the type alluded to above can occur if $\delta^{13}C$ analyses are restricted to carbon of bulk wood dry matter, since the $\delta^{13}C$ signals of cellulosic components of wood can be different by $1 - 2\text{‰}$ from those of lignin and other wood fractions. Furthermore the proportional amounts of these bulk constituents can vary appreciably, particularly between early and late wood produced during a season. Techniques for extracting specific wood fractions such as holocellulose are available should it be deemed advisable to assay specific components rather than bulk wood carbon (Leavitt and Danzer 1993, Sheu and Chiu 1995).

A recent example from the author's laboratory in which $\delta^{13}C$ signals of the bulk wood have been used to follow seasonal patterns of stress and differences between irrigated and non-irrigated treatments is shown for plantations of *Eucalyptus globulus* in Figure 3. It can be seen that the summer stress signal, corresponding to least negative $\delta^{13}C$ values for trunk wood, is replicated each summer in the rain fed plantation (Eulup site), but is almost completely suppressed once soil water stress had been combatted by a generous irrigation regime (Albany site).

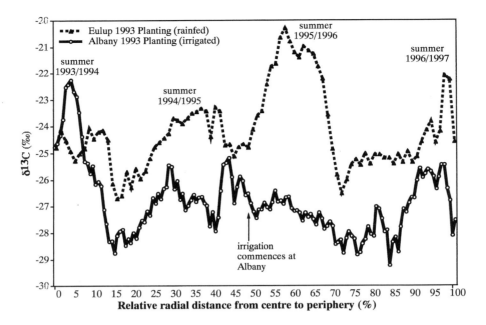

Figure 3. Comparison of $\delta^{13}C$ signals in bulk carbon of samples of wood taken progressively from the centre to the periphery of basal trunks of trees of Tasmanian bluegum (*Eucalyptus globulus*) from a rain fed (Eulup) and an irrigated (Albany) plantation. Note greater seasonal changes in $\delta^{13}C$ in the rain fed, summer droughted tree at Eulup than the irrigated one at Albany (data from Pate and Arthur 1998).

2.4   Use of $\delta^{13}C$ of phloem translocate to assess current stomatal responses and water-use efficiency of donor foliage

Application of this novel approach to $\delta^{13}C$ analysis requires access to experimental species (eg. *Lupinus, Ricinus, Banksia, Eucalyptus*) which bleed phloem sap when their phloem is cut (e.g. see Fisher 1983, Pate and Jeschke 1995, Pate and Arthur 1998). Samples of phloem exudate from petioles, stems or trunks of woody species can then be assayed for solutes and $\delta^{13}C$, and instructive inventories thereby obtained on a range of aspects of plant performance. For example in our recent studies on plantations of the Tasmanian blue gum (*Eucalyptus globulus*) in Western Australia (Pate and Arthur 1998), we have made monthly collections of phloem sap from the same dryland rain fed site (Eulup) and irrigated site (Albany) mentioned above in respect of growth ring analysis. The study provided evidence of nitrogen limitation of growth and $\delta^{13}C$ values in phloem indicative of very high levels of water stress during the dry summer at the rain fed site but not at the irrigated one. By contrast, in the winter when soil was well wetted and mineralization of N in full swing, phloem sap N levels and $\delta^{13}C$ values were closely similar for the two sites, with evidence of trees not being short of either nutrients or water.

Both sets of $\delta^{13}C$ values for phloem sap were then matched against the $\delta^{13}C$ values of recently formed xylem tissue scraped off the inner cambial face of the trunk. As shown in Figure 4, the collections of phloem sap and newly formed xylem of trunks of two age groups of trees across a season at the rain-fed Eulup site, showed close correlations between $\delta^{13}C$ values for the two entities, but evidence of seasonal peaks or troughs in $\delta^{13}C$ value of phloem being reproduced a month or so later in the recently formed wood. This is the sort of time delay one would expect between trunk phloem assimilates to pass through to the cambium and become incorporated into dry matter of new xylem (see Pate and Arthur 1998).

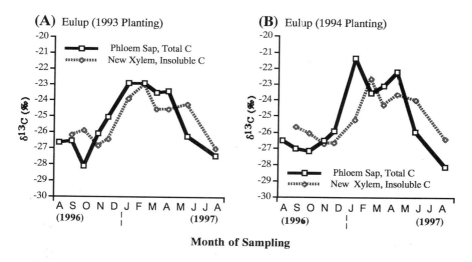

Figure 4. Relationships between $\delta^{13}C$ signals of current phloem sap total carbon and the insoluble total carbon of xylem recently scraped from the inner face of the cambium of two plantings of *Eucalyptus globulus* from a rain fed site at Eulup, Western Australia. See text and Figure 6 for rationale for this sampling approach and the results which it would be expected to generate (data from Pate and Arthur 1998).

In the above-mentioned comparison between plantations of *E. globulus* the peak summer stress value for the rain-fed plantation at Eulup averaged at -20‰ compared to -26‰ for the irrigated plantation. These values translate to c_i/c_a ratios of 0.36 and 0.63 respectively, which at the prevailing mean temperature and vapour pressure deficit conditions typically experienced at this time would suggest a much better instantaneous water-use efficiency for the rain fed (3.5 mmol C. mol H_2O) than irrigated site (2.0 mmol C. mol H_2O).

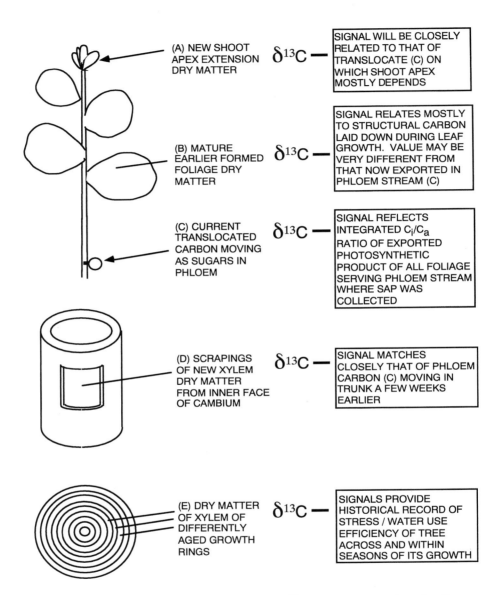

Figure 5. Protocols for sampling of a woody plant for $\delta^{13}C$ analysis of currently and earlier formed pools of assimilated carbon. The scheme depicts relationships expected between the various entities sampled. Examples of relationships between $\delta^{13}C$ of phloem and recently laid down carbon in trunk xylem are shown in Figure 4, and between phloem $\delta^{13}C$ and $\delta^{13}C$ of bulk dry matter of plant parts in Figure 6.

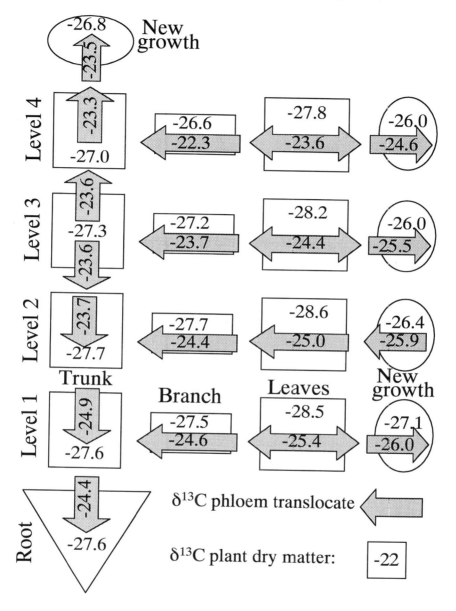

Figure 6. Interrelationships of $\delta^{13}C$ values for total carbon of phloem sap intercepted at different points in the translocatory system of 2-year old trees of *Eucalyptus globuius* and the corresponding $\delta^{13}C$ signals of dry matter total carbon of the various parts rated as donating, transporting and consuming these pools of translocate. Directions of flow of translocate are shown as derived in a recent study (Pate and Arthur 1998) describing the partitioning and utilization of carbon by the same trees during a 19 day study period in November (see text for interpretation of data).

3.   OVERVIEW OF δ ^{13}C SIGNALS IN PLANT COMPONENTS - AN INTEGRATED APPROACH TO RATIONALIZATION OF SAMPLING PROTOCOLS

The scheme presented in Figure 5 summarizes the different sampling entities which may be used when assessing the isotope discrimination status of the carbon of a woody plant such as *Eucalyptus globulus*. The type of information which each class of sample should provide is detailed, including how it might relate in isotope signal to that of other components of the system. The relationships between phloem $\delta^{13}C$ signals and those of newly formed xylem of trunk wood has already been alluded to in an earlier section (see Fig. 4). Figure 6 depicts a typical outcome when phloem $\delta^{13}C$ signals from translocate collected from different parts of a tree are compared with the $\delta^{13}C$ of the dry matter of the various sources and sinks for that translocated carbon. The data come from analysis of samples from 2-year old trees of *E. globulus* studied over a 19 day period in November 1999 (Pate and Arthur 1998). The data clearly show that phloem translocate originating from upper more-stressed parts of the trees carry $\delta^{13}C$ signals significantly less negative than those emanating from shaded older parts of the canopy. Corresponding $\delta^{13}C$ values for dry matter of root and various levels of trunk, side branches and leaves, including dry matter of most recently expanded shoot parts, are in all cases more negative than phloem carbon currently passing through or being incorporated in growth. This is because most of the carbon was laid down earlier in the year when trees were much less water stressed and in much more active growth than at the onset of summer stress when the samples of phloem translocate were collected.

4.   SUMMARY

Broadly based sampling procedures used in conjunction with one another, allow one not only to assess the current physiological status of plants of a species, but also to read back into its past history to appreciate how it has responded over time to various environmental effects, especially in relation to water use and water and nutrient status. With such refinements one can hope to gain a much more comprehensive picture than can ever be realized from indiscriminate sampling of dry matter of mixed origin and unprescribed phenology. The reader is encouraged to adopt and expand upon more definitive approaches of this kind in future investigations.

REFERENCES

Armstrong, E. L., Pate, J. S., and Tennant, D. (1994). The field pea crop in south western Australia Patterns of water use and root growth in genotypes of contrasting morphology and growth habit. *Australian Journal of Plant Physiology* 21, 517-532.

Benner, R., Fogel, M.L., Sprague, E.K., and Hodson, R.E. (1987). Depletion of $\delta^{13}C$ in lignin and its implications for stable carbon isotope studies. *Nature* 329, 708-710.

Berry, S.C., Varney, G.T., and Flanagan, L.B. (1997). Leaf $\delta^{13}C$ in *Pinus resinosa* trees and understory plants: variation associated with light and CO_2 gradients. *Oecologia* 109, 499-506.

Bert, D., Leavitt, S.W., and Dupouey, J.L. (1997). Variations of wood $\delta^{13}C$ and water-use efficiency of *Abies alba* during the last century. *Ecology* 78, 1588-1596.

Boutton, T.W. (1991) Stable carbon isotope ratios of natural materials. II. In 'Carbon Isotope Techniques.' (Eds D.C. Coleman and B. Fry.) pp. 173-185. (Academic Press: San Diego.)

Brooks, J.R., Flanagan, L.B., Buchmann, N., and Ehleringer, J.R. (1997). Carbon isotope composition of boreal plants: functional grouping of life forms. *Oecologia* 110, 301-311.

Brugnoli, E., Hubick, K.T., von Caemmerer, S., Wong, S.C., and Farquhar, G.D. (1988). Correlation between the carbon isotope discrimination in leaf starch and sugars of C_3 plants and the ratio of intercellular and atmospheric partial pressures of carbon dioxide. *Plant Physiology* 88, 1418-1424.

Buchmann, N., Guehl, J.M., Barigah, T.S., and Ehleringer, J.R. (1997). Interseasonal comparison of CO_2 concentrations, isotopic composition, and carbon dynamics in an Amazonian rainforest (French Guiana). *Oecologia* 110, 120-131.

Condon, A.G., Richards, R.A., and Farquhar, G.D. (1992). The effect of variation in soil water availability, vapour pressure deficit and nitrogen nutrition on carbon isotope discrimination in wheat. *Australian Journal of Agricultural Research* 43, 935-947.

Deines, P. (1980). The isotopic composition of reduced organic carbon. In ' Handbook of Environmental Isotope Geochemistry.' (Eds P. Fritz and J. Fontes.) pp. 329-406. (Elsevier: Amsterdam.)

Ehleringer, J.R. (1989) Carbon isotope ratios and physiological processes in arid-land plants. In 'Stable Isotopes in Ecological Research.' (Eds P.W Rundel, J.R. Ehleringer, and K.A. Nagy.) pp. 41-54. (Springer: Berlin, Heidelberg, New York.)

Farquhar, G.D., Ehleringer, J.R., and Hubick, K.T. (1989). Carbon isotope discrimination and photosynthesis. *Annual Review of Plant Physiology and Plant Molecular Biology* 40, 503-537.

Farquhar, G.D. and Richards, R.A. (1984). Isotopic composition of plant carbon correlates with water-use efficiency of wheat genotypes. *Australian Journal of Plant Physiology* 11, 539-552.

Farquhar, G.D., Hubick, K.T., Condon, A.G., and Richards, R.A. (1988). Carbon isotope fractionation and plant water use efficiency. In 'Stable Isotopes in Ecological Research.' (Eds P.W Rundel, J.R. Ehleringer, and K.A. Nagy.) pp. 21-40. (Springer: Berlin, Heidelberg, New York.)

Fisher, D.B. (1983). Year-round collection of willow sieve-tube exudate. *Planta* 159, 529-533.

Garten, C.T. Jr. and Taylor, G.E. Jr. (1992). Foliar $\delta^{13}C$ within a temperate deciduous forest: spatial, temporal, and species sources of variation. *Oecologia* 90, 1-7.

Gleixner, G., Danier, H.J., Werner, R.A., and Schmidt, H.L. (1993). Correlations between the $\delta^{13}C$ content of primary and secondary plant products in different cell compartments and that in decomposing basidiomycetes. *Plant Physiology* 102, 1287-1290.

Grigg, A.M., Pate, J.S., and Unkovich, M.J. (2000) Responses of native woody taxa in *Banksia* woodland to incursion of groundwater and nutrients from bordering agricultural land. *Australian Journal of Botany* 48, 777-792.

Hanba, Y.T., Mori, S., Lei, T.T., Koike, T., and Wada, E. (1997). Variations in leaf $\delta^{13}C$ along a vertical profile of irradiance in a temperate Japanese forest. *Oecologia* 110, 253-261.

Leavitt, S.W. (1992). Seasonal changes $^{13}C/^{12}C$ in tree rings: species and site coherence, and a possible drought influence. *Canadian Journal of Forest Research* 23, 210-218.

Leavitt, S.W. and Danzer, S.R. (1993). Method for batch processing small wood samples to holocellulose for stable-carbon isotope analysis. *Analytical Chemistry* 65, 87-89.

McNulty, S.G. and Swank, W.T. (1995). Wood $\delta^{13}C$ as a measure of annual basal area growth and soil water stress in a *Pinus strobus* forest. *Ecology* 76, 1581-1586.

Medina, E. and Minchin, P. (1980). Stratification of $\delta^{13}C$ values of leaves in Amazonian rain forests. *Oecologia* 45, 377-378.

O'Leary, M.H. (1998). Carbon isotopes in photosynthesis. *Bioscience* 38, 328-336.

Panek, J.A. (1996). Correlations between stable carbon-isotope abundance and hydraulic conductivity in Douglas-fir across a climate gradient in Oregon, USA. *Tree Physiology* 16, 747-755.

Panek, J.A. and Waring R.H. (1995). Carbon isotope variation in Douglas-fir foliage: improving the $\delta^{13}C$-climate relationship. *Tree Physiology* 15: 657-663.

Pate, J. S. (1995). Functional attributes of angiosperm hemiparasites and their host and predictions of possible effects of global climate change on such relationships. In 'Global Change and Mediterranean-type Ecosystems.' (Eds J. M. Moreno and W. C. Oechel.) pp. 161-180. (Springer Verlag: New York.)

Pate J.S. and Arthur, D. (1998). $\delta^{13}C$ analysis of phloem sap carbon: novel means of evaluating seasonal water stress and interpreting carbon isotope signatures of foliage and trunk wood of *Eucalyptus globulus*. *Oecologia* 117, 301-322.

Pate, J.S. and Dawson, T.E. (1999). Assessing the performance of woody plants in uptake and utilization of carbon, water and nutrients. *Agroforestry Systems* 45, 245-275.

Pate, J.S. and Jeschke, W.D. (1995). Role of stems in transport, storage and circulation of ions and metabolites by the whole plant. In 'Stems and Trunks: Their Roles in Plant Form and Function.' (Ed B. Gartner.) pp 177-204. (Academic Press: New York.)

Sauer, M., Siegenthaler, U. and Schweingruber, F. (1995). The climate-carbon isotope relationship in tree rings and the significance of site conditions. *Tellus* 47B, 320-330.

Schleser, G.H. and Jayasiura, R. (1985). $\delta^{13}C$ -variations of leaves in forests as an indication of reassimilated CO_2 from the soil. *Oecologia* 65, 536-542.

Sheu, D.D. and Chiu, C.H. (1995). Evaluation of cellulose extraction procedures for stable carbon isotope measurement in tree ring research. *International Journal of Environmental and Analytical Chemistry* 59, 59-67.

Teeri, J.A. (1982). Photosynthetic variation in the Crassulaceae. In ' Crassulacean Acid Metabolism.' (Eds I.P. Ting and M. Gibbs.) pp. 244-259. (*American Society of Plant Physiologists*: Rockville, Maryland.)

Walcroft, A.S., Silvester, W.B., Grace, J.C., Carson, S.D., and Waring, R.H. (1996). Effects of branch length on carbon isotope discrimination in *Pinus radiata*. *Tree Physiology* 16, 281-286.

Walcroft, A.S., Silvester, W.B., Whitehead, D., and Kelliher, F.M. (1997). Seasonal changes in stable carbon isotope ratios within annual rings of *Pinus radiata* reflect environmental regulation of growth processes. *Australian Journal of Plant Physiology* 24, 57-68.

Waring, R.H. and Silvester, W.B. (1994). Variation in foliar $\delta^{13}C$ values within the crowns of *Pinus radiata* trees. *Tree Physiology* 14, 1203-1213.

Chapter 3

Extraction and Analysis of Plant Water for Deuterium Isotope Measurement and Application to Field Experiments

J.V. Turner, P. Farrington and V. Gailitis
CSIRO Land and Water, P.O. Private Bag 5, Wembley, WA 6913 Australia.

Key words: deuterium measurement, sap water, plant water uptake, water sources

1. INTRODUCTION

Analysis of stable isotopes in water is well established as a powerful tool for investigating processes in plant-water relations such as water use efficiency and determining the source of water uptake in the soil zone. Deuterium and oxygen-18 have been used to determine the source of water taken up by plants and to study the functioning of deep root systems (White *et al.* 1985, Dawson and Ehleringer 1991, Ehleringer and Dawson 1992, Thorburn and Walker 1993 and Thorburn *et al.* 1993a and b). Provided there is sufficient difference in isotopic composition between water in the soil profile and groundwater, the stable isotopic composition of water in the xylem sap of the plant can be used to define the mixing ratio of these sources of water for the plant (see Dawson and Pate 1996). The method is based on the finding that the isotopic composition of both hydrogen and oxygen in water is not altered when taken up by roots and transported from roots to the leaves (Zimmerman *et al.* 1967, White *et al.* 1985 and Turner *et al.* 1987).

Hydrogen isotope ratios are measured by mass spectrometry after the hydrogen in water has been reduced to molecular hydrogen over zinc (Coleman *et al.* 1982). Conventional methods for δ^2H measurement require free water which must be extracted from plant tissues by squeezing (White *et al.* 1985), immiscible fluid displacement (Whelan and Barrow 1980),

37

M. Unkovich et al. (eds.),
Stable Isotope Techniques in the Study of Biological Processes and Functioning of Ecosystems, 37–55.
© 2001 *Kluwer Academic Publishers. Printed in the Netherlands.*

cryogenic vacuum distillation (Ehleringer and Osmond 1989), azeotropic distillation using toluene (Revesz and Woods 1990) or kerosene (Thorburn *et al.* 1993a) as a solvent. Each of these methods is time consuming, can present extraction difficulties where the water content is low, and may lead to isotope fractionation where phase changes occur during the extraction procedure.

This chapter describes a single-step method for analysis of the hydrogen isotope ratio of sap water in plant material and methods for preparation and storing of plant samples prior to analysis. It demonstrates the validity of the method using material from the eucalypt tree species, jarrah (*Eucalyptus marginata*), and describes a tracer study where isotopically enriched water was applied to the ground surrounding a single jarrah tree.

The standard zinc reduction technique for producing hydrogen for isotopic analysis of free water (Coleman *et al.* 1982) was modified to a single step method for δ^2H measurement on water in porous media (Turner and Gailitis 1988). The novel feature of the present method is the adaptation of the method for porous media to plant material. The method has the advantage that the extraction of water from the plant material is carried out via a microdistillation procedure and the reduction of the extracted water to hydrogen gas for isotopic analysis is achieved in a single step within a sealed reaction vessel.

Less than 30 μL of water from the plant material is required, so the method could be used to determine fine-scale isotopic variations in plant material. For example, δ^2H values can be determined in seedlings, and the movement of deuterium across an individual leaf monitored using small discs of leaf material punched out of the leaf along a transect. Because of the efficiency of the method, a greater number of replicate analyses of samples from the plant tissue are possible and excess material can be stored for later analysis if required. Analysis of $\delta^{18}O$ in plant tissue water is not possible by this method.

The method is faster and less likely than other reported methods to cause artefacts due to isotopic fractionation because phase changes (liquid-vapour-liquid) during extraction of the plant water are avoided during sample preparation. It is possible to analyse up to 30 samples per working day, with the principal limitation being the number of reaction vessels which can be accommodated in the heating block and the time required to microdistill and reduce water from the twig in the reaction vessel. Sample preparation is more straightforward than other methods and the single-step method also enables the moisture content of the plant tissue sample to be measured.

2. PROCEDURE FOR THE SINGLE-STEP MICRO-DISTILLATION METHOD

2.1 Sample weight

The microdistillation technique requires a volume of between 5 and 20µL water to be distilled from the sample of plant material and therefore its water content determines the weight of plant material required. In the example considered here, the plant material used comprises small, woody twigs (< 5mm diameter) taken from sub branches of a jarrah (*Eucalyptus marginata*) tree. After collection in the field the twigs are immediately sub-sectioned into 30mm lengths, wrapped in thin polythene film ('cling wrap') and stored frozen at -16°C to prevent water loss. Later, after removing the film the moisture content of representative material is determined by drying in an oven at 70°C for 24 hours. This moisture content then determines the weight of plant material required in the microdistillation reaction.

2.2 Deuterium analysis

Development of the microdistillation method for the analysis of δ^2H of water in plant material follows the classic method described by Turner and Gailitis (1988) for deuterium analysis of water in porous media. The sample of plant material is cut to the required weight as determined above and placed in a small, pre-weighed glass sample-holding tube that is quickly transferred to a sidearm in a reaction vessel containing zinc shot (Fig. 1). The reaction vessel is pre-prepared by evacuation on a high-vacuum manifold that allows back filling of the vessel with dry nitrogen to atmospheric pressure. Once the sample holding tube containing the sample is placed in the sidearm of the nitrogen-filled reaction tube, it is resealed via a high-vacuum stopcock and the sidearm then frozen with liquid nitrogen. Once the sample is frozen, the reaction vessel is re-evacuated on the vacuum manifold to remove the dry nitrogen. The reaction vessel is removed from the vacuum manifold and placed in an aluminium heating block such that the zinc shot in the base of the reaction vessel is heated to 450°C. The sidearm is positioned above the heating block in a position that allows convection from the hot surface of the heating block to maintain the sidearm temperature at approximately 145°C. During this stage, microdistillation and reduction of water vapour with zinc takes place with the reaction time depending on the type of plant material. Following reaction and cooling of the reaction vessel to room temperature, the hydrogen is transferred by expansion under its own pressure into a gas sample bottle. Sample bottles are attached directly to the inlet manifold of the mass spectrometer and the δ^2H of the hydrogen is determined using a normal analytical procedure.

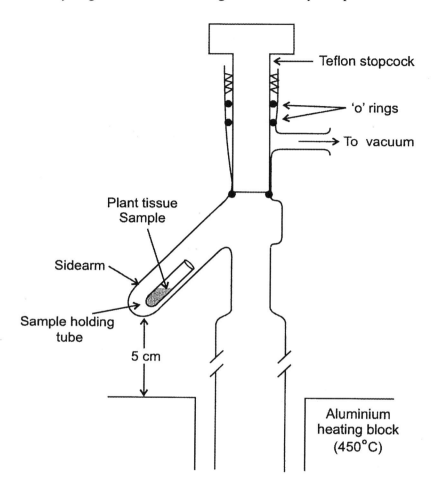

Figure 1. Design of a glass reaction tube for reduction of water to hydrogen by a single step microdistillation (after Turner and Gailitis 1988).

Isotope ratios are measured on a *VG SIRA 9* mass spectrometer. Results are expressed in standard delta notation (δ^2H) in parts per thousand (‰) relative to Vienna Standard Mean Ocean Water (*V-SMOW*) where

$$\delta^2 H = ((\frac{R_{sample}}{R_{V\text{-}SMOW}}) - 1) \times 1000 \text{‰} \tag{1}$$

and R is the ratio of the heavy and light isotopes. Precision of the analysis is estimated to be ± 1‰.

The weight of plant material placed in the glass sample tube is measured prior to reaction. After analysis the sample holding tube, containing the now dry plant material, is recovered from the sidearm and reweighed. Efficiency of water extraction is estimated by comparing the weight loss in the reaction tube to the weight loss at 70°C for 24 hours in a drying oven. A further check on the extent of reaction completion (extraction of water from the plant material and its conversion to hydrogen) can be obtained by comparing the yield of hydrogen from the reaction to the mass spectrometer ion beam strength for hydrogen in the calibrated volume of the mass spectrometer inlet system.

2.3 Determination of reaction time

The length of time required for distillation of water from samples within the sidearm of the reaction vessel depends on the type of sample material being analysed. In the case of twig material, for example, the reaction time required for quantitative extraction of water from plant material can be determined as follows. Samples are obtained from a single twig by cutting short sections (5 mm lengths) from the centre of the twig, and placing them in separate sample holding tubes. It is assumed that the isotopic composition of sap water is constant throughout a short section of twig. Replicate samples are analysed with the side arm heated by convection to 110°C from the heating block (at 450°C) for 100 minutes, 120 minutes and 140 minutes and determining the δ^2H of the hydrogen gas. After analysis, the twig samples are removed from the sidearm, reweighed and the moisture contents determined. The moisture content of the remaining sections of twigs is measured by drying in a conventional drying oven at 70°C for 24 hours. This enables comparison with moisture contents determined via the microdistillation method.

In the study highlighted here, measured moisture content of twigs averaged 48.4% so the sample weight required for analysis is about 30 mg. Table 1 shows that when the water content of twigs is distilled and reacted for 120 minutes the δ^2H is less depleted and more water is extracted than for the other times.

The effect of storage time between sampling and analysis on measured δ^2H was determined. The effect of short-term storage was assessed by collecting twigs and leaves from the same branch of the experimental jarrah tree, wrapping them in thin polythene film ('cling wrap') and storing at 0°C. One set of samples was analysed 0.5 h after collection and the second four hours later. The effect of longer-term storage was determined by reanalysing twigs that had been wrapped in polythene film in sealed vials at -16°C and stored for periods ranging from 309 to 535 days. In all analyses the samples were sectioned from the central part of the twig and, after removing the bark, placed in the sample holding tube. Analysis was carried out using a reaction time of 120 minutes.

Table 1. Effect of distillation time on amount of water extracted and δ^2H values in twigs.

Distillation Time (min)	Water content of twigs (% initial weight)	δ^2H in twig sap (‰)
100	53.6	-42.0
120	54.8	-35.3
140	54.5	-41.2

Storage of samples is shown to have little effect on the δ^2H of twig sap (Tables 2 and 3) since the δ^2H is not significantly affected when twigs are stored at 6°C for 4 hours and at -16°C for periods up to about one and a half years. Thus, collection and long-term storage of plant material under the described conditions is possible without affecting the isotopic composition of the sap water.

Table 2. Effect of storage time on amount of water extracted and δ^2H content of twigs using a distillation time of 120 minutes..

Storage time (h)	Mean water content of twigs (% initial weight)	Mean δ^2H in twig sap (‰)
0.5	49.8 (1.5)	-35.8 (-0.9)
4	53.8 (0.8)	-36.7 (-2.2)

Values in brackets are standard errors of mean (n = 4)

The methods for collecting and storing twigs have been developed to prevent evaporative loss and associated isotopic artefacts. Immediately after collection, samples should be wrapped tightly in several layers of plastic film and placed in an airtight vial. As soon as practicable, samples should be frozen as low temperature storage ensures that the sample is preserved without evaporative water loss. Before analysis, the sample must be thawed progressively while remaining wrapped in plastic film to prevent contamination from condensation of atmospheric water vapour on the sample.

Table 3. Effect of long-term storage time on δ^2H of sap in twigs using a distillation time of 120 minutes and a reaction temperature of 145°C.

Time between collection and analysis (Days)	δ^2H in fresh twig (‰)	δ^2H in stored twig (‰)	Difference between twigs
535	-20.90	-20.85	0.05
493	-19.20	-19.71	0.51
456	-22.50	-21.47	1.03
342	-23.56*	-20.61*	2.95
309	-22.86*	-23.73*	0.87

* Mean of 4 twigs

2.4 Method validation: isotopic recovery experiments

The analytical method was tested by imbibing small, live jarrah shoots from the experimental tree with water of known δ^2H and then measuring the δ^2H of the sap water as described above. Two experiments were carried out. In the first, water of two different δ^2H values was used; one a tap water (δ^2H = -6.3 ‰) and the other an isotopically depleted water (δ^2H = -88.0 ‰). In the second experiment, water with three different δ^2H values was used; an isotopically enriched water (δ^2H = +90.7‰), tap water (δ^2H = -10.8‰) and an isotopically depleted water (δ^2H = -112.5‰).

Jarrah shoots from a single location in the experimental tree with similar numbers of leaves were cut from the tree under water to prevent air entrapment. Four shoots were retained for analysis. The remaining shoots were placed in individual polycarbonate vials (140mL capacity) containing water of known δ^2H. After sealing to prevent evaporation, the vials and shoots were weighed and placed in a glasshouse at 0900h. Twenty-four hours later the vials were reweighed to determine the water use of the shoots and one set of shoots was harvested for analysis. A twig and leaf were sampled from each shoot using the procedure described above. Make-up water with the same δ^2H was added to the vials of the remaining set of shoots, the vials were reweighed and returned to the glasshouse. The same sampling procedure was repeated after 24 and 48 hours (Exp. 1 only).

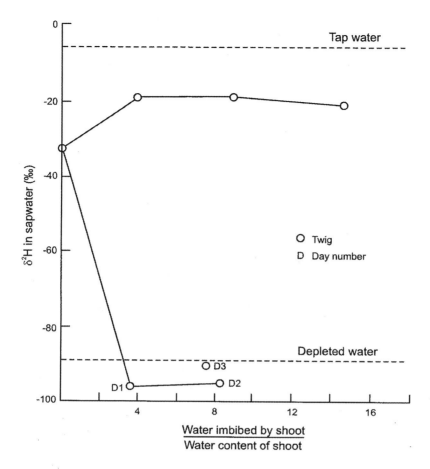

Figure 2. Changes in δ^2H values of sapwater in stems of shoots following imbibition of water of two δ^2H values plotted against the water turnover in the shoot for the first imbibing experiment.

The initial δ^2H of sap water within the shoots in the first imbibing experiment averaged -33.6 ‰ ($\sigma = 0.83$, n = 4). Once the experiments commenced, the δ^2H of the shoots changed rapidly (Fig. 2). Within 24 hours, the δ^2H values were close to that of the feed water whereupon they remained constant for the next two days. The rate of water throughput in the shoot amounted to about one pore volume per half day (Fig. 2). By the third day water uptake by the shoots was highly variable, possibly due to blockage in the xylem at the cut end of the shoot. On both days the δ^2H values of the sap water within the shoots was lower than that of the feed water. For tap water, the sap water within the shoots was about 13‰ lower and for the depleted water about 8.3‰ lower. The probable explanation for this was that the amount of water extracted from the shoots during the reaction was less than that from oven drying (Fig. 4). This suggests that the temperature of the sidearm of the reaction vessel was too low to obtain complete distillation of the water from the plant material. Incomplete extraction of water from the sample by distillation leads to δ^2H values less than the true value in twigs (Thorburn *et al.* 1993a) and soils (Turner and Gailitis 1988, Walker *et al.* 1991). Therefore, in the second experiment the temperature of the sidearm during the distillation and reaction period was increased from 110°C to 145°C by shielding the sidearm of the reaction vessel within a metal enclosure, thereby increasing the heating effectiveness of convected heat from the heating block (Fig. 1).

Before conducting the second imbibing experiment, the δ^2H compositions of leaves and twigs were +35.4‰ and -26.1‰ respectively, showing the expected enrichment in δ^2H in the leaves due to evaporation (Leaney *et al.* 1985). Two days after the imbibing experiments began; δ^2H values approached their respective feed waters (Fig. 3). The greatest difference was for the enriched water where the twigs were about 10‰ lighter than the feed water. Differences between the other treatments were insignificant (3.8‰ for tap water and 1.2‰ for depleted water). Leaf water showed less response to the imbibing experiments than did sap water of twigs. The δ^2H values of the leaves were more enriched for the tap and depleted water, but more depleted for the enriched water treatments (Fig. 3).

The amount of water extracted by distillation was in close agreement to oven drying (Fig. 4), indicating that the modification used to heat the sidearm of the reaction achieved a more complete distillation.

These experiments demonstrate that reliable δ^2H compositions of sap water in twigs of eucalypts can be obtained using the single step method provided the sample weights are less than 30 mg, the reaction time is 2 hours and the sidearm temperature is 145°C.

In the study reported here the procedure for analysing δ^2H in sap has been developed for eucalypt twigs. The specification endorsed may not necessarily be suited to other plant material of interest, so tests should be conducted on each type of plant material to ensure that the reaction time and the temperature required to heat the sidearm are sufficient for complete extraction of water from plant tissues.

Other methods for extracting free water from plant tissue such as azeotropic distillation; vacuum distillation and sap expression are discussed in the chapter in this volume by Walker *et al.*

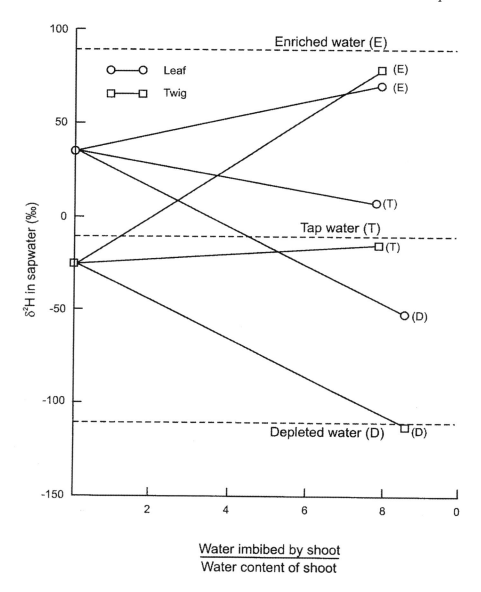

Figure 3. Changes in δ^2H values of sapwater in leaves and stems of shoots following imbibition of water at three δ^2H values plotted against the water turnover in the shoot for the second imbibing experiment.

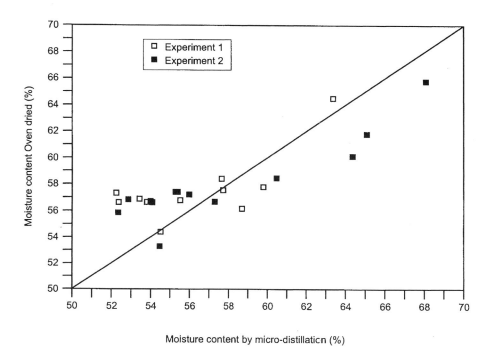

Figure 4. Scatter plot showing (a) the moisture content of twigs determined by oven drying at 70°C for 24 hours and by microdistillation drying under vacuum for 2 hours for the first imbibing experiment (Experiment 1), and (b) relationship between moisture content of twigs determined by oven drying at 70°C for 24 hours and by microdistillation drying under vacuum at 145°C for 2 hours for the second imbibing experiment (Experiment 2).

3. INTERPRETATIONAL APPROACHES TO DETERMINE PLANT WATER SOURCES

Isotopic studies of plant water sources require selection of an appropriate methodology that will provide solutions to a particular issue. For example, studies of ecosystem dependence on groundwater have become prevalent in recent years and thus determining whether trees are dependent on groundwater resources has become a common and practical application of isotope techniques. Such applications require characterisation of both temporal and spatial variation of the isotopic composition of potential source waters. Thus in designing a field program some *a priori* knowledge of potential source waters and their isotopic composition is required and the temporal and spatial sampling frequency at which they should be sampled. Information from a single tracer such as δ^2H is of course essential, but data on a second and independent tracer will increase the level of discrimination between water sources that can be achieved. Because δ^2H and $\delta^{18}O$ are strongly correlated one with the other via the well-known meteoric water line, even when source waters have undergone evaporation, the additional information provided by $\delta^{18}O$ is limited. Lin *et al.* (1996) for example, circumvent this limitation by use of tracer water that was isotopically enriched in deuterium but not oxygen-18. A similar approach is described below in relation to water uptake experiments by *Eucalyptus marginata*. Two and three component source identification models that can be used to analyse isotopic water uptake results are presented below.

It is clear from previous studies that isotopic data at varying degrees of spatial and temporal resolution are needed to achieve useful results using isotope techniques. It is unlikely that a single reconnaissance-level sampling of plant material and potential water source candidates at one time step will yield interpretable results. Such data are more likely to be of use in broader scale design and methodologies for such studies.

3.1   Two and three component isotope mixing relations.

Two or three component source mixing models can be applied to stable isotope data in plant tissue with the objective of determining the relative contributions of the potential water sources to the xylem-water of the study species. Two component source models have been used by White *et al.* (1985) and Thorburn and Walker (1993).

A three-component source model can be developed as follows. Firstly, the well-known two-component isotope source model can be written as:

$$V_1/V_2 = \frac{(\delta_3 - \delta_2)}{(\delta_1 - \delta_3)} \qquad (2)$$

Where V_1 and V_2 are the source volumes of components *1* and *2* respectively making up the mixed component *3*. In this case component *3* corresponds to the xylem twig water and V_1/V_2 is the ratio of volumetric components making up V_3. The δ values of components 1, 2 and 3 are the measured δ^2H values of the two source components *1* and *2* and the xylem twig water *3*. This simple model can be used where the isotopic composition of the xylem twig water can be resolved in terms of two components, for example soil water and groundwater. In some circumstances this may be an oversimplification and consideration of a third source component may be required. In cases where a second independent tracer is available, the mixing equation can be solved for three components (see e.g. Lin *et al.* 1996, where water artificially enriched in deuterium was introduced to generate a second and independent tracer). Natural abundance oxygen-18 can potentially be used as this second tracer. However, in many instances it is expected that δ^2H and $\delta^{18}O$ will be highly correlated via the relation between ^2H and ^{18}O. Thus because $\delta^{18}O$ is generally not independent of δ^2H due to the meteoric water line relationship between δ^2H and $\delta^{18}O$, it cannot be used to solve the three component mixing model. An alternative approach can be adopted using only δ^2H as follows. Firstly we write a three component tracer mass balance expression as:

$$\delta_1 V_1 + \delta_2 V_2 + \delta_3 V_3 = \delta_4 V_4 \tag{3}$$

with

$$\delta_1' V_1 + \delta_2' V_2 + \delta_3' V_3 = \delta_4' V_4 \tag{4}$$

where δ' is an independent tracer and

$$V_1 + V_2 + V_3 = V_4 \tag{5}$$

Where volumetric components (V) and corresponding isotopic compositions (δ and δ' where δ' is the isotopic composition of a second tracer, e.g. ^{18}O) of the three components *1, 2* and *3* contributes to the mixed component *4*. Component *4* corresponds to the measured xylem twig water. Because only the ratios of volumetric contributions to V_4 are required Equation 4 can be divided through by V_4 to obtain $R_1 = V_1/V_4$, $R_2 = V_2/V_4$ and $R_3 = V_3/V_4$, so that $R_1 + R_2 + R_3 = 1$.

R_2 is given by

$$R_2 = \left\{ \frac{\delta_4 - \delta_1 - R_3\left(\delta_3 - \delta_1\right)}{\left(\delta_2 - \delta_1\right)} \right\} \tag{6}$$

where R_3 is

$$R_3 = \left\{ \frac{\delta_2'\left(\delta_1 - \delta_4\right) + \delta_1'\left(\delta_4 - \delta_1\right) + \delta_4'\left(\delta_2 - \delta_1\right) - \delta_1'\left(\delta_2 - \delta_1\right)}{\delta_1'\left(\delta_3 - \delta_1\right) - \delta_2'\left(\delta_3 - \delta_1\right) - \delta_1'\left(\delta_2 - \delta_1\right) + \delta_3'\left(\delta_2 - \delta_1\right)} \right\} \tag{7}$$

and solution is obtained by back substitution of R_3 in Equation 6 and R_2 and R_3 in $R_1 + R_2 + R_3 = 1$.

Because a second *independent* tracer (δ' in Equation 4) is generally not available, the three component model can be solved by assuming values for one of the ratios R_1, R_2 or R_3 ranging between 0 and 1 in increments of say 0.01 and using the measured δ^2H data. Because of this assumption, specific fractional values for R_1, R_2 or R_3 cannot be obtained, with a single tracer, but rather a range of values that provide plausible solutions for R, i.e R_1, R_2 or R_3 all between 0 and 1. Thus inspection of the range of resulting values that are physically possible for the other two ratios can be used to determine the possible range of values for these other two ratios. In analysis of the results, measured δ^2H for each of groundwater, soil water, rainfall and twig xylem water are substituted for δ_1, δ_2, δ_3 and δ_4.

A time series of δ^2H of water in the xylem sap of twigs is generally required in order to determine the temporal variation in water source or mixture of sources being used by a plant. The possible sources of water for trees can be considered to be groundwater (saturated region of the soil), soil water (the unsaturated region of the soil) and rainfall as surface water. The surface soil compartment can be distinguishable from the remainder of the profile as it has the most dynamic evaporation and infiltration processes. In the intermediate soil profile, the bulk soil compartment, isotopic composition, and moisture levels are often constant over the time frame of a particular experiment, however, each field experiment will require the selection of appropriate, physically reasonable water source components and measurement of the isotopic composition of these source components.

3.2   Method application to water uptake by *Eucalyptus marginata* using enriched deuterium

An example application of the above methods is given in this section. The single step method for measurement of δ^2H in soils and plant material was used and analysis of the experimental results using the two-component source model (Equation 2) was applied. The experiment concerns the final one of three experiments where water of artificially different δ^2H was applied to a single jarrah tree. Natural abundance isotopic studies on the same tree are given in Farrington *et al.* (1996). The experimental jarrah tree had shown limited response to the deuterium applied to the unsaturated zone in two previous experiments. The third and final experiment was conducted during the following summer, 9 months after the second experiment. In this experiment isotopically enriched water was added to the ground surface over an annular area around the tree. The tracer remaining from the two previous experiments was assumed to have leached out of the profile by winter rains during 1993. Before the tracer was applied, the soil water content at depths down to at least 75 cm was at levels unavailable to plants as it was below 0.015 g g^{-1}, but below 2 m depth, water contents rose steadily to 0.03 g g^{-1}. The δ^2H values of soil moisture ranged from -10.0‰ in the upper profile to below -20‰ in the deeper profile. The annulus had an outer radius of 6m and an inner radius of 3.5m. The volume of water (3600L) was calculated to be sufficient to wet the profile to a depth of 0.75 m over the annular area. To achieve an even application, the volume of enriched water was divided into four amounts of 900L each and applied evenly using a hand held hose and spray over each quarter of the annulus after the soil surface was treated with a wetting agent. After application, the annulus was covered by polythene sheeting to reduce evaporation during the experiment.

During the 42 days of measurement less than 2 mm of rain fell and the daily air temperatures averaged 28°C maximum and 16°C minimum.

The δ^2H composition of the enriched water applied to the annulus averaged +412‰. One day after watering the δ^2H values in the water in the upper soil layers had been depleted to about two thirds that of the applied water, possibly through dilution. The enriched water had infiltrated to 1.5 m with most of the applied water being present in the upper 75 cm of the profile. During the period of measurement, the δ^2H values in the soil profile above 75cm fell steadily but still remained highly enriched. However the soil water content in the same depth fell steadily from 0.044 to 0.019 g g^{-1}. Some deuterium enrichment occurred in the profile above 1.5 m because infiltration of the enriched water down the profile was variable. Below a depth of 1.5 m the soil water was slightly affected by the tracer application, with δ^2H values less than - 10‰.

Prior to application of the enriched water, the δ^2H values in the twig sap sampled from the five branches generally fell within the range of -18 and -24‰. After the tracer application the δ^2H values in the tree first showed signs of enrichment within 24 hours. The δ^2H values in twig sap water peaked three to four days after the enriched water was applied (Fig. 5). During the next five weeks the level of enrichment in the twig sap declined steadily to reach values similar to that present before the enriched water was applied.

The response in δ^2H values in twig sap varied between the trunks (Fig. 5). The response was greatest in trunks 3 and 4 where maximum δ^2H values of +41.7 and +53.5‰ respectively were recorded. By contrast trunk 2 gave limited response with maximum δ^2H values reaching only +6.6‰. In the largest trunk (trunk 1) the extent of enrichment differed between the branches. The δ^2H level in the upper branch peaked at only +6.0‰, but in the lower branch the δ^2H level in the sap became much more enriched, reaching +25‰.

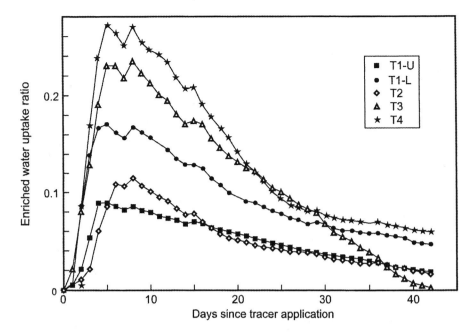

Figure 5. Changes in δ²H of sap water of twigs from five locations in the study tree following application of isotopically enriched water. Locations are: T1-U, trunk 1 upper branch; T1-L, trunk 1 lower branch; T2, trunk 2; T3, trunk 3; T4, trunk 4. Tracer experiment 3.

    Daily transpiration rates were measured in trunks 1 and 3 and showed a similar pattern with time. The fluctuations in the rate were mainly responses to the weather conditions. The transpiration rate followed closely an indicator of evaporative demand i.e. daily maximum air temperature. The rate in trunk 3 was consistently 60% that of trunk 1, but following watering, the transpiration rate of trunk 3 increased for some days relative to trunk 1.

    Uptake of isotopically-enriched water by the tree relative to other sources of soil water can be estimated using the two component source equation (Equation 2) in which $\delta^2H_n^{twig}$ is equivalent to δ_3 and is the deuterium composition of twig sap n days after enriched water was applied; $\delta^2H_n^{soil}$ is equivalent to δ_1 and is the deuterium composition of soil water between 0 and 75cm depth n days after the enriched water was applied; and δ_2 is the deuterium composition of soil water taken up by the tree from sources other than enriched water. Thus V_1/V_2 is the enriched water uptake ratio.

    Daily changes in the δ²H values of twig sap following the application of enriched water are determined from Fig. 5. The δ²H values of soil water can be obtained by averaging the composition between 0 and 75cm depths for each sampling. Figure 6 shows that enriched water uptake ratio was highest between five and eight days after tracer application and then declined steadily. The highest ratio occurred in trunk 4 (0.27) and the lowest was in the upper branch of trunk 1 (0.09).

Daily uptake of enriched water for each trunk can be estimated using the daily values for the enriched water uptake ratio and current transpirational flux. Daily transpirational fluxes for trunks 1 and 3 should be obtained from sap flow meters based on heat pulse methods. Fluxes for trunks 2 and 4 were estimated from trunk 1 using the assumption that the fluxes were proportional to the sapwood area ratio relative to trunk 1, which were 0.18 and 0.36 respectively for trunks 2 and 4. Since uptake of enriched water in trunk 1 differed between branches, the uptake by trunk 1 was assumed to be the mean of the upper and lower branches.

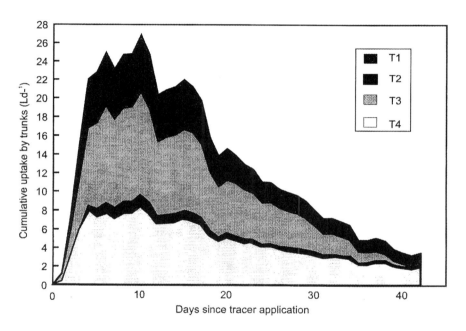

Figure 6. Uptake ratio of isotopically enriched water from five locations in the study tree following application of isotopically enriched water. Locations are: T1-U, trunk 1 upper branch; T1-L, trunk 1 lower branch; T2, trunk 2; T3, trunk 3; T4, trunk 4. Tracer experiment 3.

Uptake of enriched water reached a peak of $26.9 Ld^{-1}$ for the tree nine days after the enriched water was applied. Trunk 3 obtained more enriched water during the first three weeks. The tracer study shows that between five and eight days after application of the tracer, the tree extracted up to 18% of the daily water requirements from the annulus where the tracer had been applied, after which the contribution from enriched water fell steadily. As a result, out of a total transpiration by the tree of about 5900L during the seven weeks following irrigation, 9.8% (equivalent to 560L) of tree water used was extracted from the enriched water in the annulus.

4.   CONCLUSIONS

The pattern in which enriched water is taken up by trees reflects the spatial distribution of the root system and this should be considered in both the design of experiments and the interpretation of results. The root system of jarrah is three tiered (Abbot *et al.* 1989) consisting of an extensive horizontal root network from which two classes of roots i.e. riser and sinker roots develop. Riser roots ramify close to the soil surface and sinker roots grow straight downwards for several metres, the depth depending on the soil physical characteristics. The rapid response by the tree to the enriched water application and the daily contribution of 18% enriched water to the total water uptake five and eight days later was evidence of the high density of the riser and horizontal roots in the annulus.

This chapter shows three elements of the approach to water uptake studies in vegetation using isotopic tracers. Firstly, development of an efficient, single-step method for the measurement of δ^2H values in plant material has been demonstrated. Basic source component analysis equations that can be adapted and applied to the results of isotopic water uptake studies are also provided, and finally application of these methods to an enriched isotopic tracer study involving a *Eucalyptus marginata* tree. The results clearly showed the rapid uptake of tracer water when it is applied to the ground surface, illustrating how effectively uptake of available soil moisture can occur from the upper soil zone in response to an 'artificial' rainfall event. The purpose of the paper has thus been to demonstrate a range of practical methodologies that are required in order implement water uptake studies using isotopic methods. Each individual case study will require its own set of applicable methodologies and adaptation of some of the methodologies presented here to the particular case will be required.

REFERENCES

Abbott, I., Dell, B. and Loneragen, O. (1989). The jarrah plant. In 'The jarrah forest. A complex Mediterranean ecosystem.' (Eds B. Dell, J J Havel, and N. Malacjzuk.) pp. 41-51. (Kluwer Academic Publishers: Dordrecht, Netherlands.)

Coleman, M.L., Shepherd, T.J., Durham, J.J., Rouse, J.E. and Moore, G.R. (1982). Reduction of water with zinc for hydrogen isotope analysis. *Analytical Chemistry* 54, 993-995.

Dawson, T.E. and Ehleringer, J.R. (1991). Streamside trees that do not use stream water. *Nature* 350, 355-357.

Dawson, T. E. and Pate, J. S. (1996). Seasonal water uptake and movement in root systems of Australian phreatophytic plants of dimorphic root morphology: a stable isotope investigation. *Oecologia* 107, 13-20.

Ehleringer, J.R. and Dawson, T.E. (1992). Water uptake by plants: perspectives from stable isotopes composition. *Plant, Cell and Environment* 15, 1073-1082.

Ehleringer, J.R. and Osmond, C.B. (1989). Stable isotopes. In 'Plant Physiological Ecology Field Methods and Instrumentation.' (Eds R.W. Pearcy, J.R. Ehleringer, H.A. Mooney and P.R. Rundell.) pp. 281-300. (Chapman and Hall Ltd.: London.)

Farrington, P., Turner, J.V. and Gailitis, V. (1996). Tracing water uptake by jarrah (*Eucalyptus marginata*) trees using natural abundances of deuterium. *Trees, Structure and Function* 11, 9-15.

Leaney, F.W., Osmond, C.B., Allison, G.B. and Ziegler, H. (1985). Hydrogen-isotope composition of leaf water in C3 and C4 plants: its relationship to the hydrogen-isotope composition of plant matter. *Planta* 164, 215-220.

Lin, G., Phillips, S.L., and Ehleringer, J.R. (1996). Monsoonal precipitation responses of shrubs in a cold desert community on the Colorado Plateau *Oecologia* 106, 8-17.

Revesz, K. and Woods, P.H. (1990). A method to extract soil water for stable isotope analysis. *Journal of Hydrology* 115, 397-406.

Thorburn P.J. and Walker, G.R. (1993). The source of water transpired by *Eucalyptus camaldulensis*: soil, groundwater or streams? In 'Stable isotopes in Plant Carbon - Water Relations.' (Eds J.R. Ehleringer, A.E. Hall and G.D.Farquhar.) pp. 511-527. (Academic Press: San Diego, Ca.)

Thorburn, P.J., Walker, G.R. and Brunel, J.P. (1993a). Extraction of water from Eucalyptus trees for analysis of deuterium and oxygen-18: Laboratory and field techniques. *Plant, Cell and Environment* 16, 269-277.

Thorburn, P.J., Hatton, T.J. and Walker, G.R. (1993b). Combining measurements of transpiration and stable isotopes of water to determine groundwater discharge from forests. *Journal of Hydrology* 150, 563-587.

Turner, J.V., Arad, A. and Johnston, C.D. (1987). Environmental isotope hydrology of salinized experimental catchments. In 'Hydrology and Salinity in the Collie River Basin, Western Australia.' (Eds A.J. Peck and D.R. Williamson.) Special edition. *Journal of Hydrology* 94, 89-107.

Turner, J.V. and Gailitis, V. (1988). Single-step method for hydrogen isotope ratio measurement of water in porous media. *Analytical Chemistry* 60, 1244-1246.

Walker, G.R., Woods, P.H. and Allison G.B. (1991). Inter-laboratory comparison of extraction methods to determine the stable isotope composition of soil water. In 'Isotope Techniques in Water Resources Development.' *1991 Proceedings International Symposium Vienna, IAEA-SM-319/40.* pp. 509-517. (International Atomic Energy Agency: Vienna.)

White, J.W.C., Cook, E.R., Lawrence, J.R. and Broecker, W.S. (1985). The D/H ratios of sap in trees: Implications for water sources and tree ring D/H ratios. *Geochimica et Cosmochimica Acta* 49, 237-246.

Whelan, B.R. & Barrow, N.J. (1980). A study of a method for displacing soil solution by centrifuging with an immiscible liquid. *Journal of Environmental Quality* 9, 315-320.

Zimmerman, U., Ehhalt, D., and Munnich, K.O. (1967). Soil-water movement and evapotranspiration changes in the isotopic composition of water, In 'Isotopes in Hydrology.' *1967 Proceedings International Symposium Vienna* IAEA-IUGS. pp. 567-586 (International Atomic Energy Agency: Vienna.)

Chapter 4

The Use of Stable Isotopes of Water for Determining Sources of Water for Plant Transpiration

Glen Walker[1], Jean-Pierre Brunel, John Dighton[1], Kate Holland[1], Fred Leaney[1], Kerryn McEwan[1], Lisa Mensforth, Peter Thorburn and Colin Walker
[1]*CSIRO Land and Water, PMB 2, Glen Osmond, SA 5064 Australia. Email: glen.walker@adl.clw.csiro.au*

Key words: stable isotopes of water, plant water sources, groundwater, modelling, errors, inverse techniques

1.   INTRODUCTION

Over the last ten years, there has been a large increase in the number of vegetation studies that have incorporated measurements of the stable isotopic composition of water. There are many methods for measuring the amount of water being used by plants, but until recently it has been difficult to determine from where plants obtained their water. This has been particularly difficult where there is more than one available water source (e.g. where groundwater is shallow or streams are nearby).

There are several reasons why it is necessary to know from where plants source their water:

1. The community has increasingly acknowledged the need to understand more fully the water relations of natural areas with phreatophytic (deep-rooted plants that obtain water from the water table or the layer of soil just above it) or wetland vegetation;

2. Water use by vegetation may conflict with demands for extraction of groundwater for industry, or decrease base flow to streams at times of important environmental requirement;

3. Manipulation of streams for salinity mitigation may affect plants' reliance on stream water; and

M. Unkovich et al. (eds.),
Stable Isotope Techniques in the Study of Biological Processes and Functioning of Ecosystems, 57–89.
© 2001 *Kluwer Academic Publishers. Printed in the Netherlands.*

4. Where plantations are being suggested as a means to lower water tables in areas of salinity.

In all of the above, there is a need to better understand how vegetation responds to change in environment or alternatively how vegetation may modify the environment.

Isotopes are increasingly being used where more standard physical methods are limited, since:

a) measurement of zones of root activity is most often destructive, intrusive and labour-intensive. Furthermore, the mere presence of roots is not of itself evidence of water uptake (Ehleringer and Dawson 1992);

b) recent advances in soil moisture monitoring means that it is now possible to delineate changes in soil moisture on a fine temporal scale. However, this is problematic for determining plant water use in the presence of shallow water tables, streams or in areas of lateral water movement, and there are calibration problems when employing such techniques with some soils;

c) piezometric methods for determination of rates of groundwater use only work well if groundwater use exceeds aquifer transmissivity; and

d) lysimeters have been used for determining water use strategies but they are difficult to use in natural field conditions, and do not replicate natural groundwater conditions.

Obviously, isotopic methods complement these approaches, and provide information that simply can not be obtained by other means.

In this chapter, we aim to highlight the important considerations that apply when using isotopes for determining plant water sources. By means of a set of case studies, we will show many of the limitations, as well as the strengths, of the methods employed.

The isotope methods that have been developed have the advantage of being applicable to natural systems in remote locations. Although they may require reasonably frequent visits to the field for samples, they do not require sensitive equipment to be maintained in the field and in most cases involve little disturbance to the field site. Furthermore, isotope methods can be integrated easily into other water balance work, they cost only a fraction of other water balance methods and require little in the way of specialist facilities (except in already established analytical laboratories). On the negative side, the isotope methodology does not of itself guarantee unequivocal results with a high degree of resolution. In every instance substantial supporting corroborative evidence is required before a clearly defined picture can be obtained.

The method depends on sources of water having different endogenous isotopic compositions. Such natural variations in isotopic composition arise because of isotopic fractionation, caused principally by transport processes

(Gat 1981) and phase transitions through the atmosphere, lithosphere and biosphere (Barnes and Allison 1983). Since the relative proportion of the fractionating processes are likely to be different for groundwater, surface stream water and soil, different sources of water for plant use will often, but not always, have different isotope values. It is essential that the variation in isotopic compositions of sources exceeds the errors.

Isotopic methods have been successfully used to differentiate water sources for plants with access to both ocean and freshwater (Sternberg and Swart 1987). Dawson (1993) used isotopes to demonstrate hydraulic lift by sugar maples during drought periods to corroborate the observed fluctuations in soil water potentials. Again incorporating other plant physiological methods, Flanagan *et al.* (1992) combined measurements of δ^2H with predawn plant water potential to compare the relative uptake of summer precipitation by four co-occurring trees and shrubs in the semiarid region of the south west United States. A study by Dawson and Ehleringer (1991) used isotopes to show that streamside trees were not using stream water, despite apparent availability.

Isotopes have also been used to demonstrate temporal changes in water sources, e.g. White *et al.* (1985) who showed that water extraction by white pine (*Pinus strobus*) in the eastern United States changed over time between surface soils to the deeper groundwater. Similarly, Ehleringer *et al.* (1991) investigated the seasonal differences in water sources within a scrub community in southern Utah. They found that some species were completely dependent on summer rainfall, other species used summer and winter rains; while another species showed no response to summer rainfall, indicating that it used deeper water resources. More recently, Dawson and Pate (1996) compared xylem sap water of deep and shallow roots of *Banksia* with that of water in trunk xylem to follow seasonal dependence on soil and groundwater.

The work described in this chapter should not be confused with isotopic methods for transpiration measurement (e.g. Calder 1991 and Calder 1992). These involve artificial tracers, such as deuterium oxide, to be injected into the base of the tree, and collected as leaf condensate in bags attached at different levels in the trees. The rate at which the tracer moves through the tree can then be used to measure the transpiration rate.

2.   TESTING THE ISOTOPE APPROACH FOR SOURCING PLANT WATER

2.1   Brief description of field methods

The basis of the isotope method is to compare the isotopic composition of the plant material to that of the various sources and then use the data to infer from where the plant has obtained its water. It is assumed that the isotopic composition of source water is unchanged as it is extracted by the plant. Hence, if the plant obtained all of its water from one source, the isotopic composition of the plant water should be identical to that of the source.

2.1.1   Evaluation of the isotopic composition of defined potential sources of water to plants

Sources of plant water may include: soil water at various depths; water from the capillary fringe from the top of a body of groundwater; stream water and episodic flood water.

Soil can be sampled destructively using hand-held augers or obtained mechanically with hollow-stemmed augers in which no drilling fluid is used. Groundwater can be sampled from shallow piezometers (ensuring minimal evaporative losses), while stream waters can be sampled with grab samples. It is often advisable to collect samples of current rainfall as well. This can be collected in gauges, but a less dense fluid such as paraffin oil needs to be added to prevent evaporation. Evaporation would of course result in isotopically enriched samples, so it is important that water is placed quickly into containers - glass pickle jars with an additional seal afforded by wrapping electrical tape around the lid.

2.1.2   Isotopic composition of the water from within plants

Sampling should be designed to minimise evaporation from the plant material. The sampling point within the plant needs to be chosen carefully. This is discussed in Section 2.3.

2.2   Extraction techniques and potential errors

The largest source of error in the overall methodology is likely to be in the extraction of isotopically unaltered water from both soils and plants, once samples are taken back to the laboratory. Indeed, in a comparison of laboratory methods (see Case Study 1) for soil water extraction, the error in extraction has been shown in some cases to exceed the natural variation sources.

Many of the water extraction methods apply to both soil and plant material; the most commonly used methods being vacuum distillation (Jusserand 1980, Saxena and Dressie 1984) and azeotropic distillation (Woods 1989)

Extraction methods for soil water should only extract water that is freely available to plants while excluding that associated with the water of crystallisation (e.g. from gypsum); or those molecules associated with the physical structure of clays; or chemically bound in organic matter. The best methods involve no change in phase of the water concerned. An example of this is centrifugation of soils with heavy organic immiscible liquids, such as tri-chloro-ethylene (Whelan and Barrow 1980, Reynolds 1984). However, most of such techniques fail to extract water from anything other than the wettest of soils, and they are unsuitable for sampling over a range of hydrological conditions.

In order to increase the applicability of the method to most field conditions, a number of techniques involving a change to the gaseous phase have been developed. Any technique involving a phase change from liquid to gaseous form is subject to a Rayleigh distillation process (Clark and Fritz 1997, and see chapter in this volume by Dawson and Brooks) in which the heavier isotopes of water change phase more slowly than the lighter isotopes. Hence, if all the water is not collected, the water remaining will be enriched in heavier isotopes, while the collected water sample will be depleted in the same heavier isotopes. The deviation is given by

$$\delta = -\frac{1000(f^{\alpha-1} - 1)f}{1-f} \tag{1}$$

where f is the fraction of water collected and α is the fractionation factor for a given temperature (Fritz and Fontes 1980). A plot of this is shown in Figure 1.

2.2.1 Vacuum distillation

The simplest of the techniques involving a phase change is vacuum distillation. During vacuum distillation, the sample is cooled down to approximately -70°C, the system is closed and a vacuum created above the sample. The sample is gradually heated and vapour collected using traps cooled with dry ice/ethanol (~ -60°C). The amount of time needed for the process to be completed depends on the temperature. If the temperature is too high, there is the danger that water bound in organic material might be produced, thereby biasing the sample. Also, if the sample contains high quantities of gypsum, or any other hydrated mineral salt component, water of crystallisation may be released. On the other hand, when opting for a 'safe' low temperature, the amount of time needed to collect the sample may be protracted and thus complete extraction maybe well nigh impossible.

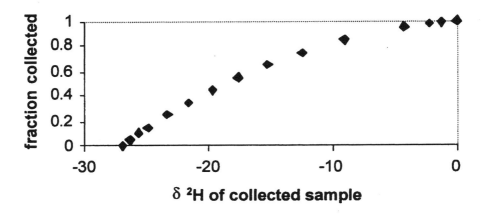

Figure 1. Hydrogen isotopic composition of the collected water as a function of the fraction of water collected (at 100°C).

2.2.2   Azeotropic distillation

In an alternative distillation technique, the sample containing water (e.g. plant or soil material) and an organic solvent, such as toluene or kerosene, are heated in a round bottom boiling flask (Fig. 2). At the boiling point (a temperature always less than that of the pure components), the azeotrope is formed, in which water has become completely miscible with the organic solvent. The resultant vapour is then condensed, with the water separating from the heavier liquid. The former collects in the receiver, while the solvent returns to the flask. The sample is collected by drainage through the stopcock and heated with paraffin wax to remove contaminants. The rate at which water is distilled depends on the azeotropic temperature of the solvent-water mixture and the type of materials from which water is extracted (i.e. sand<<clay).

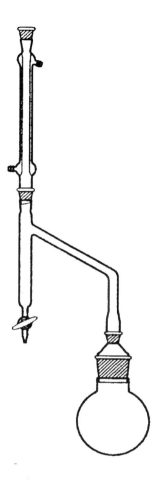

Figure 2. Dean and Stark distillation apparatus.

There are also some novel one step techniques that have been developed to measure the δ^2H ratio of soil water, in which an aliquot of soil is reacted with zinc at 400 ^{0} C in special glass vessels in which the evolved hydrogen gas is collected (see chapter in this volume by Turner *et al.*). Oxygen isotopes are also determined similarly (i.e. one step) by equilibrating CO_2 of known isotopic composition directly with the soil sample in suitable containers. These methods are not discussed here.

CASE STUDY 1: COMPARISON BETWEEN DIFFERENT LABORATORIES

There is no single way to validate with confidence all the various extraction methods used in the analysis of soil samples. However, the most commonly used methods to test such methodologies are to:

a) dry the sample, add water of known isotopic content and then extract the water and measure its isotope composition, or;

b) compare results between methods involving and not involving a phase change.

In the first of these protocols, after oven drying of the sample, some water remains adsorbed to the clay particles. Hence as this water diffuses into the added water, the isotopic composition of the soil sample becomes obscured. The second method can only be used for wet soils and therefore does not provide a test over a reasonable range of soil water potentials, and indeed problems with extraction and fractionation are likely to be less important for all methods where wetter soils are used.

An international comparison designed to investigate the influence of different extraction techniques for stable isotopes of water from soil was conducted and reported on by Walker *et al.* (1994). Samples of four soil types (sand, gypseous sand, high and low water content clays) were sent to fourteen laboratories around the world. The extraction techniques employed by the laboratories included azeotropic, vacuum and microdistillation methods.

The results from two soil types (sand and dry clay) are shown in Figure 3. The deviation for the sand is seen to be about 20‰ for δ^2H and 3‰ for $\delta^{18}O$, while for the dry clay the values varied by 45‰ for δ^2H and 5‰ for $\delta^{18}O$. The lines in Figure 3 represent the Rayleigh distillation line at 35°C, using the mass balance of the air-dry sample plus water as a reference. The proximity of the results about this line suggests that the main source of error is due to incomplete extraction. The variation increased with decreasing soil water content and was greater for clays than sand at comparable soil matric suction.

The spread of results is unfortunately greater than the natural variability of plant sources and this would therefore preclude use of the isotope methodology for determining sources of plant water.

Walker *et al.* (1994) showed that lower temperature methods had the greatest deviations, suggesting that heat is necessary for complete extraction. The low temperature methods had been developed in order to avoid extraction of interstitial water from clays since this component is not thought to play any significant role in hydrological processes. The study concluded that fractionation during the extraction process was insignificant provided the soil was heated above 100°C (Walker *et al.* 1994).

In terms of using the isotope method to source plant water, the likely zones of plant water extraction are the wetter areas of the soil, which are less prone to error. Consequently, it is difficult to define an absolute error for soil water extraction.

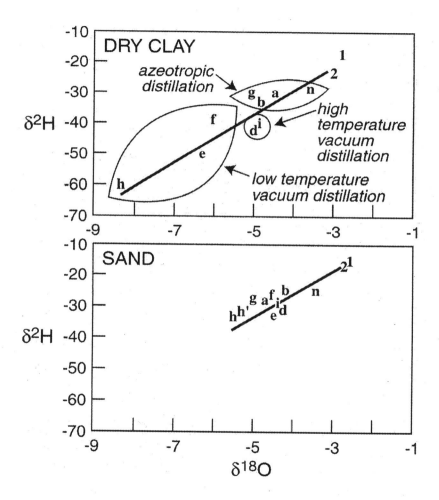

Figure 3. Sand and dry clay δ^2H vs δ^{18}O plots for different laboratories (Figure 1, Walker *et al.* 1994).

Tests of extractions of plant material have been carried out using seedlings growing in water of known isotopic composition. Such tests have been conducted on lucerne (*Medicago sativa*), black box (*Eucalyptus largiflorens*) and melaleucas (*Melaleuca halmaturorum*). For example, Thorburn *et al.* (1993b) investigated the extraction of water of known isotopic composition from *Eucalyptus* seedlings by azeotropic distillation using three solvents (hexane, toluene and kerosene) with different azeotrope and boiling point temperatures. They found that kerosene and toluene gave the most accurate extractions, with a small negative bias (~-2‰ for δ^2H and ~ -0.4‰ for $\delta^{18}O$) due to incomplete extraction.

2.3   Errors due to plant sampling and plant processes

In addition to errors due to the extraction of bulk water from plant material, errors may be associated with plant sampling. One of these is a mixing of xylem water with the enriched, evaporated water moving in from leaves in the phloem. This can be minimised by avoiding any green parts of a plant as these are potential regions where enriched water would abound. Similarly, leaves should be removed and bark scraped from woody twigs or shoots. For herbaceous species such as lucerne / alfalfa (*Medicago sativa*), samples need to be taken from near the crown, that is farthest away from transpiring surfaces. Whole roots have also been used for sampling stable isotopes of water from alfalfa (Thorburn and Mensforth 1993). Obvious advantages are to be gained when assessing plant sources of water by collecting the xylem fluid directly from roots, stems and trunks by the mild vacuum extraction technique advocated in Dawson and Pate (1996). In such cases the voided water should be collected and sealed immediately on extraction to avoid isotope changes through evaporation.

Case study 2 shows that there is no significant change in sap δ^2H values between the trunk and twigs adjacent to leaves in *Eucalyptus* trees. Therefore, the sampling of small twigs with bark removed is a valid methodology for *Eucalyptus* spp., thus avoiding destructive sampling and allowing repeated measurements on the same tree.

CASE STUDY 2: INTRA TREE VARIATION IN ISOTOPIC COMPOSITION

Variations in δ^2H within the canopy of *Eucalyptus* trees, and between samples of trunks, branches and twigs from trees are reported in Brunel *et al.* (1991) (site: Ouyen) and Thorburn *et al.* (1993b) (site: Chowilla).

2.3.1 Comparison between different tree tissues

Trees at each site were sampled at various points (leaves, small twigs, big twigs, primary branch, secondary branch and trunk) and separate heartwood and sap wood samples obtained. The results from the analysis of leaves, small twigs, big twigs, primary branch, secondary branch and trunk are shown in Figure 4. For sapwood, there was no trend in δ^2H values between the bottom of the trunk and the extremity of the twigs. The greatest difference within any one tree was 6‰ (-28 to -34‰, site 2HTD, sampling 9), most differences were within 2‰, i.e. within expected analytical precision. The differences in δ^2H values between trunks, branches and twigs were ≤2‰ at Chowilla and ≤5‰ at Ouyen. The analysis of 18 paired samples of twig and sapwood found variations of ~3‰, with a mean difference of 0.3‰ over a wide range of δ^2H values. These differences were not systematic, and reflected analytical error, rather than enrichment processes within the tree. Leaf δ^2H values at Chowilla were ~40‰ greater than in other tree parts. Unfortunately in this study, δ^2H values for samples of extracted xylem fluid were not available for comparison with the above samples.

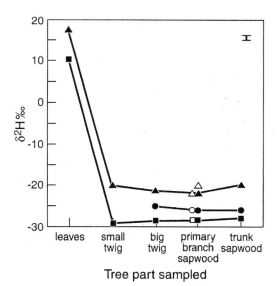

Figure 4. Variation in δ^2H of water extracted from sapwood of different parts of *Eucalyptus camaldulensis* (■ and ●) and *E. largiflorens* (▲) trees at Chowilla, South Australia. Open symbols show heartwood δ^2H values of primary branch (of two trees), and half filled symbols show bark δ^2H values from those trees. Vertical bar indicates the precision of analyses (Figure 3, Thorburn *et al.* 1993b).

The close match within 5‰ between all tree parts (apart from leaves which displayed evaporative enrichment in δ^2H) contrasts with results from several other species. For example, Dawson and Ehleringer (1991) found significant ^2H enrichment (up to 55‰) in small stems of *Acer grandidentatum, A. negundo* and *Quercus gambelli,* which they attributed to cuticular water loss. Bariac *et al.* (1983) also observed a significant trend in δ^2H values in the alfalfa plant.

2.3.2   Comparison between root and twig isotopic composition

Samples of roots and surrounding soil materials were also taken and analysed for δ^2H (Thorburn *et al.* 1993b). Root δ^2H values ranged from -9‰ to -32‰, with greater variation in shallow (<0.2 m depth) than deeper samples. The twig samples of the five trees studied were -23‰ to -27‰, indicating that the roots sampled did not provide a good indication of sapwood δ^2H values, but represented local soil isotopic composition. This is shown in Figure 5, by the relationship between the plot of root δ^2H‰ and soil δ^2H‰. The twig δ^2H values were similar to the groundwater (at ~ 3 m depth) and deeper soil δ^2H values at the sites. The poor correlation between root and twig δ^2H values might well indicate that many of the roots sampled were not significantly involved in transporting water to the tree canopy. The relationship between root δ^2H values and soil δ^2H values, particularly for the surface roots may reflect the roots being in equilibrium with the surrounding soil water rather than groundwater.

Another source of error may arise where there is a lack of mixing of water from different sources and parts of the root system. Dye studies have shown continuity of flow between certain roots and certain stems with little mixing. If this was the case, the isotopic composition of the plant could be strongly dependent on the sampling technique. Tests done on intra-canopy variations found these variations were not significant during sampling of mallee-form eucalypts near Ouyen (Brunel *et al.* 1995) (refer Case Study 2). This also needs to be examined at each study site and sampling over the whole canopy should take place.

A further possible source of error is a lack of consideration of travel times within the plant. It may take some days, or weeks for larger trees, for water in xylem conducting vessels to travel throughout a tree, exchanging all the time with tissue water and phloem water. A large rainfall event between the time of uptake and when the plant water was sampled may considerably change the isotopic composition of the soil water. This change in soil water isotopic composition would be reflected in the plant water isotopic composition following uptake and transport to the sampling site. This problem was highlighted in a study by Thorburn *et al.* (1993b), who found that surface roots had sapwater δ^2H values in excess of the possible soil water and groundwater sources. They concluded that this was probably due to rain (with δ^2H of -40‰) which had fallen in the 2 days prior to sampling. Hence, one should attempt to collect soil water and plant water after a period of 1 – 2 weeks with relatively constant climatic conditions, i.e. without large rainfall events or sudden weather changes in the intervening period.

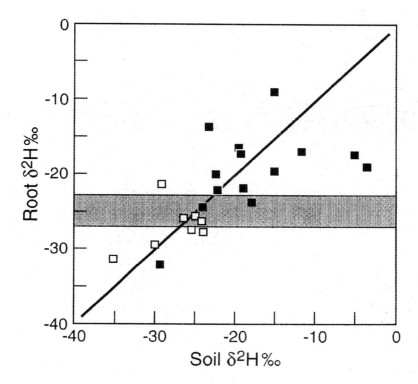

Figure 5. Comparison of root δ^2H and soil δ^2H values. Woody roots (15 – 25 mm diameter) were sampled from five *Eucalyptus camaldulensis* and *E. largiflorens* trees from the surface to 2 m depth (from <0.2 m depth and from >0.2 m depth). The hatched area shows the range of δ^2H values of twigs from the trees sampled (twig δ^2H values matched groundwater values), and the solid line is the 1:1 line (Figure 6, Thorburn *et al.* 1993b).

2.4 Validation of the Isotope Approach for Sourcing Plant Water

In sections 2.2 and 2.3, we have described some of the possible errors associated with obtaining isotopic values for plant water sources and plant water. Other possible errors inherent in the methodology for identifying plant water sources using stable isotopes are that:

a) There is no significant fractionation when water is taken up by roots;
b) The plant water source samples are representative of all the sources. Of most concern here is the spatial variability of isotopes in soil, as studies have shown variability over short distances. Thus, different sampling sites can lead to different soil water isotope values. The time-consuming nature of isotope analyses makes a sampling routine to account for variability impossible. There is probably less variability in the wettest parts of soils as these represent the most recent rainfall or groundwater which is the most available to the plant; and
c) Inter-canopy variations are likely to occur if tree to tree variation is significant. Different outcomes would be reached depending on number of trees sampled.

The first of these sources of error has been investigated in many studies involving a wide range of plants from various environments (Zimmermann *et al.* 1966, Barnes and Allison 1983, White *et al.* 1985, Sternberg and Swart 1987, Brunel *et al.* 1991, Thorburn *et al.* 1993b). The most commonly used method for testing extraction of plant water is to use seedlings under well-watered conditions or aquarium-cultured plants (e.g. Thorburn *et al.* 1993b). However, it is not clear how data from such tests might apply to the situation in large trees under stress, say for example a mature 20 – 30m tree growing over saline groundwater, sampled in the middle of summer.

It would be almost impossible to derive relative magnitudes of each of the above sources of error, as relating to specific trees in a particular water use study. As mentioned, errors associated with soil water extraction and spatial variability in soils may be biased towards dry soils, whereas plant water use would tend to be biased towards wet soils. Hence, a simple arithmetic addition of errors might well lead to an exaggerated measure of error.

In the following example (Case Study 3A), an alternative approach was taken in which the isotopic composition of the presumed 'plant available water' of the soil was matched with that obtained from the plant. This involved matching the isotopic composition of plant available water from the soil and / or groundwater, i.e. soil which is neither too salty, nor too dry with the isotopic composition of the plant water. This reduces the need to analyse all samples for isotopic composition, improving sampling efficiency and reducing costs of analysis.

CASE STUDY 3A: STUDY AT OUYEN, VICTORIA

The validity of the field sampling techniques used for the isotope approach for sourcing plant water was analysed by Brunel *et al.* (1995) in a study of mallee-form eucalypts in a semi-arid environment. This study was able to provide an estimate of the error due to field techniques for predicting plant water sources.

The field study was conducted near Ouyen, Victoria in south eastern Australia (Brunel *et al.* 1995). The climate is semi-arid, with mean annual rainfall of 340 mm yr^{-1} and and pan evaporation of 1350 mm yr^{-1}. Mallee-form eucalypts grow on a 7 m aeolian sand dune, which overlies a 1 m thick layer of lacustrine clay. This, in turn, separates the sand dune from the underlying regional sand aquifer. A perched water table above the clay layer was present throughout the sampling period. There is a saline discharge area at the base of the sand dune.

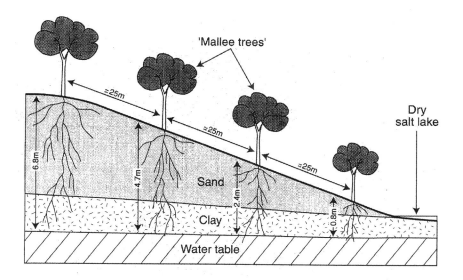

Figure 6. Schematic cross-section of field site, showing trees on dune and depth to groundwater beneath the trees at the four sites (Figure 3, Brunel *et al.* 1995).

The trees were sampled from four sites, growing at 0.8 m to 6.8 m depth over the shallow saline (~40 dS m^{-1}) groundwater (refer Fig. 6). Samples of plant and soil materials were taken on 10 occasions over 2 years, encompassing both very wet and dry conditions. Details are given in (Brunel *et al.* 1991). Soil matric potential was determined using the 'filter paper method' described in Greacen *et al.* (1989). The respective depths to groundwater and groundwater salinity were measured in piezometers installed at each of the four sites. Plant water sources were evaluated using two methodologies - the plant water availability and compartmental models.

2.4.1   Plant water availability model

A simple model of plant water availability was used to interpret the soil water potential and osmotic potential measurements obtained in the study. It was postulated that:

Soil water suctions (the sum of matric and osmotic potentials) less than -3.5 MPa would be too extreme for the mallee to extract significant amounts of water (a soil water chloride concentration of 20 g L^{-1} was used as a surrogate for -3.5 MPa). The values postulated are consistent with other values reported for arid to semi-arid native vegetation; and

Zones of high soil water suction (matric potential greater than -1.0 MPa; soil water chloride concentrations less than 10 g L^{-1}) would be the preferred source of water for the plant to access. It was assumed that essentially all water would be obtained from such zones.

The above assumptions led us to believe that there would have been a wide zone of available soil water with a wide range of isotopic compositions. However, the aridity of the area and salinity of the underlying groundwater meant that only a narrow depth interval contained water available for the plant. Using the soil osmotic and matric potential criteria discussed above, a predicted range of isotopic compositions was thus determined for each soil profile, indicating the zones from which the plant was most likely to be obtaining water.

These predicted isotopic compositions were then compared to the isotopic composition of twig water. For example, at Site 2HBD for trip 4 (Fig. 7a), the tree would be expected to be extracting water from 0.1 – 0.4 m depth, as the matric potential for 0 – 0.1 m depth is too high and the osmotic potential below 0.4 m depth would be too high. The δ^2H values in the 0.1 – 0.4 m depth range vary from -1.7 ‰ to 2.8 ‰, compared with -1.3 ‰ to -1.7 ‰ for the twigs. Only one sample had enough water for ^{18}O analysis, this was -0.88 ‰ (0.4 m depth) compared to the twig values of -1.6 ‰ and 0.6 ‰.

Similar matches were found at the other sites and sampling times. Of special interest was site 2HTD at sampling time 2 (Fig. 7b), when two zones of relatively fresh and plentiful soil water were presumably available to the plants. The surface 0 – 0.1 m interval was moist and fresh, as was the zone between 1.5 – 1.75 m. The isotopic values were intermediate between the two possible sources, indicating partial extraction from each.

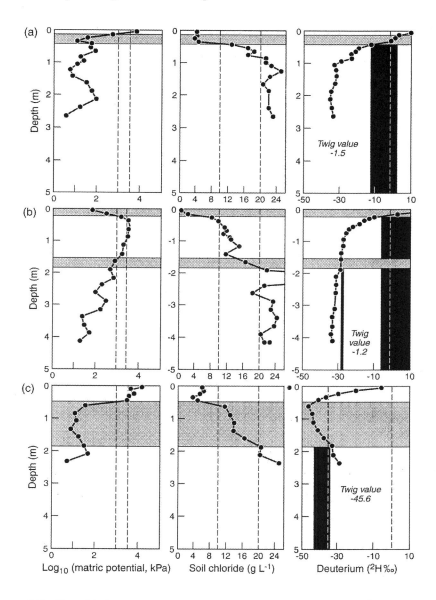

Figure 7. Profiles with depth of soil matric suction, chloride concentration and ²H composition in soil water for: (a) Site 2HBD, Sampling 4; (b) Site 2HTD, Sampling 2; (c) Site 2HBD, Sampling 8. The zone represented by light shading is the depth interval in which the soil is not drier than -3.5 MPa and more saline than 20 g L⁻¹ chloride. The ²H composition of this zone is represented by dark shading. This can be compared with the measured ²H composition of sap water in twigs shown on the plot by a dashed line (Figure 5, Brunel *et al.* 1995).

For sampling time 8 at Site 2HBD (Fig 7c), the soil matric potential turned out to be low from the surface to 0.3 m depth and then high below 0.5 m depth. The chloride concentration is low to a depth of 0.5 m, intermediate from 0.5 to 1.75 m depth and high below this. The soil water δ^2H values in the $0.5 - 1.75$ m depth range varied from -48 ‰ to -38‰ compared with the twig values of -49.4 ‰ and -41.7 ‰, again in principle agreeing with the plant water availability model.

Hence, the plant water availability model using soil isotope values in conjunction with soil water potential values (osmotic and matric potentials) allowed the comparison with twig isotope values. The results of this comparison over all sites is shown in Figure 8.

2.4.2   Estimation of total error and validation of overall methodology

Given the range of conditions from the 10 sampling trips and 4 field sites, which covered a large range of soil matric and osmotic potentials, the overall validity of the technique can be evaluated. The range predicted by the plant water availability model varied from $3 - 4$ ‰ to a range so large as to make the comparison meaningless. The compilation of all of the predicted and observed ranges for all sites and sampling times showed that the values lie close to the 1:1 line. A comparison of the δ^2H results (Fig. 8) found that the predicted values overlapped with the observed values within an error range of ±5 ‰, even though the ^2H composition varied from -48 ‰ to -1 ‰. A similar analysis for ^{18}O yielded a range of ±1 ‰.

The deviation from the 1:1 line gives an indication of the total errors associated with sampling, water extraction, analysis and model assumptions (Brunel *et al.* 1995). The error appears to be less than 5‰ for δ^2H, and so for the method to be useful, differences in the δ^2H of the water sources must be > 5‰. The above authors further concluded that their study validated the assumption that "there is no significant fractionation of water during uptake of soil water by the plant" thus agreeing in principle with numerous other studies. However, we regard the data reported in this study as the strongest evidence that this assumption holds for a wide range of soil conditions, including those of high total potential.

A study reported by Mensforth and Walker (1996) investigated the response of the roots of *Melaleuca halmaturorum* to fluctuating saline groundwater. This combined three separate approaches to determine from where the trees were sourcing water. New root growth was recorded via an observation window, which showed that all new root growth was confined to the surface $0 - 0.1$ m depth and there was no new root growth at greater depth. The tree and soil water potentials were higher (less negative) at the end of winter and the results from the stable isotopes of water analysis indicated that the twig water was similar to a mixture of water sources from the 0 – 0.1 m depth and the 0.1 – 0.2 m depth (or groundwater).

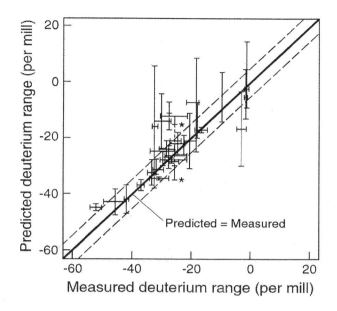

Figure 8. Measured twig δ^2H vs twig δ^2H predicted from soil matric potential and chloride concentration. The predicted data for the sample marked with an asterisk are for two distinct ranges, one from near the soil surface and one from closer to the water table. The twig data for this sample lie midway between the two predicted ranges. The dashed lines on either site of the 'predicted=measured' line mark the 5‰ interval (Figure 6, Brunel *et al.* 1995).

This combination of measurements indicated that the trees were sourcing their water from the surface 0 – 0.2 m depth interval even when the groundwater was within 0.3 m of the surface at the end of winter.

3.   INVERSE TECHNIQUES FOR INFERRING SOURCE WATER

The isotope method is a type of inverse technique, i.e. one uses a measured output to infer something about the system. The errors discussed in the previous section are errors associated with a forward process; that is, if the source has a certain isotopic composition, what are the likely errors when trying to measure the same isotopic composition in the plant? However, our endeavour is to use the isotope method for the opposite purpose, that is, to infer the sources of water once we know the isotopic composition of the xylem water. This protocol introduces its own errors.

The isotope technique should not be thought of as a precise 'fingerprint type method' providing unique tracer-based information on the sources of water transpired by plants. In fact, there is limited information in any single data point. With both $\delta^{18}O$ and δ^2H, there are potentially two pieces of information against which to match all possible sources of water in a system. With no spatial variability or other sources of water, there are still numerous depths from which plants may extract water within their root zone. Even so, δ^2H is highly correlated with $\delta^{18}O$ so that in many cases, $\delta^{18}O$ provides essentially identical information to that from δ^2H alone (see chapter in the volume by Turner *et al.*). Furthermore, with only one data point, one can fit only one parameter, usually the fraction of water from each of two sources. With two pieces of information, one can either fit two parameters with no redundancy in the data or one parameter with some check on the value.

Thus, in the absence of any other data, dual isotope analysis can be used to estimate the relative fraction of each of either two or three sources of water. It is clear that the error in the fraction is linked to the difference in isotopic composition of the sources, the correlation between the δ^2H and $\delta^{18}O$ and the degree to which the sources can be compartmentalised into 2 or 3 sources. The method is robust provided the difference in isotopic composition of sources exceeds measurement errors and 'systematic' errors (i.e. errors due to assumptions). The latter is difficult to quantify, although some sensitivity analysis provides an indication. Case Study 3B helps show how a compartmental inverse technique might work and where errors may enter.

CASE STUDY 3B: STUDY AT OUYEN, VICTORIA - COMPARTMENTAL MODEL

The simplifying assumptions made when using a compartmental model were tested using results from a detailed sampling programme at the Ouyen site. The soil profile was considered as three compartments - the surface soil (0 – 0.4 m depth), an intermediate depth (from 0.4 – 0.8 m for 2HBD to 0.4 – 1.6 m for 2HTD) and the capillary fringe / groundwater. The following assumptions were made:

1. the surface soil isotopic compositions would oscillate widely in response to evaporation and infiltration;
2. intermediate depth soil isotopic compositions would remain relatively constant; and

3. the capillary fringe / groundwater would be too saline to be used by the trees.

To reproduce these layers for this model, the individual measurements in each layer were weighted by the volumetric water content. The estimated fraction of water taken from each compartment using the inverse method is compared to the zone of extractable water using the criteria described in Case Study 3A.

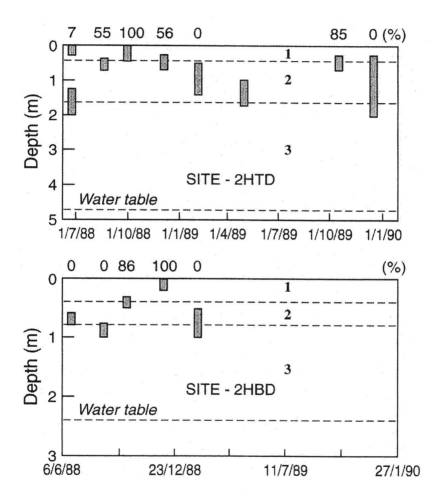

Figure 9. Expected depth of water extraction based on water availability (shaded zones). Zones marked 1 and 2 are the upper and lower compartments, respectively, and zone marked 3 represents the zone too saline for water extraction. Also shown are the percentages of water extracted from the upper compartment as calculated from the isotopic composition (Figure 7, Brunel *et al.* 1995).

The isotopic composition of the sapwater was assumed to be a mixture of water from the different compartments. Using the same examples as discussed in Case Study 3A (Fig. 7) the compartmental model provides further information relating to the proportion of water taken from each defined compartment.

At site 2HBD (Sampling 4), the compartmental model indicates that 70 – 90% of the water was being extracted from the surface 40 cm, depending on the weighting given to the δ^2H and δ^{18}O results. For Sampling 6 at Site 2HBD, the compartment model indicates that no water was being taken from the upper layer. This is similar to the result from Sampling 2 at Site 2HTD, where 7% of plant water was apparently being obtained from the upper layer. The results for sites 2HBD and 2HTD are shown in Figure 9. A sensitivity analysis showed that the proportion of water derived from either layer could be changed by up to 20% depending on the weighting given to the δ^2H and δ^{18}O values.

The following example (Case Study 4) shows quite a different approach, although the same considerations apply. This example uses a δ^2H – δ^{18}O plot, with several sources to infer degree of use of stream water.

CASE STUDY 4: ARE TREES USING GROUNDWATER OR STREAM WATER?

A field study by Thorburn and Walker (1993) reported that changes to the flow regime resulting from river regulation led to a decline in the health of the Eucalyptus woodland species on the Chowilla Floodplain. River red gums (*Eucalyptus camaldulensis*) growing over shallow saline (~40 dS m^{-1}) groundwater were sampled at various distances (between 0.5 and 40 m) from a permanent freshwater creek.

Four sites, A, B, C and M, were chosen on the Chowilla floodplain. The creek water was of low salinity (~0.8 dS m^{-1}), while the groundwater salinity ranged from 8 dS m^{-1} in winter at site M (Fig. 10) to 50 dS m^{-1} at site C in summer. Samples from soil profiles at the inland sites were taken and analysed for matric and osmotic potential, and δ^2H and δ^{18}O isotopes. The soil profiles at the creek-side sites were saturated and, accordingly, effective sampling was not possible. Creek water and groundwater samples were taken on five occasions and analysed for δ^2H and δ^{18}O isotopes.

The δ^2H and δ^{18}O data from all of the tree sap, creek and groundwater samples lie to the right of the global meteoric water line (as shown in Fig. 11), indicating enrichment by evaporation relative to rainfall. The four groups of samples, i.e. soil, creek water and groundwater data form a straight line with δ^2H = 3.1δ^{18}O – 16.1 (r^2 = 0.87). This regression was used to calculate the proportion of sapwater derived from the possible tree water sources.

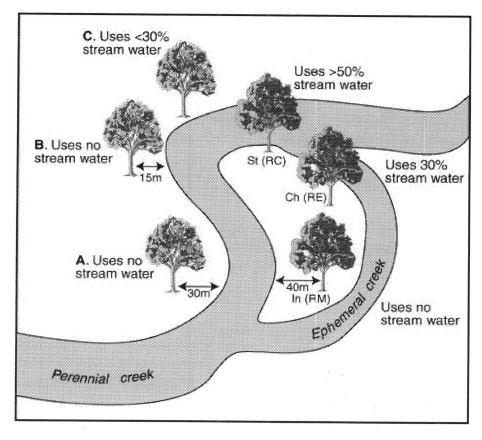

Figure 10. Schematic diagram of stream water use by red gums (adapted from Figure 17, Jolly and Walker 1995).

Sampling of the soil profile, stream and groundwater found seasonal differences in isotopic composition. The twig samples from sites 15 to 40 m from the nearest stream (sites A, B and M) matched the soil water δ^2H and $\delta^{18}O$ values from 0.2 m depth and below 1.4 m depth, including the groundwater. A comparison of the total soil water potential (matric and osmotic potentials) and predawn xylem water potentials showed that the trees at sites A, B and M were extracting their water from $1 - 1.5$ m depth, with soils less than 1 m depth being too dry. The soil moisture from $1 - 1.5$ m depth is derived from capillary rise through the clay from the groundwater.

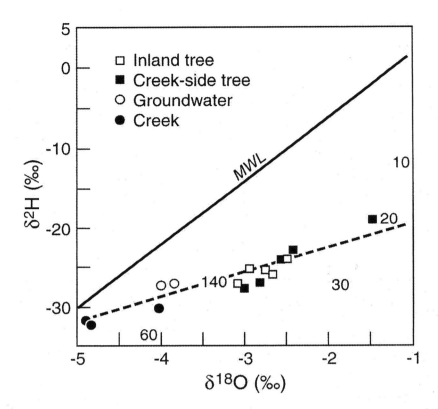

Figure 11. $\delta^2H - \delta^{18}O$ plot for all tree sap and water source samples collected during winter and early spring 1991 (the numbers indicate the depth in cm of the soil water sample). The meteoric water line and the regression line (dashed line) for all samples (excluding soil samples) are also shown (Figure 5, Thorburn and Walker 1993).

The isotopic composition of twig samples from site C showed evidence of mixing of streamwater and soil water from 0.05 – 0.25 m depth within the tree during summer. The isotopic composition of the twig samples was displaced towards that of the stream water from the soil water at 0.05 – 0.25 m depth. However, the soil water potential measurements indicated that the soil from 0.05 – 0.25 m depth was too dry for the tree to extract water while the soil water from 0.5 m depth to the groundwater was available for the trees. Mensforth *et al.* (1994) used the simple mixing model of Thorburn and Walker (1993) to determine the proportion of groundwater used by the trees at each site and found that < 30% of the twig water was derived from the stream water in summer.

The study concluded water uptake by red gums is strongly dependent on the long-term availability of surface water. In all cases, mixed usage of soil water and groundwater was significant (even when saline) as these are a reliable source of water and could provide a supply of nutrients.

4. INTEGRATING ISOTOPE DATA INTO WATER BALANCE STUDIES

The best way to overcome the limited capability of isotope methods is to integrate them with other measurements. Useful auxiliary data includes soil and leaf water potential, transpiration flux, soil salinities, piezometric data and stomatal conductance. However, integration is not straightforward. In this section, we illustrate three different approaches:
1. Using isotopes to partition the transpiration flux (Case Study 5);
2. Comparing isotope methods to piezometric data to substantiate groundwater use over a limited period of time (Case Study 6);
3. Using the data to test other conceptual or quantitative models that are not based on the isotope data i.e. an independent appraisal of the technique used (Case Study 7).

CASE STUDY 5: MEASUREMENT OF GROUNDWATER DISCHARGE

The groundwater discharge flux from the semi-arid infrequently flooded Chowilla floodplain on the River Murray was measured. A shallow (2 – 4 m depth), saline (11 – 30 dS m^{-1}) unconfined aquifer underlies the floodplain. Thorburn *et al.* (1993a) combined measurements of tree water use and stable isotopes of water to determine both rates and sources of water used by the *Eucalyptus* communities on this floodplain. The groundwater was the only reliable source of water for the forests, however the high groundwater salinity levels were thought to limit groundwater discharge fluxes. It was felt that rainfall stored in the soil profile would further complicate estimates of groundwater discharge, thus making traditional measurements of discharge of limited use.

Transpiration rates were determined from heat pulse measurements of tree sap velocity that were empirically scaled based on relationships between sap flux density and tree size (Hatton and Hsin-I Wu 1995). Transpiration of *E. camaldulensis* was approximately six times that of *E. largiflorens* on a tree basis. Thorburn *et al.* (1993a) postulated that the difference might be related to the lower leaf area index and / or the higher soil and groundwater salinities beneath the trees of *E. largiflorens*.

Tree water sources were determined using a method similar to that in Case Study 3. Maximum total soil water potentials of -3.5 MPa for *E. largiflorens* and -2.4 MPa for *E. camaldulensis* were used (Eldridge *et al.* 1993, Mensforth *et al.* 1994).

Table 1. Depth in the soil profile from which water was extracted, percentage of transpired water that originated from the groundwater, and groundwater discharge fluxes (mm d^{-1}) by *E. largiflorens* (sites BH, BM and BT) and *E. camaldulensis* (Site RM) trees in different seasons (Tables 3 and 4, Thorburn *et al.* 1993a).

| | *E. largiflorens* | | | | *E. camaldulensis* | |
	BH		BM	BT	RM	
Summer 1991						
Source (m)	1.7 – 3.3		0.2 – 0.6	0.2 / 3.0	0.1 / 2.8	
% Groundwater	100		100	65 (± 18)	79 (± 8)	
Discharge (mm d^{-1})	0.3		nd[a]	nd[a]	1.0	
Winter 1991	(a)[c]	(b)[c]				
Source (m)	0.2 / 4.0[b]	0.9 – 1.3	0.2 – 0.9	0.1 / 4.0	0.3 / 2.4	
% Groundwater	44 (± 21)	100	100	51 (± 17)	58 (± 10)	
Discharge (mm d^{-1})	0.13	0.3	0.2	0.1	1.0	
Autumn 1992					(a)[c]	(b)[c]
Source (m)	0.8 – 1.3		0.3 – 0.7	0.5 – 0.6	1.3 – 3.2	0.3 / 3.2
% Groundwater	100		100	100	100	76 (± 9)
Discharge (mm d^{-1})	0.3		0.2	0.2	2.0	2.0

[a] not determined, as transpiration measurements were not made
[b] tree water source was a mixture of water from 0.2 m and 0.4 m depth
[c] measurements (a) and (b) groundwater transpiration rates changed twice during a season

Groundwater discharge fluxes were determined from the transpiration fluxes and the proportion of groundwater taken by the trees. Transpiration rates turned out to be lower than the measured rates at sites where 'stored' rainfall was available from the shallower parts of the soil profile. The groundwater discharge flux varied through time, as the relative proportions of water uptake from the available sources changed.

The groundwater discharge fluxes determined at these sites appear to be small from a groundwater balance perspective. Potential water table fluctuations of 1 – 10 mm d^{-1} or less with lateral replenishment are at or below the limit of automated water level recorders and changes in atmospheric pressure may affect water table levels if the aquifer is partially confined. The integration of the isotope approach into traditional water balance study work should allow calculation of groundwater discharge fluxes which have been possible previously by more conventional methods.

CASE STUDY 6: TALL WHEATGRASS GROWING OVER SHALLOW GROUNDWATER

Tall wheatgrass (*Agropyron elongatum*) is a salt-tolerant pasture species that has been recommended for use in areas affected by dryland salinity. It is a perennial grass with an extensive root system, prolific near the soil surface, and extending to a depth of 3.5 m. Bleby *et al.* (1997) investigated the water use characteristics of this species in the field, with the aim of determining its suitability for stabilising water table levels and reducing the potential for dryland salinity in areas of groundwater discharge.

The study concentrated on tall wheatgrass growing in the Upper South East catchment of South Australia. The area receives an annual rainfall of 600 mm yr^{-1}. The study was conducted on established pastures of *A. elongatum* on a low-lying flat with marginally saline shallow groundwater (0 − 1.5 m depth; 0 − 10 dS m^{-1}). Groundwater levels beneath the two pastures were monitored continuously over the summer months, rising from 1.5 m depth to the surface during particularly wet winters.

Root density and distribution measurements beneath tall wheat grass showed that root density declined rapidly below 0.2 m depth though excavation found roots present to >1.5 m depth. Evapotranspiration from the tall wheatgrass pastures calculated using the ventilated chamber technique was 4.15 mm day^{-1} in summer, and declined to 0.34 mm day^{-1} in Autumn, (Bleby *et al.* 1997).

The higher water level in winter resulted in a relatively homogenous chloride and isotope soil profile in winter and spring. The isotope measurements indicated that the plants were taking water from 0.15 − 0.25 m depth in autumn and from 0 − 0.1 m depth in spring and from 0.15 − 0.45 m depth in summer. However, only towards the early autumn, when the soil moisture was depleted, did the isotopes show that the wheatgrass was directly using groundwater. This was matched with fluctuations in the water table. This continued for about a month before the wheatgrass senesced. The example shows the importance of matching sampling with temporal patterns of water use. In this case, the water strategy of the wheatgrass clearly varied over the year and frequent sampling was needed to authenticate this in detail. In a similar study, McEwan and Leaney (1996) found that during a wet year *Pinus radiata* was not using groundwater, but hypothesised that during a dry year, this may change.

CASE STUDY 7: IMPACT OF SALINITY ON LUCERNE GROWN OVER SHALLOW GROUNDWATER

The aim of this case study was to show how isotopes could be integrated with other field measurements through modelling studies. In many cases where vegetation is growing over shallow water tables, we become interested in answering a number of challenging questions such as:

If the water table is changed (i.e. in its depth or salinity), is it possible to predict what would happen to soil salinity, the plants, and the amount of groundwater used?

This was tested by Zhang *et al.* (1999), using a comprehensive data set from a lysimeter, for which the groundwater was spiked with δ^2H enriched water, and simulated with a soil-vegetation-atmosphere model (WAVES). Only two plant parameters were allowed to change in the calibration period, in the first year of the experiment.

Table 2. Effect of saline groundwater table on evapotranspiration, upflow, and leaf area index (Table 2, Zhang *et al.* 1999)

	Before ET (mm/d)	Upflow (mm/d)	LAI	After ET (mm/d)	Upflow (mm/d)	LAI
Lysimeter 1	5.9	3.3	2.2	3.8	1.1	1.3
WAVES	5.3	3.7	2.4	3.7	1.3	1.7
Lysimeter 2	4.4	2.2	1.9	3.1	0.7	1.4
WAVES	4.1	2.3	2.3	3.0	0.8	1.8

At the experimental site at Griffith, NSW, leaf area index, evapotranspiration, groundwater table movement and isotope composition were measured concurrently. A fresh water table was maintained at 60 cm in 1990 and dropped to 100 cm in 1991. A saline water table with isotopically-enriched water was introduced in March 1992 and maintained at 100 cm until the end of the experiment (Zhang *et al.* 1999).

The plots of Figure 12 show that the lucerne plant under stress is using the isotopically enriched water in early 1993, some 10 months after the introduction of the saline water table. This delay and the proportion of isotopically-enriched water being used by the plant were correctly simulated in the model, suggesting that some of the assumptions of water uptake and soil moisture movement had remained consistent (Zhang *et al.* 1999). This gave confidence in other aspects, as calibration of the model was independent of the measurement and also of the structure of the model. While the model matched other measured data, such as transpiration and growth, its structure simulated these in a manner which appeared to be closely related to the measurement techniques. Hence these would have provided only a partial check. The understanding of the processes and the ability to model this enabled Zhang *et al.* (1999) to predict the impacts of changes in groundwater on salinity and associated parameters. Zhang *et al.* (1999) found in their study that lucerne used surprisingly little groundwater and died back with increased salinity.

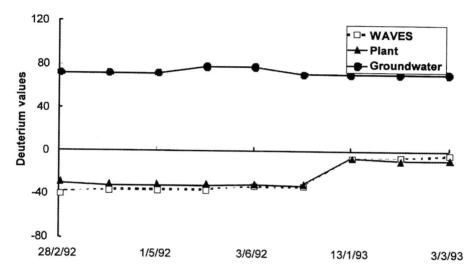

Figure 12. Comparison of the δ^2H values (δ‰) of groundwater, water extracted from lucerne, and calculated from the WAVES model during the period of 1992 – 1993. The δ‰ values of the groundwater were artificially high, thus making the stable isotope technique very sensitive to groundwater used by plants. This technique was found to be valuable for determining plant water extraction patterns and data suggested that plants were using only small (i.e. 20%) proportions of the saline groundwater. (Figure 5a, Zhang *et al.* 1999).

5.   SUMMARY AND CONCLUSIONS

Over the past decade there has been an escalating increase in the use of isotopes for determining the sources of water available to and used by plants. Studies have encompassed a variety of plants including agronomically-important species (wheat, alfalfa, wheatgrass), mangroves, halophytes, deciduous trees and evergreen woody species. Increasingly, these studies are being used in an integrated manner towards evaluating water balances, of which the isotope component is only one part.

Perhaps, the greatest potential source of the error with the isotopic technique is that associated with the extraction of water from soils. This can exceed the variation in δ^2H and $\delta^{18}O$ of water sources. It is accordingly important that extraction procedures are properly verified, and inexperienced personnel receive thorough training in specialist laboratories or alternatively have analyses done in those laboratories. Our intensive field study in SE Australia has found that the difference between expected isotopic values of sources and those found in plant xylem was less than 5 ‰ for δ^2H and 1 ‰ for ^{18}O. Since this experiment was conducted over a range of extreme conditions, we believe the method to be robust.

There are several issues relating to sampling where caution should be exercised. The sampling site from within a plant should be checked to avoid any enriched evaporated water. Direct sampling of water moving in xylem is to be encouraged, wherever possible. Roots are generally unsuitable for sampling unless specifically studying root processes. If twigs are used, samples should be taken from the whole canopy to minimise mixing errors. Sampling should be avoided within a few days after rain as the travel times through the plant may mean that this confuses the interpretation. Isotope measurements represent points in time and therefore temporal changes in plant water sources in the sampling regime may need to be considered. For example, the use of groundwater may only occur after other sources are used during a lengthy dry period. Rainfall should generally be collected as this often helps interpretation. It is critical to the method that all materials are collected in sealed glass jars as rapidly as possible, to avoid the potential of isotopic enrichment of the sample water due to evaporation.

It must be remembered that the isotope method is an inverse method in that the isotopic composition of the plant is used to infer information about the sources and vice versa. There is limited information contained in isotope measurements and hence for the compartmental model, the possible number of water sources is limited to 2, or at most 3. The error in the fraction of water derived from these sources is dependent on:
a) the difference in isotope values of these sources;
b) the degree of correlation in δ^2H and $\delta^{18}O$ values of different sources;
c) the errors in measurement; and
d) the degree to which sources can be compartmentalised into 2 or 3 sources.

The traditional $\delta^2H - \delta^{18}O$ plot is useful where there are several possible sources.

When isotopes are used by themselves the data provide a relatively blunt instrument of analysis. It is only where it is possible to integrate isotope data with other information such as from piezometric data, transpiration fluxes, soil and leaf water potential that one can obtain a more sophisticated interpretation of plant water use strategies. In particular, the use of isotope data to test the calibration and assumptions in a model that has been calibrated independently of the isotope data brings forth many possibilities. Models, whether conceptual or quantitative, offer a means of bringing prior knowledge into the interpretation of plant processes.

With the experimental basis described in this chapter, there should be no reason why isotopes can not be used much more widely in agronomic, forestry and ecological studies. Isotopes allow measurements of processes that cannot be obtained in any other way, provided some of the difficulties with the method are not underestimated. The advances in soil moisture monitoring and transpiration measurements, as well as vegetation-soil models, together with isotopes, should see large advances in our understanding of plant water strategies over the next ten years.

REFERENCES

Bariac, T., Ferhi, A., Jusserand, C. and Letolle, R. (1983). Soil-Plant-Atmosphere: A contribution to the study of isotopic composition of the water of various components of this system. In 'Isotope and Radiation Techniques in Soil Physics and Irrigation Studies.' Proceedings of an International Symposium, Vienna. (International Atomic Energy Agency: Vienna.)

Barnes, C.J. and Allison, G.B. (1983). The distribution of deuterium and oxygen-18 in dry soil. Theory. *Journal of Hydrology* 60, 141-56.

Bleby, T.M., Aucote, M., Kennett-Smith, A.K., Walker, G.R. and Schachtmann, D.P. (1997). Seasonal water use responses of tall wheatgrass [*Agropyron elongatum* (host) Beauv.] in a saline environment. *Plant, Cell and Environment* 20, 1361-71.

Brunel, J.-P., Walker, G.R. and Kennett-Smith, A.K. (1995). Field validation of isotopic procedures for determining water used by plants in a semi-arid environment. *Journal of Hydrology* 167, 351-68.

Brunel, J.-P., Walker, G.R., Walker, C.D., Dighton, J.C. and Kennett-Smith, A.K. (1991). Using stable isotopes of water to trace plant water uptake. In 'Stable Isotopes in Plant Nutrition, Soil Fertility and Environmental Studies.' (Ed. G.R. Stewart.) pp. 543-51. (International Atomic Energy Agency and Food and Agriculture Organization of the United Nations: Vienna.)

Calder, I.R. (1991). Technical Note: Implications and assumptions in using the 'total counts' and convection-dispersion equations for tracer measurements - with particular reference to transpiration measurements in trees. *Journal of Hydrology* 125, 149-58.

Calder, I.R. (1992). Deuterium tracing for the estimation of transpiration from trees Part 2. Estimation of transpiration parameters using a time-averaged deuterium tracing method. *Journal of Hydrology* 130, 27-35.

Clark, I.D. and Fritz, P. (1997) Environmental Isotopes in Hydrogeology. (Lewis: Boca Raton, Florida.)

Dawson, T.E. (1993). Hydraulic lift and water use by plants: implications for water balance, performance and plant-plant interactions. *Oecologia* 95, 565-74.

Dawson, T.E. (1996). Determining water use by trees and forests from isotopic, energy balance and transpiration analysis: the role of tree size and hydraulic lift. *Tree Physiology* 16, 263-72.

Dawson, T.E. and Ehleringer, J.R. (1991). Streamside trees that do not use stream water. *Nature* 350, 335-37.

Dawson, T.E. and Pate, J.S. (1996). Seasonal water uptake and movement in root systems of Australian phreatophytic plants of dimorphic root morphology: a stable isotope investigation. *Oecologia* 107, 13-20.

Ehleringer, J.R. and Dawson, T.E. (1992). Water uptake by plants: perspectives from stable isotope composition. *Plant, Cell and Environment* 15, 1073-82.

Ehleringer, J.R., and Rundel, P.W. (1989). Stable Isotopes: History, Units and Instrumentation. In 'Stable Isotopes in Ecological Research.' Ecological Studies, Vol. 68. (Eds P.W. Rundel, J.R. Ehleringer, and K.A. Nagy.) pp. 1-15. (Springer-Verlag: New York.)

Ehleringer, J.R., Phillips, S.L., Schuster, W.F.S. and Sandquist, D.R. (1991). Differential utilisation of summer rains by desert plants: implications for in competition and climate change. *Oecologia* 88, 430-434.

Eldridge, S.R., Thorburn, P.J., McEwan, K.L. and Hatton, T.J. (1993). Health and structure of Eucalyptus communities on Chowilla and Monoman Islands of the River Murray floodplain, South Australia. CSIRO Division of Water Resources, Divisional Report No. 93/3, Adelaide, SA.

Flanagan, L.B., Ehleringer, J.R. and Marshall, J.D. (1992). Diferential uptake of summer precipitation among co-occuring trees and shrubs in a pinyon-juniper woodland. *Plant, Cell and Environment* 15, 831-36.

Fritz, F., and Fontes, J.-C. (1980). Introduction. In 'Handbook of Environmental Isotope Geochemistry. Volume 1. The Terrestrial Environment.' (Eds F. Fritz and J.-C. Fontes.) pp. 1-19. (Elsevier Scientific Publishing Company: Amsterdam, The Netherlands.)

Gat, J.R. (1981). Isotopic Fractionation. In 'Stable Isotope Hydrology: Deuterium and Oxygen-18 in the Water Cycle.' (Eds J.R. Gat and R. Gonfiantini.) (International Atomic Energy Agency: Vienna, Austria.)

Greacen, E.L., Walker, G.R. and Cook, P.G. (1989). Evaluation of the filter paper method for measuring soil water suction. International Meeting at University of Utah on Measurement of Soil and Plant Water Status. pp. 137-45. (Utah State University: Utah, U.S.A.)

Hatton, T.J. and Hsin-I Wu (1995). Scaling theory to extrapolate individual tree water use to stand water use. *Hydrological Processes* 9, 527-40.

Jolly, I.D. and Walker, G.R. (1995). A sketch of salt and water movement in the Chowilla floodplain. (CSIRO Division of Water Resources: Adelaide, S.A.)

Jusserand, C. (1980). Extraction de l'eau interstitielle des sediments et des sols - Comparison des valeurs de l'oxygene 18 par differentes methodes, premiers resultats. *Catena* 7, 87-96.

McEwan, K.L. and Leaney, F.W. (1996). Using the natural abundance composition of stable isotopes in sap water, groundwater and soil water to estimate wateruse by pine plantations near Mt Gambier, South Australia . CSIRO Division of Water Resources, Consultancy Report for Department of Environment and Natural Resources No. 96-34.

Mensforth, L.J. and Walker, G.R. (1996). Root dynamics of *Melaleuca halmatuorum* in response to fluctuating saline groundwater. *Plant and Soil* 184, 75-84.

Mensforth, L.J., Thorburn, P.J., Tyerman, S.D. and Walker, G.R. (1994). Sources of water used by riparian *Eucalyptus camaldulensis* overlying highly saline groundwater. *Oecologia* 100, 21-28.

Reynolds, B. 1984. A simple method for the extraction of soil solution by high speed centrifugation. *Plant and Soil* 78, 437-440.

Saxena, K., and Dressie, Z. (1984). Estimation of Groundwater Recharge and Moisture Movement in Sandy Formations by Tracing Natural Oxygen-18 and Injected Tritium Profiles in the Unsaturated Zone. In 'Isotope Hydrology.' pp. 139-50. (International Atomic Energy Agency: Vienna, Austria.)

Sternberg, L.S.L. and Swart, P.K. (1987). Utilization of freshwater and ocean water by coastal plants of southern Florida. *Ecology* 68, 1898-905.

Thorburn, P.J. and Mensforth, L.J. (1993). Sampling water from alfalfa (*Medicago sativa*) for analysis of stable isotopes of water. *Communications in Soil Science and Plant Analysis* 24, 549-57.

Thorburn, P.J. and Walker, G.R. (1994). Variations in stream water uptake by *Eucalyptus camaldulensis* with differing access to stream water. *Oecologia* 100, 293-301.

Thorburn, P.J., and Walker, G.R. (1993). The Source of Water Transpired by *Eucalyptus Camaldulensis*: Soil, Groundwater or Streams? In 'Stable Isotopes and Plant Carbon - Water Relations.' (Eds J.R. Ehleringer, A.E. Hall, and G.D. Farquhar) pp. 511-527. (Academic Press: San Diego, U.S.A.)

Thorburn, P.J., Hatton, T.J. and Walker, G.R. (1993a). Combining measurements of transpiration and stable isotopes of water to determine groundwater discharge from forests. *Journal of Hydrology* 150, 563-87.

Thorburn, P.J., Walker, G.R. and Brunel, J.-P. (1993b). Extraction of water from *Eucalyptus* trees for analysis of deuterium and oxygen-18: laboratory and field techniques. *Plant, Cell and Environment* 16, 269-77.

Walker, G.R., Woods, P.H. and Allison, G.B. (1994). Interlaboratory comparison of methods to determine stable isotope composition of soil water. *Chemical Geology (Isotope Geoscience Section)* 111, 297-306.

Whelan, B.R. and Barrow, N.J. (1980.) A study of a method for displacing soil solution by centrifuging with an immiscible liquid. *Journal of Environmental Quality*. 9, 315-319.

White, J.W.C., Cook, E.R., Lawrence, J.R. and Broeker, W.S. (1985). The D/H ratios of sap in trees: Implications for water sources and tree ring D/H ratios. *Geochimica at Cosmochimica Acta* 49, 237-46.

Woods, P.H. (1989). The use of azeotropic distillation to obtain water from natural materials for isotopic determinations. Third Australian Stable Isotope Conference, Adelaide. (CSIRO: Australia.)

Zhang, L., Dawes, W.R.J., Slavich, P.G., Meyer, W.S., Thorburn, P.J., Smith, D.J. and Walker, G.R. (1999). Growth and groundwater uptake response of lucerne to changes in groundwater levels and salinity: lysimeter, isotope and modelling studies. *Agricultural Water Management* 39, 265-82.

Zimmermann, U., Ehhalt, D. and Munnich, K.O. (1966). Soil-water movement and evapotranspiration: Changes in the isotopic composition of the water. In 'Isotopes in Hydrology.' Proceeding of the International Symposium on Isotopes in Hydrology. pp. 567-85. (International Atomic Energy Agency: Vienna.)

Chapter 5

What do $\delta^{15}N$ Signatures tell Us about Nitrogen Relations in Natural Ecosystems?

George R. Stewart
Faculty of Science, The University of Western Australia, Crawley, WA 6009 Australia. Email:
gstewart@science.uwa.edu.au

Key words: nitrogen isotope fractionation, N cycling, N nutrition, N_2 fixation

1.   INTRODUCTION

There is a persistent and perhaps widespread view that the behaviour of ^{15}N in soils and plants is too complex to permit variations in its natural abundance to be used as a tracer or even as a probe to explore plant-soil relationships in natural ecosystems (Högberg 1997, Handley and Scrimgeour, 1997, Griffith *et al.* 1999). It is well established that fractionation in the content of ^{15}N in plant and soil material may occur as a consequence of both biological and physico-chemical processes (Handley and Raven 1992, Högberg 1997). However it is argued that the processes determining $\delta^{15}N$ signatures of plants and soils are little understood and interpretations of $\delta^{15}N$ are generally empirical (Handley and Raven 1992). This may well apply despite the number of studies that have reported distinct patterns and differences in nitrogen isotope signatures between plant communities and between species within those communities. The interpretation of such results is somewhat controversial. Many such studies are based on what I would regard as rather young communities (post-glacial) and subject to very significant and poorly quantified anthropogenic inputs of nitrogen (see Pearson and Stewart 1993).

Here I will examine some case studies in which natural abundance ^{15}N has been used to explore the nitrogen sources assimilated in natural ecosystems and its use as a tracer. Before doing so it is worthwhile

M. Unkovich et al. (eds.),
Stable Isotope Techniques in the Study of Biological Processes and Functioning of Ecosystems, 91–101.
© 2001 *Kluwer Academic Publishers. Printed in the Netherlands.*

considering the factors known to affect the amount of ^{15}N in soil and hence in the nitrogen sources assimilated by plants.

2.   FACTORS AFFECTING ^{15}N FRACTIONATION IN SOILS AND PLANTS

The δ^{15}N of nitrogen available to plants in soils may be influenced by soil age (Vitousek *et al.* 1989), with older soils tending to have more ^{15}N than younger soils. ^{15}N accumulation as soils mature probably relates to the ^{15}N enrichment brought about by processes such as leaching, volatilisation, denitrification and uptake of soil nitrogen by plants and other organisms. Various processes of disturbance including the frequency of fire (Schulze *et al.* 1991), extent of soil wetting (Selles *et al.* 1986, Garten 1993, Sutherland *et al.* 1993) and flooding (Yoneyama 1994) are reported to modify soil ^{15}N content. Soil δ^{15}N may be influenced by the presence of nitrogen-fixing species or nitrogen-rich plant parts in the soil profile (Pate and Unkovich 1999) and by point sources of isotopically distinct nitrogen such as animal excreta (Mizutani *et al.* 1986, Erskine *et al.* 1998). Furthermore it has been suggested that community averaged values of foliar δ^{15}N for diverse kinds of vegetation are strongly related to rainfall (Handley *et al.* 1999).

A logical starting point for understanding the basis of differences in δ^{15}N signal between plants is that the nitrogen source assimilated by a plant has a strong influence on its subsequent ^{15}N signature. Thus plants relying primarily on NH_4^+-N are commonly reported to be ^{15}N-depleted (Nadelhoffer and Fry 1994). Furthermore, since soil nitrogen species may carry different δ^{15}N signatures one might expect that plants specializing in the utilisation of one source would exhibit different δ^{15}N values from those utilising other sources. However there may also be physiological and biochemical processes which lead to isotope fractionation at whole plant and plant organ levels during or after a particular nitrogen source has been absorbed (Robinson *et al.* 1998).

The type and extent of mycorrhization may influence plant δ^{15}N values (Högberg 1990). Depleted δ^{15}N values in many temperate woody perennials have been attributed to the dependence of such species on ectomycorrhizas (Nadelhoffer and Fry 1994). Experimental work involving VA mycorrhizal *Ricinus communis* has found that presence of the fungal symbiont changed the δ^{15}N value of the host by as much as 2‰ (Handley *et al.* 1993). In a parallel experiment, ectomycorrhizal seedlings of *Eucalyptus globulus* had δ^{15}N values not significantly different from non-mycorrhizal seedlings (Handley *et al.* 1993). The nitrogen isotope signature carried by particular species of plant are likely to reflect the interaction of a combination of possibly many soil and plant processes.

3.   VARIATION IN NITROGEN ISOTOPE SIGNATURES IN NATURAL COMMUNITIES

Over a period of several years we have examined the variation in $\delta^{15}N$ in species of diverse Australian plant communities (Pate *et al.* 1993, Stewart *et al.* 1995, Schmidt and Stewart 1997, Erskine *et al.* 1998, Pate *et al.* 1998). Several general points emerge from these investigations (Table 1).

Table 1. Foliar $\delta^{15}N$ values of Australian Plant Communities. Plants were sampled and analysed as described in Erskine *et al.* (1998). Community mean and standard deviations are shown (adapted from Stewart and Schmidt 1999).

Plant Community	Foliar $\delta^{15}N$ (‰)
Alpine-subalpine Heathland	-4.5 (1.7)
Sub-antarctic Feldmark	-2.4 (1.7)
Sub-tropical Wallum Heathland	-1.7 (2.9)
Subtropical Woodland	0.2 (1.4)
Tropical Woodland	0.5 (1.3)
Deciduous Monsoon Forest	4.9 (1.3)
Semi-arid Mulga Woodland	8.6 (0.6)
Tropical Coral Cay	10.6 (1.1)
Sub-antarctic Bird Colonies	13.1 (2.5)

Community averaged values show a considerable range, the most ^{15}N enriched values are associated with sites where there is a marked input of animal-derived nitrogen such as the penguin colonies on Macquarie Island and noddy colonies on Heron Island (see below). The component species of mulga also showed high $\delta^{15}N$ values (Pate *et al.* 1998) and were shown to be predominantly nitrate-assimilating (Erskine *et al.* 1998). In contrast plants from sites where nitrogen sources other than nitrate predominated were shown to have low $\delta^{15}N$ values (Schmidt and Stewart 1997) with the most depleted ^{15}N values at remote sites in the alpine and subalpine vegetation of Tasmania and the upland vegetation of Macquarie Island (Erskine *et al.* 1998). The range of values within these sites was largest in the alpine/sub-alpine vegetation (15‰) and least in the guano-dominated vegetation of Heron Island (2‰). The range of $\delta^{15}N$ values observed for putative nitrogen fixing species was extremely wide, -3 to 10.5‰. In the *Acacia*-dominated mulga vegetation (Pate *et al.* 1998) we found the eight common *Acacia* species had a mean $\delta^{15}N$ value of 9.1 ± 1‰. This value was the same as that for non-fixing shrubs and trees, concurring with the non-nodulated status of the acacias. In mulga soils symbiotic nitrogen fixation by acacias would thus appear insignificant. Evidence was obtained of nitrogen fixation by lichens and termites, and, in a leached soil fringing a lake, legumes were nodulated and fixing N. More generally, however, we have encountered considerable variation in $\delta^{15}N$ values of potential reference plants, thus making it difficult to accurately quantify the magnitude of symbiotic nitrogen fixation in many natural plant communities.

4.   MYCORRHIZAL-STATUS AND NITROGEN ISOTOPE SIGNATURES

Within all of the communities that we have studied there are consistent differences in $\delta^{15}N$ between component species. Our general impression so far is that, irrespective of the community average value, certain groups of species tend to have higher and others lower $\delta^{15}N$ values. Of particular interest are those species that consistently have the highest $\delta^{15}N$ values. These are almost always species belonging to plant families that are regarded as being non-mycorrhizal or are weakly mycorrhizal. The best examples of this in Australia are members of the Proteaceae, possibly the only major family of woody plants that is consistently non-mycorrhizal. In the alpine/subalpine vegetation, the northern eucalypt woodland, coastal heathlands and mulga woodland, members of the Proteaceae have $\delta^{15}N$ values higher than the community average (Table 2). Other families which have higher $\delta^{15}N$ values are the generally non-mycorrhizal Cyperaceae and Restionaceae (G. R. Stewart and J. S. Pate, unpubl.).

Table 2. Foliar $\delta^{15}N$ values of Proteaceae (see Table 1 for details).

Plant Community	Proteaceae $\delta^{15}N$ (‰)	Community $\delta^{15}N$ (‰)
Wallum Heathland	0.57 (1.50)	-1.7 (2.90)
Banksia Woodland	1.09 (1.14)	0.66 (1.32)
Sub-tropical Heathland	1.50 (1.86)	-0.66 (2.71)
Savanna Woodland	1.07 (0.81)	-0.12 (1.10)
Mulga Woodland	10.44 (1.99)	8.6 (0.61)
Alpine-subalpine Heathland	1.81 (1.15)	-4.5 (1.7)

Stock *et al.* (1995) also showed that Proteaceous species in two Cape ecosystems in South Africa had enriched ^{15}N signatures relative to other species. Comparable results have been obtained for subarctic communities where non-mycorrhizal species, or those with arbuscular mycorrhizae, were enriched in ^{15}N relative to ectomycorrhizal species that were in turn enriched relative to ericoid mycorrhizal species (Michelsen *et al.* 1966, Nadelhoffer *et al.* 1996). Michelsen *et al.* (1996) suggest that in the subarctic, mycorrhizal species specialize in using organic nitrogen in the litter and that this recalcitrant nitrogen source is depleted in ^{15}N relative to soil mineral nitrogen.

Evidence is now emerging that associated mycorrhizal fungi have the potential to influence plant δ^{15}N in a number of ways. They can assimilate and transport large amounts of nitrogen and thereby possibly enhance nitrogen uptake by host plants (Arnebrant *et al.* 1993, Frey and Schlepp 1993). Increased substrate availability is likely to result in greater discrimination against ^{15}N. In addition, ectomycorrhizal fungi may enable host plants to utilise organic forms of nitrogen that the host cannot assimilate (Abuzinadah and Read 1986a and b, Abuzinadah and Read 1988, Abuzinadah *et al.* 1986, Finlay *et al.* 1992), and their hyphae may be concentrated in pockets rich in such organic nitrogen (Read 1991). Our own studies with *Eucalyptus* (Turnbull *et al.* 1995) have shown that colonisation of *E. maculata* and *E. grandis* roots with ectomycorrhizal fungi confers ability to utilise amino acids and protein nitrogen sources inaccessible to non-mycorrhizal seedlings. Mycorrhizal infection may therefore broaden the nitrogen source of the host to include a range of inorganic and organic compounds, possibly carrying δ^{15}N signatures different from other more generally used sources of nitrogen.

Despite what is said above there is still little concrete evidence that the nitrogen sources accessed by different mycorrhizal plants carry uniquely different isotope signatures. Thus, we were unable to relate the nitrogen isotope signature of soil nitrogen fractions to those of component species of a wallum heathland (Schmidt and Stewart 1997). Rather we showed that the fine roots of ericoid and ectomycorrhizal species were enriched in ^{15}N and suggested that there was fractionation at the point of nitrogen transfer from fungal to plant partner with the plant receiving ^{15}N depleted nitrogen. We hypothesise that species differences in nitrogen isotope signatures relate more to physiological differences in root functioning rather than being indicative of differences in the exploitation of soil nitrogen sources. In these studies ^{15}N natural abundance measurements indicate either differences in physiological processes among species that lead to differences in nitrogen isotope signatures or differential utilisation of soil nitrogen sources carrying different isotope signatures.

5. ^{15}N NATURAL ABUNDANCE OF RAINFOREST EPIPHYTES

A study of the foliar abundance of ^{15}N of vascular rainforest epiphytes and associated soil-rooted trees revealed marked differences (Stewart *et al.* 1995). Leaves of epiphytes were depleted relative to the trees in six rainforest communities in Australia, Brazil and the Solomon Islands and the data generally interpreted as indicating that epiphytes utilised nitrogen sources depleted in ^{15}N. The epiphytes could be separated into those having both low δ^{15}N values (-4 to -2‰) and low leaf-nitrogen contents and those having somewhat higher δ^{15}N values (-2 to 0‰) and higher nitrogen contents. It was suggested that species in the second group might have assimilated nitrogen derived from associated free-living nitrogen fixers, while those in the first group were probably relying on depleted nitrogen from atmospheric deposition. In this study ^{15}N natural abundance values clearly indicated that epiphytes were accessing nitrogen sources distinct from those of their supporting phorophytes.

6. ^{15}N AS A TRACER OF NITROGEN FLUXES IN NATURAL ECOSYSTEMS

There has been considerable criticism of attempts to employ ^{15}N as a tracer of nitrogen fluxes in natural environments (Handley and Scrimegour 1997). In an attempt to offset this I have selected three examples that illustrate how one may employ a highly distinctive signature of a specific nitrogen source to see the extent to which this can be traced through the plant species at a particular site.

The first example is a study conducted on the subantarctic, Macquarie Island (Erskine *et al.* 1998). Here there is a massive input of nitrogen associated with the deposition of bird guano and where the guano-nitrogen carries a distinctive, highly enriched ^{15}N signature. Royal penguin colonies contribute guano-based nitrogen equivalent to around 150,000 kg per year. As shown in Table 3, the ^{15}N natural abundance of plants and soils in the vicinity of royal penguin colonies suggest that most of the nitrogen used by plants at these sites is derived from the excreted guano whereas plants growing in upland plateau sites exhibited very low δ^{15}N values (-10‰). It was further suggested that a major portion of the nitrogen acquired by plants at the upland sites was derived from the volatilisation of ammonia released by hydrolysis of uric acid in the guano since volatilised ammonia collected above the royal penguin colonies gave δ^{15}N values of -10.0 ± 3.1‰ and plants upwind of the colony had δ^{15}N values of -10‰. Samples of the subtidal kelp, *Durvillaea antarctica* collected from the coast adjacent to the site of a penguin colony showed considerable ^{15}N enrichment ($\delta^{15}N$ 11.9‰). The same species collected at a site distant from any bird colonies had a $\delta^{15}N$ value of 1.6‰. The results obtained in this study made it possible to formulate a model to describe the ^{15}N enrichment and the flow of nitrogen for the Macquarie Island ecosystem (Erskine *et al.* 1998). In this instance the input of nitrogen with a distinctive $\delta^{15}N$ value makes it relatively simple to interpret source-sink relationships on Macquarie Island.

Table 3. Foliar $\delta^{15}N$ values of plant communities on Macquarie Island. Sites were characterised in relation to the major features influencing nitrogen relations. Different letters indicate significant differences at the $P < 0.05$ level (ANOVA and post-hoc LSD test). Standard deviations are in parentheses, n-number of samples (see Erskine *et al.* 1998 for details).

Site Characteristics	Foliar $\delta^{15}N$ (‰)	n
Giant petrels, humans, seals	12.9 (4.8)[a]	40
Albatross, burrowing petrels, penguins	7.0 (4.6)[b]	74
Particulate/gaseous nitrogen deposition, rabbits	1.3 (1.8)[c]	26
Deposition of volatilised ammonia	-5.2 (2.5)[d]	66
Remote from animal influence	-2.4 (1.7)[e]	21

The second example is again from sites dominated by a high input of guano, in this case from roosting and nesting seabirds on islands of the Great Barrier Reef. Mean $\delta^{15}N$ values of plants from the Low Isles (no seabird colonies) were 2.0 and 2.4‰ for cay and mangrove sites respectively (Table 4). Mean values for plants growing at sites associated with seabird colonies on Raine, Heron and One Tree Islands were 13.4, 9.3 and 12.6‰. The $\delta^{15}N$ values of freshly collected guano ranged from 8.6 to 10.9‰. Soil $\delta^{15}N$ values at seabird sites on Heron and One Tree Islands were in the range 10.8 – 14.9‰. Enrichment of soil and plant relative to guano is explainable in terms of volatilisation of ^{15}N-depleted ammonia derived from hydrolysis of uric acid in the guano. The impact of guano inputs is quite localised, plant and soil samples from beach sites on Heron that were removed from the direct impact of nesting and roosting birds were much less enriched in ^{15}N. Soil $\delta^{15}N$ values were 4 – 5‰ while those of the plants were 5 – 7‰. Examination of macro-algae in the Heron reef showed that they had $\delta^{15}N$ values in the range 2.0 – 4.5‰, similar to the values found for macro-algae from Wistari Reef (a reef with no cay). Again in this study natural abundance ^{15}N could be used to trace the flow of N from source to sinks.

The third example is the use of $\delta^{15}N$ signals to trace the movement of nitrogen from putative hosts to the root hemiparasite *Santalum acuminatum* growing in coastal heathlands of SW Australia (Tennakoon *et al.* 1997a and b). At each of three sites the parasite was found to parasitise most intensely common nitrogen fixing trees or shrubs (*Allocasuarina campestris, Acacia pulchella* and *Acacia rostellifera*). These legumes were well nodulated and showed $\delta^{15}N$ values close to atmospheric (0 to -1.5‰). As expected the associated *Santalum* displayed $\delta^{15}N$ values close to those of the major hosts and significantly different from those of other, non-fixing species.

Table 4. Foliar $\delta^{15}N$ values of plant communities on Great Barrier Reef Islands

Sampling Site	Foliar $\delta^{15}N$ (‰)
Low Isles Cay	1.6 (3.2)
Low Isles Mangrove	2.2 (2.1)
Raine Island	13.4 (4.4)
One Tree Island	11.8 (3.0)
Heron Island *Pisonia* Forest	10.6 (1.1)
Heron Island Beach	6.9 (2.6)

7.   CONCLUDING REMARKS

Should we be pessimistic about the quality of the information that can be derived from measurements of ^{15}N natural abundance in plants and soils? It may be that we need more precision in defining the questions or hypotheses that measurements of $\delta^{15}N$ values are going to answer or support. Mere collection of $\delta^{15}N$ values in the hope that they will provide both the question and answer is not the way forward. Some of the examples illustrated here show the successful use of variation in $\delta^{15}N$ values to trace the flow of nitrogen within plant communities. These same studies show that there may well be problems associated with the use of the natural abundance technique to quantify nitrogen fixation particularly in woody plant communities. However this arises from observations that some of what would be regarded as control non-fixing species exhibit $\delta^{15}N$ values that are similar to, or less than those expected from plants deriving most of their nitrogen from fixation. We should regard this as an interesting scientific challenge, rather than a cause for pessimism. The challenge is how to explain why some species have highly depleted ^{15}N contents. Generalised more widely, the problem is to account for the diversity of $\delta^{15}N$ values found among the component species of different ecosystems. The explanation that this reflects the selective use of nitrogen sources carrying a different nitrogen isotope signature, while seductively attractive, is as yet mostly unproven. It is however consistent with a growing body of evidence that within a particular plant community, species differ in their capacity to assimilate inorganic and organic nitrogen sources. The challenge then is to link the plant $\delta^{15}N$ value to that of specific soil nitrogen sources as was done by Pate *et al.* (1993) for the nitrophilous species *Ptilotus polystachyus* in which leaf and xylem sap $\delta^{15}N$ values were similar to that of soil nitrate. I suggest that studies based solely on the measurement of $\delta^{15}N$ values of plant tissue are likely to prove of limited use. If we wish to study nitrogen relations at the plant, community or ecosystem level ^{15}N profiles are best deployed as part of a suite of measurements. In most of our studies we have combined the determination of $\delta^{15}N$ values with metabolic profiling of leaves and xylem sap, enzymic profiling, measurements of soil nitrogen sources (Pate *et al.* 1993, Schmidt and Stewart 1997, Erskine *et al.* 1998) and more recently we have added to this, measurement of the capacity of root systems to take up and assimilate different nitrogen sources (Schmidt and Stewart 1999). My final conclusion would be that the diversity of plant $\delta^{15}N$ values is a cause for optimism, a source of further study and one likely to inform us better about plant nitrogen relations.

REFERENCES

Abuzinadah R. A. and Read D. J. (1986a). The role of proteins in the nitrogen nutrition of ectomycorrhizal plants. I. Utilisation of peptides and proteins by ectomycorrhizal fungi. *New Phytologist* 103, 481-493.

Abuzinadah R. A. and Read D. J. (1986b). The role of proteins in the nitrogen nutrition of ectomycorrhizal plants. III. Protein utilisation by *Betula, Picea* and *Pinus* in mycorrhizal association with *Hebeloma crustuliniforme. New Phytologist* 103, 507-514.

Abuzinadah R. A. and Read D. J. (1988). Amino acids as nitrogen sources for ectomycorrhizal fungi: utilisation of individual amino acids. *Transactions of the British Mycological Society* 91, 473-479.

Abuzinadah R. A., Finlay R. D., and Read D. J. (1986). The role of proteins in the nitrogen nutrition of ectomycorrhizal plants. II. Utilisation of protein by mycorrhizal plants of *Pinus contorta. New Phytologist* 103, 495-506.

Arnebrant K., Ek H., Finlay R. D., and Söderström B. (1993). Nitrogen translocation between *Alnus glutinosa* (L.) Gaertn. seedlings inoculated with *Frankia* sp. and *Pinus contorta* Doug. ex Loud seedlings connected by a common ectomycorrhizal mycelium. *New Phytologist* 124, 231-242.

Erskine, P. D., Bergstrom, D. M., Schmidt, S., Stewart, G. R., Tweedie, C. E., and Shaw, J. D. (1998). Subantarctic Macquarie Island - a model ecosystem for studying animal derived nitrogen sources using ^{15}N natural abundance. *Oecologia,* 117, 187-193

Finlay R. D., Frostegård C., and Sonnerfeldt A.-M. (1992). Utilisation of organic and inorganic nitrogen sources by ectomycorrhizal fungi in pure culture and in symbiosis with *Pinus contorta* Dougl. ex Loud. *New Phytologist* 120, 105-115.

Frey B. and Schlepp H. (1993). Acquisition of nitrogen by external hyphae of arbuscular mycorrhizal fungi associated with *Zea mays* L. *New Phytologist* 124, 221-230.

Garten C. T., Jr. (1993). Variation in foliar ^{15}N abundance and the availability of soil nitrogen on Walker Branch watershed. *Ecology* 74, 2098-2113.

Griffith, H., Borland, A., Gillon, J., Harwood, K., Maxwell, K., and Wilson, J. (1999). Stable isotopes reveal exchanges between soil, plants and the atmosphere. In 'Physiological Plant Ecology.' (Eds. M.C. Press, J.D. Scholes and M.G. Baker) pp 415-441. (Blackwell Science Ltd: Oxford.)

Handley, L. L. and Raven, J. A. (1992). The use of natural abundance of nitrogen isotopes in plant physiology and ecology. *Plant Cell and Environment* 15, 965-985.

Handley, L. L. and Scrimgeour, C. M. (1997). Terrestrial plant ecology and 15N natural abundance: the present limits to interpretation for uncultivated systems with original data from a Scottish old field. *Advances in Ecological Research* 27, 133-212.

Handley, L. L., Daft, M. J., Wilson, J., Scrimgeour, C. M., Ingelby, K., and Sattar, M. A. (1993). Effects of the ecto- and va-mycorrhizal fungi *Hydnagium* and *Glomus clarum* on the δ^{15}N and δ^{13}C values of *Eucalyptus globulus* and *Ricinus communis*. *Plant Cell and Environment* 16, 375-382.

Handley, L. L., Austin, A. T., Robinson, D., Scrimgeour, C. M., Raven, J. A., Heaton, T. H. E., Schmidt, S., and Stewart, G. R. (1999). The ^{15}N natural abundance (δ^{15}N) of ecosystem samples reflects measures of water availability. *Australian Journal of Plant Physiology* 26, 185-199.

Högberg, P. (1990). [15]N natural abundance as a possible marker of the ectomycorrhizal habit of trees in mixed African woodlands. *New Phytologist* 115, 483-486.

Högberg, P. (1997). [15]N natural abundance in soil-plant systems. *New Phytologist* 137, 179-203.

Michelson, A., Schmidt, I. K., Jonasson, S., Quarmby, C., and Sleep, D. (1996), Leaf ^{15}N abundance of subarctic plants provides evidence that ericoid, ectomycorrhizal and non- and arbuscular mycorrhizal species access different sources of soil nitrogen. *Oecologia* 105, 53-63.

Mizutani H., Hasegawa H., and Wada E. (1986). High nitrogen isotope ratio for soils of seabird rookeries. *Biogeochemistry* 2, 221-247.

Nadelhoffer, K. J. and Fry, B. (1994). Nitrogen isotope studies in forest ecosystems. In 'Stable Isotopes in Ecology and Environmental Science.' (Eds K. Lathja and R. H. Michener.) pp. 22-44. (Blackwell Scientific Publication: London.)

Nadelhoffer, K. J., Shaver, G., Fry, B., Giblin, A., Johnson, L, and McKane, R. (1996). [15]N natural abundance and N use by tundra plants. *Oecologia* 107, 386-394.

Pate, J. S. and Unkovich, M. J. (1999). Measuring symbiotic nitrogen fixation: case studies of natural and agricultural ecosytems in a Western Australian setting. . In 'Physiological Plant Ecology.' (Eds. M. C. Press, J. D. Scholes and M. G. Baker) pp 53-173. (Blackwell Science Ltd: Oxford.)

Pate, J. S., Stewart, G. R., and Unkovich, M. J. (1993). [15]N natural abundance of plant and soil components of a *Banksia* woodland ecosystem in relation to nitrate utilization, life form, mycorrhizal status and N_2-fixing abilities of component species. *Plant, Cell and Environment* 16, 365-373.

Pate, J. S., Unkovich, M. J., Erskine, P. D., and Stewart, G. R. (1998). Australian mulga ecosystems δ^{13}C and ^{15}N abundance of biota components and their ecophysiological significance. *Plant Cell and Environment* 21, 1231-1242.

Pearson, J. and Stewart, G.R. (1993). The deposition of atmospheric ammonia and its effects on plants. (Tansley Review 56). *New Phytologist* 125, 283-305.

Read, D. J. (1991). Mycorrhizas in ecosystems. *Experientia* 47, 376-391.

Robinson, D., Handley, L. L., and Scrimgeour, C. M. (1998). A theory for $^{15}N/^{14}N$ fractionation in nitrate-grown vascular plants. *Planta* 205, 397-406.

Schmidt, S. and Stewart, G. R. (1997). Waterlogging and fire impacts on nitrogen availability and utization in a subtropical wet heathland (wallum). *Plant, Cell and Environment* 20, 1231-1241.

Schmidt, S. and Stewart, G. R. (1999). Glycine metabolism by plant roots and its occurrence in Australian plant communities. *Australian Journal of Plant Physiology* 26, 253-264.

Schulze E.-D., Gebauer G., Ziegler H., and Lange O. L. (1991). Estimates of nitrogen fixation by trees on an aridity gradient in Namibia. *Oecologia* 88, 451-455.

Selles F., Karamanos R. E., and Kachanoski R. G. (1986). The spatial variability of nitrogen-15 and its relation to the variability of other soil properties. *Soil Science Society of America Journal* 50, 105-110.

Stewart, G. R., Schmidt, S., Handley, L. L., Turnbull, M. H., Erskine, P. D., and Joly, C. A. (1995). 15N natural abundance of vascular rainforest epiphytes: implications for nitrogen source and acquisition. *Plant, Cell and Environment* 18, 85-90.

Stock, W. D., Wienland, K. T., and Baker, A. C. (1995). Impacts of invading N_2-fixing Acacia species on patterns of nutrient cycling in two Cape ecosystems: evience from soil incubation studies and ^{15}N natural abundance values. *Oecologia* 101, 375-382.

Sutherland R. A., van Kessel, C., Farrell R. E., and Pennock D. J. (1993). Landscape-scale variations in plant and soil nitrogen-15 natural abundance. *Soil Science Society of America Journal* 57, 169-178.

Tennakoon, K. U., Pate, J. S. and Arthur, D. (1997a) Ecophysiological aspects of the woody root hemiparasite *Santalum acuminatum* (R. Br.) A. DC and its common hosts in south western Australia. *Annals of Botany* 80, 245-256.

Tennakoon, K. U., Pate, J. S., and Stewart, G. R. (1997b). Haustorium-related uptake and metabolism of host xylem solutes by the root hemiparasitic shrub *Santalum acuminatum* (R. Br.) A. DC (Santalaceae). *Annals of Botany* 80, 257-264.

Turnbull, M. H., Goodall, R., and Stewart, G. R. (1995). The impact of mycorrhization on nitrogen source utilisation in *Eucalyptus grandis* and *Eucalyptus maculata*. *Plant, Cell and Environment* 18 1386-1394.

Vitousek, P. M., Shearer, G., and Kohl, D. H. (1989). Foliar ^{15}N natural abundance in Hawaiian rainforest: patterns and possible mechanisms. *Oecologia* 78, 383-388.

Yoneyama T. (1994). Nitrogen metabolism and fractionation of nitrogen isotopes in plants. In 'Stable Isotopes in the Biosphere.' (Eds E. Wada, T. Yoneyama, M. Minagawa, T. Ando and B.D. Fry) pp 92-102. (Kyoto University Press: Kyoto.)

Chapter 6

Assessing N_2 Fixation in Annual Legumes using ^{15}N Natural Abundance

Murray Unkovich[1,2] and John S. Pate[1]

[1]Centre for Legumes in Mediterranean Agriculture (CLIMA) and Department of Botany, The University of Western Australia, Crawley WA 6009 Australia: [2]Victorian Institute of Dryland Agriculture, Mallee Research Station, Walpeup Victoria 3507 Australia. Email: murray.unkovich@nre.vic.gov.au

Key words: annual legumes, symbiotic N_2 fixation, $^{15}N/^{14}N$ isotope ratios, nitrogen cycling

1.   INTRODUCTION

This chapter serves as an introduction to the use of natural variations in ^{15}N for estimating N_2 fixation by field-grown annual legumes. While all of the examples cited come from Australian studies, the principles illustrated can be applied effectively in any region. Much of the material is extracted from that presented in Unkovich *et al.* (1994b, 1997) and Pate *et al.* (1994) and complements the chapter in this volume by Peoples *et al.* dealing with perennial woody and herbaceous legumes.

The ability of legumes to acquire nitrogen (N) through the agency of symbiotic N_2-fixing root nodule bacteria sets them apart from virtually all other groups of plants. In addition to providing a cash crop, symbiotically competent legumes have been long used in farming systems throughout the world to provide N-rich above- and below- ground residues whose decomposition eventually supplies mineral N for growth of crops and pastures. In order to design strategies for optimising management of legume N for maximal production with minimal N pollution of water resources, it is essential that N inputs by legumes and the subsequent fate of this N be quantitatively assessed.

M. Unkovich et al. (eds.),
Stable Isotope Techniques in the Study of Biological Processes and Functioning of Ecosystems, 103–118.
© 2001 *Kluwer Academic Publishers. Printed in the Netherlands.*

Measurement of the amount of N fixed by a legume crop or pasture requires measurements of the percentage of legume nitrogen derived from the atmosphere (%Ndfa) and parallel measurements of total N in legume biomass. Although assessments of N yield are usually less costly in practical terms than those involved in estimating %Ndfa, considerable effort is still required when obtaining accurate biomass yields, particularly in grazed pastures (see for example Bolger *et al.* 1995 and Unkovich *et al.* 1998).

Bergersen (1980), Peoples *et al.* (1989), Pate and Unkovich (1999), and Unkovich and Pate (2000) have discussed in detail the practicalities, strengths and weaknesses of the various methods used for estimating N_2 fixation and the reader is referred to these articles for comprehensive discussions of available methodologies. The more reliable of the techniques currently available for field estimates of N_2 fixation are (a) xylem solute analysis, (b) ^{15}N isotope dilution, and (c) ^{15}N natural abundance (NA). The usefulness of the first of these three techniques is restricted principally to ureide-producing semi-tropical legume species (Ledgard and Peoples 1988), with the notable exception of the work of Herridge (1988) and Herridge and Doyle (1988) on the amide-producing species *Lupinus angustifolius*. The most reliable of the techniques currently available for measuring N_2 fixation are based on the use of nitrogen stable isotopes.

2.   [15]N ISOTOPE DILUTION THEORY

There are two stable isotopes of nitrogen, ^{14}N and ^{15}N, with the vast bulk (>99.5%) of this N comprised of the lighter isotope, and only 0.3630 – 0.3700 % of atoms as [15]N (Letolle 1980). Natural variations in ^{15}N abundance are normally expressed as parts per thousand (‰) deviation from atmospheric N_2 (0.3663⁻ atom %[15]N, Junk and Svec 1958), which by definition is given a delta (δ)[15]N value of 0‰ (see chapter in this volume by Dawson and Brooks).

Since atmospheric N_2 and derived natural soil nitrogen contains very little ^{15}N it is possible to use ^{15}N-enriched tracers to assess N_2 fixation. Indeed following the pioneering work of Burris *et al.* (1942), Norman and Krampitz (1945) and Allos and Bartholomew (1955), this so-called ^{15}N isotope dilution technique had been widely employed as a means of assessing the extent of dependence of a range of legumes on fixation of atmospheric N_2 (see reviews of Chalk 1985, Chalk and Ladha 1999, Warembourg 1993). However, over the past decade or so technological advances in isotope ratio mass spectrometry have made it easier to assess accurately the small natural differences that exist in the abundance of ^{15}N in soil and plant materials. This has revealed that soils are often slightly enriched in ^{15}N compared to the atmosphere. It has therefore been possible in many situations to measure symbiotic N$_2$ fixation in field settings simply by applying isotope dilution principles relating to ^{15}N natural abundance without recourse to additions of ^{15}N-labelled materials. It is this approach that is selected for detailed discussion in the rest of this chapter.

The proportional dependence of a legume on atmospheric and soil N can be estimated by comparing the natural δ^{15}N content of legume biomass with that of an adjacent 'reference' non-legume or, in certain situations, against the δ^{15}N value of the non-nodulated legume subsisting solely on soil N.

The following formula can thus be used to calculate the percentage of plant nitrogen derived from the atmosphere (%Ndfa):

$$\%\text{Ndfa} = \frac{\delta^{15}\text{N reference plant} - \delta^{15}\text{N legume}}{\delta^{15}\text{N reference plant} - \text{B}} \times \frac{100}{1} \tag{1}$$

where 'reference plant' refers to a non N$_2$-fixing plant (or non-nodulated legume) selected to match as close as is practical the study legume in temporal and spatial uptake of soil sources of N, and the factor 'B' refers to the δ^{15}N value of the effectively nodulated legume grown in media totally lacking combined N and thus totally dependent on symbiotic N fixation for its N requirements. This B value essentially corrects for any isotopic discrimination during distribution of symbiotically fixed N. (The significance of the requirement for a B value in the protocol of measurement is discussed in Section 3.3.)

2.1   Sensitivity of estimates of N₂ fixation using ^{15}N natural abundances

Since the δ value for atmospheric N$_2$ is essentially constant and set arbitrarily at 0‰ (Mariotti 1983), soil δ^{15}N needs to be significantly different from 0‰ before the NA approach can be applied successfully. The minimum significant enrichment (or depletion) of soil-derived N necessary to detect N$_2$ fixing activity, or a change therein (i.e. the sensitivity of the technique) can be theoretically predicted, given the analytical precision of the δ^{15}N measurement and the desired resolution of %Ndfa (see Unkovich and Pate 1993). For example if a sensitivity of 10% Ndfa is required, the error of the mass spectrometric analysis would simply need to be one tenth of that of the reference plant (soil-available N) enrichment. For example, given an analytical precision of δ^{15}N measurement of ± 0.2‰, a soil enrichment of at least 2‰ would be required to detect a change of 10%Ndfa. Thus, since

$$\%Ndfa = \frac{\delta^{15}N \text{ reference plant} - \delta^{15}N \text{ legume}}{\delta^{15}N \text{ reference plant} - B} \times \frac{100}{1}$$

$$\%Ndfa = ((2.0 - 1.8)/(2.0)) \times 100$$

$$= 10\%Ndfa$$

(B value assumed to be 0‰)

Following this line of reasoning further, the relationship between legume $\delta^{15}N$ and %Ndfa for a series of non-fixing reference plant enrichments would be as illustrated in Figure 1.

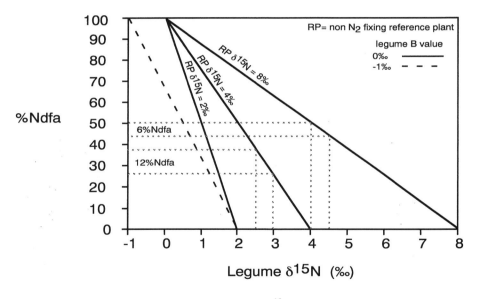

Figure 1. Hypothetical relationships between legume $\delta^{15}N$ and %Ndfa for a series of non-fixing reference plant (RP) $\delta^{15}N$ values. The scheme predicts how the sensitivity of %Ndfa estimates would be affected by a 0.5‰ change in legume $\delta^{15}N$ at reference plant $\delta^{15}N$ values of 2, 4 and 8‰. The relationship shown by the broken line (reference plant = 2‰) is for a B value of -1‰. This would give a higher precision of %Ndfa than a B of 0‰.

The relationship between $\delta^{15}N$ and %Ndfa is depicted in Figure 1 as linear for a given reference plant $\delta^{15}N$ value, so the sensitivity of estimates of %Ndfa are always directly proportional to that of the non-fixing reference plant component. For example, given a reference plant $\delta^{15}N$ of 4‰, a change in legume $\delta^{15}N$ of 0.5‰ would give a change of 12%Ndfa, whereas for a reference plant $\delta^{15}N$ of 8‰ the change would be only half of this (6%Ndfa). In effect the potential precision of estimates of %Ndfa is determined by both the reference plant enrichment and the analytical resolution of $\delta^{15}N$. This theoretical index of precision does not imply a similar degree of accuracy in actual estimates of %Ndfa within a given legume crop, since the efficacy of the technique is conditioned by other factors, such as variations across the site in legume N_2 fixation and soil $\delta^{15}N$, and the rigour of the plant sampling regime in accommodating such variations.

It should be appreciated from the above analysis that at high N_2 fixation rates, when legume $\delta^{15}N$ approaches 0‰, the methodology will become relatively insensitive to changes in reference plant $\delta^{15}N$. Hence, when reference plant species with substantially different $\delta^{15}N$ values are compared in turn with the legume, closely similar estimates of %Ndfa will result. Conversely at low fixation rates (high legume $\delta^{15}N$ values), choice of reference plants which match the legume N uptake as precisely as possible become paramount to accuracy. For example, in the extreme case of a poorly fixing legume with a $\delta^{15}N$ value of 3.5‰ compared to reference plants with values of 4 and 6‰, the %Ndfa estimates would be 13 and 40% respectively.

3.　　APPLICATION OF THE METHODOLOGY

The above rationale was used by Unkovich *et al.* (1994b) to ascertain the overall effectiveness of the NA method for estimating N_2 fixation in annual crops and pastures across south western Australia. In their survey of non-fixing plants from 243 clover and medic pastures 88% of plants had $\delta^{15}N$ values of 2‰ or greater (Fig. 2) and this enrichment was shown to be sufficient to estimate %Ndfa of these pastures with high (± 6%Ndfa) precision (Fig. 3). By comparision $\delta^{15}N$ values of reference plants in lupin and field pea crops in south-west Australia also mostly exceed 2‰, though rarely reach 5‰ (Fig. 4, and Pate *et al.* 1994). In eastern Australia, soil total N δ values over a 400km^2 catchment were shown to range widely from 2.5‰ – 6.8‰ (Ledgard *et al.* 1984).

The existence of predominantly positive, as opposed to negative, $\delta^{15}N$ values for the above and most other ecosystems (see chapter in this volume by Stewart) raises the question of how soil mineral N becomes enriched in ^{15}N relative to atmospheric N_2. Since the lighter ^{14}N reacts more quickly than ^{15}N (see chapter in this volume by Dawson and Brooks) any processes in which N is lost from an ecosystem (e.g. as gaseous NH_3, N_2, NOx, or through leaching) would result in more ^{14}N being lost from the soil than ^{15}N.

As a consequence the remaining N in the soil will become relatively enriched in [15]N. One would therefore expect that in ecosystems prone to denitrification or ammonia volatilisation, the soil NH_4^+ and NO_3^- would show elevated ^{15}N levels (see for example Evans and Ehleringer 1993, Hogberg 1990, Mizutani *et al.* 1985). Conversely in ecosystems with 'tight' N cycles, $\delta^{15}N$ of soil is likely to be close to atmospheric N_2 (0‰).

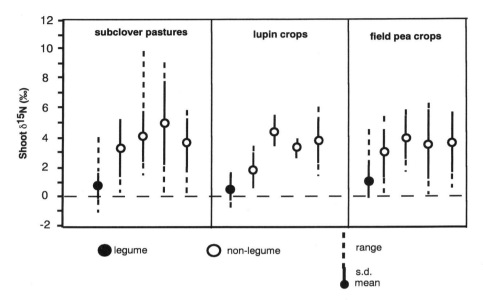

Figure 2. Variations in shoot $\delta^{15}N$ value of potential reference plants and study legumes across a range of agricultural ecosystems in south west Australia. δ values are given as means, ranges and standard deviations of means for each plant species.

3.1   Variation in $\delta^{15}N$ enrichment of soil and non N_2-fixing reference plants

It must be stressed that the methodology employing ^{15}N natural abundance values can be applied with a high degree of confidence only where the forms of soil N currently available to plants at a study site are relatively uniform in $\delta^{15}N$ throughout a rooting zone across a site. Where these prerequisites are not met, large spatial variations in soil $\delta^{15}N$ are likely to be evident between matched legume-reference plant pairings sampled across the study site, and reliability of assessment of %Ndfa accordingly reduced. Effects attributable to spatial and temporal variations in $\delta^{15}N$ have been investigated in several Australian studies in both crops and pastures (Bergersen *et al.* 1985 and Turner *et al.* 1987) and the general conclusion reached that such variations are not usually substantial enough to invalidate use of the NA method (see Table 1). We have investigated differences between the $\delta^{15}N$ of soil NO_3^- and NH_4^+ (Fig. 4) and in this instance concluded that the levels of variation present did not negate efficaceous use of the method.

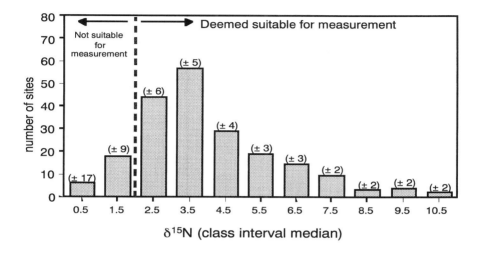

Figure 3. Histogram showing distribution of $\delta^{15}N$ values of the recommended reference plant, capeweed, across 240 subterranean clover-based pastures. Numbers in parentheses indicate predicted precision of %Ndfa estimates at centre of each class interval based on an analytical precision of ±0.2‰ (see Unkovich *et al.* 1993) and assuming a B value of -0.58‰ (see Table 4). The arbitrary line drawn at a reference plant value of 2‰ was chosen as a cut off point for use of the NA methodology based on the error of estimate being too high (±9%Ndfa) for the class interval below this value (see Unkovich *et al.* 1994b and Sanford *et al.* 1994).

3.2   Selection of non N₂-fixing reference plants

Where legume and reference plant are grown in close association (e.g. in typical pastures) obvious complications will arise were there to be direct transfer of fixed N of near zero delta value from legume to close-companion reference species. Attendant lowering of the reference plant delta below that of the soil available N would then result in %Ndfa being correspondingly underestimated. Pate *et al.* (1994) investigated this possibility in pea and lupin crops and subterranean clover pasture and found that, while such transfer appeared to complicate the NA methodology when grass reference species were used in mixed pastures, this did not appear to apply to broadleaf reference species. The authors therefore concluded that dicotyledonous species were preferable reference plants for pasture legume studies. In application of the NA technique to legume crops, there was no evidence for N transfer between lupin and non-leguminous grass or broadleaf reference species, but some evidence was obtained of N transfer from field pea to monocotyledonous reference species (Pate *et al.* 1994). Thus, it was suggested that reference plants should be sampled in pair-wise fashion as close as possible (= 2m) to the legume, but always distinctly outside the rooting catchment of field pea.

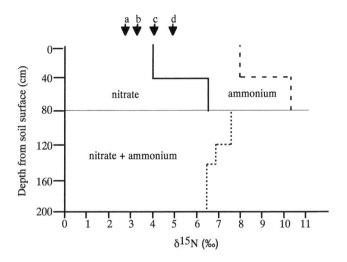

Figure 4. $\delta^{15}N$ profile for soil NO_3^- and NH_4^+ in 40 cm increments under a lupin crop 5 weeks after germination of the legume (lupin). Potential reference species $\delta^{15}N$ are shown as (a) *Bromus mollis*, (b) *Hordeum vulgare*, (c) *Emex australis* and (d) *Arctotheca calendula*. Concentrations of NH_4^+ were very low below 80 cm, necessitating analysis of the NH_4^+ and NO_3^- fractions together.

It is of course paramount that the reference plants utilised should have accessed pools of soil N of closely similar $\delta^{15}N$ value to that taken up by the study legume. Non-nodulating isolines of the legume might be ideal for such a purpose (Bergersen *et al.* 1985, Herridge *et al.* 1990), but these are currently available for very few of the more common grain crops (e.g. soybean) and pasture legumes. Another option used by Pate *et al.* (1994) for field peas is to trial reference species in the field by determining which one(s) acquire $\delta^{15}N$ values closest to those of adjacent non-nodulated (uninoculated) peas growing in the same soil. This approach is only possible, of course, where indigenous rhizobia capable of effectively nodulating the legume are absent from the study site.

Table 1. $\delta^{15}N$ values of legume and reference plants and derived value for proportional dependence on N_2 fixation (%Ndfa) using a 10-point sampling procedure for (A) a subterranean clover based pasture and a field pea crop, and (B) a lupin crop. (From Unkovich *et al.* 1994b)

(A) Subclover and Field Pea

Sampling point	Subclover $\delta^{15}N$	Capeweed $\delta^{15}N$	%Ndfa	Pea $\delta^{15}N$	Radish $\delta^{15}N$	%Ndfa
1	-0.21	3.85	88	0.52	3.28	77
2	-0.31	2.27	87	0.30	2.64	80
3	-0.20	2.52	84	0.61	2.46	67
4	-0.39	3.99	93	0.38	2.77	78
5	-0.37	6.79	96	0.05	4.25	93
6	-0.01	6.88	91	0.66	2.26	63
7	-0.20	3.36	88	0.32	2.80	80
8	-0.89	1.82	Note [A]	0.47	2.54	73
9	-0.57	3.80	97	0.78	3.05	68
10	-0.42	4.51	95	0.36	2.70	78
Mean	-0.36	3.99	91	0.45	2.88	76
± s.e.	0.08	0.54	1.3	0.06	0.16	2.44

[A] %Ndfa not calculated since reference plant $\delta^{15}N$ value less than 2‰.

(B) Lupin

Sampling point	Lupin $\delta^{15}N$	Barley $\delta^{15}N$	%Ndfa	Brome-grass $\delta^{15}N$	%Ndfa	Cape-weed $\delta^{15}N$	%Ndfa
1	0.0	3.6	94.7	3.6	94.7	3.0	93.8
2	0.4	2.8	80.0	2.1	73.9	4.0	85.7
3	0.0	4.0	95.2	-	-	3.3	94.3
4	0.0	4.1	95.3	2.9	93.5	4.0	95.2
5	0.0	4.2	95.4	3.4	94.4	3.4	94.4
6	0.0	4.0	95.2	3.8	95.0	3.8	95.0
7	-0.2	4.2	100.0	3.5	100.0	4.1	100.0
8	-0.2	3.0	100.0	3.0	100.0	-	-
9	0.2	2.8	86.7	4.6	91.7	3.7	89.7
10	0.1	4.0	92.9	2.8	90.0	1.8	85.0
Mean	0.03	3.67	93.5	3.30	92.6	3.45	92.6
± s.e.	0.02	0.06	0.60	0.08	0.86	0.07	0.54

Table 2. $\delta^{15}N$ values (‰) of peak biomass shoot dry matter of reference plants harvested within (in) grazed subterranean clover pasture, lupin or pea crop or adjacent but at least 100cm distant from (out) legume areas.

Legume	Site	Reference plant	In	Out	s.e.d	sig. level[A]
(A) Pea	Mt Barker	Radish	2.9	3.7	0.11	***
	Research Station	Barley	1.5	2.5	0.08	***
	Avondale	Barley	4.5	5.1	0.25	ns
(B) Lupin	Mt Barker	Grass	3.4	3.1	0.20	ns
	Research Station	Capeweed	3.3	3.3	0.13	ns
	Gingin	Ryegrass	0.5	0.3	0.23	ns
		Barleygrass	1.7	1.2	0.70	ns
		Capeweed	1.5	1.3	0.27	ns
	Reagan's Ford	Ryegrass	1.1	2.1	0.27	ns
		Bromegrass	3.1	2.5	0.52	ns
		Capeweed	3.9	3.4	0.10	*
(C) Subclover	Mt Barker	Grass	3.7	5.5	0.17	**
	Research Station		6.0	6.9	0.13	**
			5.9	6.4	0.14	*
		Capeweed	5.0	5.8	0.43	ns
			4.2	4.1	0.11	ns
			4.6	4.2	0.14	ns

[A] ns, not significant; * P<0.05; ** P<0.01; ***P<0.001. s.e.d., standard error of difference between two means.

3.3 ^{15}N isotope fractionation, 'B' values and timing of sampling

N_2-fixing legumes without access to soil N have been often shown to have shoot $\delta^{15}N$ ('B') values slightly different from atmospheric N_2 (Ledgard 1989, Bergersen *et al.* 1988). In virtually all cases the B values recorded for shoots of a legume are negative, whereas those of corresponding nodules are highly positive, and those of roots somewhere between the two (Table 3).

Although it has been suggested that the above mentioned fractionation relates to isotopic discrimination against ^{15}N during N_2 fixation (e.g. Yoneyama *et al.* 1986) we present here substantial evidence that only minimal, if any, fractionation occurs during the actual process of N_2 fixation (Tables 4 and 5). However, as indicated from the data of Table 3, appreciable isotope fractionation can take place during subsequent transport and biochemical transformations within the host plant (Unkovich and Pate

2000, Handley and Scrimgeour 1997, Peoples *et al.* 1989). Accordingly, when N_2 fixation estimates are performed on a shoot-only basis, a correction needs to be made for 'within-plant' discrimination effects. The extent to which such internal fractionation occurs can be assessed using $\delta^{15}N$ data (B values) for effectively nodulated plants grown in minus N pot culture. When deriving such B values, one should use the same microsymbiont as is infecting the study legume at the site in question. This is deemed necessary since the bacterial strain that infects a legume may in certain instances have a significant influence on the fractionation of fixed N which subsequently takes place between plant parts (Table 4, and Bergersen *et al.* 1986, Ledgard 1989, Unkovich *et al.* 1994b, 1998).

Table 3. $\delta^{15}N$ (B values) of shoots and nodulated roots of fully symbiotic sand-cultured plants of subterranean clover, lupin and field pea inoculated respectively with *Rhizobium* WU95, *Bradyrhizobium* WU425 and *Rhizobium* SU391 (From Unkovich *et al.* 1994b).

Species		Shoot	Nodulated root
Trifolium subterraneum	cv. Mt Barker	-0.9	0.9
	cv. Daliak	-0.4	1.1
	cv. Esperance	-0.7	3.3
Trifolium arvense		-0.3	0.8
Trifolium balansa		-0.3	1.6
Trifolium campestris		-0.6	0.1
Trifolium fragriferum		-0.4	1.0
Lupinus angustifolius	cv. Ilyarrie	-0.2	2.4
	cv. Danja	-0.2	2.0
	line 75A/333	-0.2	2.1
Pisum sativum	cv. Dundale	-0.4	1.4
	cv. Wirrega. 4.4	-0.3	1.6
	cv. Progretta	-0.2	2.3
	cv. Dinkum	-0.2	1.4
	line 82-14 p44	-0.4	1.9
	line 80-58 p12	-0.4	1.8

B values of the minus N-cultured glasshouse plants should also be assessed at the same stages in plant development as those at which the N_2 fixation estimates are made in the field, since fractionation effects can change substantially with plant age (Unkovich *et al.* 1994a and b). An adjustment may also be required to correct for any N supplied through the cotyledons of the sown seed (Bergersen *et al.* 1988), especially where the size of this seed reserve is large compared to that accumulated within the plant at the time of harvest.

Table 4. $\delta^{15}N$ (B values) of shoots, nodulated roots and whole plants of sand-cultured fully symbiotic subterranean clover (*Trifolium subterraneum* cv. Trikkala) partnered with mixed rhizobial inoculants prepared from soils from a range of pasture sites or with WU95, the recommended commercial inoculum for this clover.

Inoculum origin	Shoot δ	Nodulated Root δ	Whole plant δ
Airport	0.06	1.06	0.30
Mt Shadforth	-0.09	0.70	0.13
Porongurups	-0.21	1.59	0.27
Kangaroo Valley	-0.23	1.65	0.10
WU95	-0.38	0.47	-0.16
Stirling South	-0.41	1.65	0.09
Manypeaks	-0.54	1.79	0.20
Torbay 5	-1.22	2.69	0.15
Torbay 4	-1.27	2.27	-0.06
Redmond	-1.37	2.71	-0.16
Mean	-0.58	1.64	0.09
(± s.d.)	(± 0.50)	(± 0.73)	(± 0.20)

Table 5. $\delta^{15}N$ values for nodulated legumes grown in glasshouse culture in the absence of combined N and harvested at around mid-flowering. Whole plant values include a correction for seed N (Bergersen *et al.* 1988), though the 'uncorrected' values were < 0.1 ‰ different in each case (from Unkovich and Pate 2000).

Species		Shoot $\delta^{15}N$ (‰)	Whole plant $\delta^{15}N$ (‰)
Cicer arietinum	Chickpea	-1.34	0.28
Lupinus luteus	Yellow lupin	-0.88	0.18
Lens culinaris	Lentil	-0.51	-0.09
Medicago polymorpha	Burr medic	0.02	0.15

The timing of sampling of field crops or pastures is also important since there may be changes in $\delta^{15}N$ during senescence which do not relate to changes in the relative contributions of N sources for plant growth. For instance, amines are produced within the plant during remobilisation and decay, and these may be lost as gaseous NH_3 (Wetselaar and Farquhar 1980) which is much depleted in ^{15}N (Turner *et al.* 1983, O'Deen 1989). These losses will result in an increase in the enrichment of the maturing plant that is not due to changes in the balance between symbiotic N_2 fixation and soil N uptake by the plant. This is illustrated for a field-grown lupin crop in Figure 5 where it can be seen (5A) that the total crop N declines over the course of pod fill, even after fallen leaf N is accounted for. During this period (119 – 177 days after sowing) there is an increase of about 1‰ in shoot δ^{15}N (5B) while nodule δ^{15}N had reached a plateau at 119 days after sowing, indicating that N₂ fixation had probably ceased. We therefore advocate that for annual crops, sampling for N₂ fixation estimates using ^{15}N natural abundance be conducted a peak crop N accumulation, typically around mid-flowering, and before significant leaf drop. At this time complications arising from fractionation during remobilisation of N will be minimal and the sampling should integrate the whole time course of N₂ fixation for the crop.

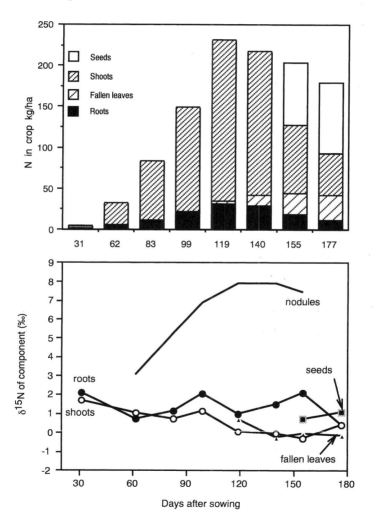

Figure 5. (A - above) Accumulation of N in a broadacre crop of lupin (*Lupinus angustifolius*) showing the decline in total crop N that occurs during remobilisation of N and plant senescence, and (B - below) changes in δ^{15}N values with time for the same crop components.

4.   SUMMARY

The NA method has the potential to provide reliable and relatively easily performed assessments of N_2 fixation by field-grown annual legumes, provided that plant available soil N is sufficiently enriched, 'B' values for the association are assessed, and appropriate reference plants are readily available at the study site. The technique is particularly useful at farm and broad ecosystem levels, and with suitable precautions, also appears reasonably well suited to pastures subjected to a range of grazing pressures (e.g. Bolger *et al.* 1995, Peoples *et al.* 1995, Sanford *et al.* 1995, Unkovich *et al.* 1997).

REFERENCES

Allos, H. F. and Bartholomew, W. V. (1955). Effect of available nitrogen on symbiotic fixation. *Soil Science Society America Proceedings* 19, 182-184.

Bergersen, F. J. (1980). Measurement of nitrogen fixation by direct means. In 'Methods for Evaluating Biological Nitrogen Fixation.' (Ed. F. J. Bergersen.) (John Wiley & Sons: Chichester.)

Bergersen, F. J., Peoples, M. B., and Turner, G. L. (1988). Isotopic discrimination during the accumulation of nitrogen by soybeans. *Australian Journal of Plant Physiology* 15, 407-420.

Bergersen, F. J., Turner, G. L., Gault, R. R., Chase, D. L., and Brockwell, J. (1985). The natural abundance of ^{15}N in an irrigated soybean crop and its use for the calculation of nitrogen fixation. *Australian Journal of Agricultural Research* 36, 411-423.

Bergersen, F. J., Turner, G. L., Amarger, N., Mariotti, F., and Mariotti, A. (1986). Strain of Rhizobium lupini determines natural abundance of ^{15}N in root nodules of *Lupinus* species. *Soil Biology and Biochemistry* 18, 97-101.

Bolger, T. P., Pate, J. S., Unkovich, M. J., and Turner, N. C. (1995). Estimates of seasonal nitrogen fixation of annual subterranean clover-based pastures using the ^{15}N natural abundance technique. *Plant and Soil* 175, 57-66.

Burris, R. H., Eppling, F. J., Wahlin, H. B., and Wilson, P. W. (1942). Studies of biological nitrogen fixation with isotopic nitrogen. *Soil Science Society of America Proceedings* 7, 258-262.

Chalk, P. M. (1985). Estimation of N_2 fixation by isotope dilution: An appraisal of techniques involving ^{15}N enrichment and their application. *Soil Biology and Biochemistry* 17, 389-410.

Chalk, P.M. and Ladha, J.K. (1999). Estimation of legume symbiotic dependence: an evaluation of techniques based on ^{15}N dilution. *Soil Biology and Biochemistry* 31, 1901-1917.

Evans, R. D. and Ehleringer, J. R. (1993). A break in the nitrogen cycle in aridlands? Evidence from $\delta^{15}N$ of soils. *Oecologia* 94, 314-317.

Handley, L. L. and Scrimgeour, C. M. (1997). Terrestrial plant ecology and ^{15}N natural abundance: The present limits to interpretation for uncultivated systems with original data from a Scottish old field. *Advances in Ecological Research* 27, 133-212.

Herridge, D. F. (1988). The narrow leaved lupin (*Lupinus angustifolius*) as a nitrogen-fixing rotation crop for cereal production. I. Indices of nitrogen fixation. *Australian Journal of Agricultural Research* 39, 1003-1015.

Herridge, D. F. and Doyle, A. D. (1988). The narrow leaved lupin (*Lupinus angustifolius*) as a nitrogen-fixing rotation crop for cereal production. II. Estimates of nitrogen fixation by field-grown crops. *Australian Journal of Agricultural Research* 39, 1017-1028.

Herridge, D. F., Bergersen, F. J., and Peoples, M. B. (1990). Measurement of nitrogen fixation by soybean in the field using the ureide and natural ^{15}N abundance methods. *Plant Physiology* 93, 708-716.

Hogberg, P. (1990). Forests losing large quantities of nitrogen have elevated N-^{15}:N-^{14} ratios. *Oecologia* 84, 229-231.

Junk, G. and Svec, H. J. (1958). The absolute abundance of the nitrogen isotopes in the atmosphere and compressed gas from various sources. *Geochimica et Cosmochimica Acta* 14, 234-243.

Ledgard, S. F. (1989). Nutrition, moisture and rhizobial strain influence isotopic fractionation during N₂ fixation in pasture legumes. *Soil Biology and Biochemistry* 21, 65-68.

Ledgard, S. F. and Peoples, M. B. (1988). Measurement of nitrogen fixation in the field. In 'Advances in Nitrogen Cycling in Agricultural Ecosystems.' (Ed. J. R. Wilson.) pp. 351-367. (C.A.B. International: Wallingford, U.K.)

Ledgard, S. F., Freney, J. R., and Simpson, J. R. (1984). Variations in natural enrichment of ^{15}N in the profiles of some Australian pasture soils. *Australian Journal of Soil Research* 22, 155-164.

Letolle, R. (1980). Nitrogen-15 in the natural environment. In 'Handbook of Environmental Isotope Geochemistry.' (Eds. P. Fritz and J.C. Fontes.) (Elsevier Scientific: New York)

Mariotti, A. (1983). Atmospheric nitrogen is a reliable standard for ^{15}N natural abundance measurements. *Nature* 303, 685-687.

Mizutani, H., Kabaya, Y., and Wada, E. (1985). Ammonia volatilization and high ^{15}N/^{14}N ratio in a penguin rookery in Antartica. *Geochemical Journal* 19, 323-327.

Norman, A. G. and Krampitz, L. O. (1945). The nitrogen nutrition of soybeans: II. Effect of available soil nitrogen on growth and nitrogen fixation. *Soil Science Society America Proceedings* 10, 191-196.

O'Deen, W. A. (1989). Wheat volatilised ammonia and resulting isotope fractionation. *Agronomy Journal* 81, 980-985.

Pate, J. S. and Unkovich, M. J. (1999). Measuring symbiotic nitrogen fixation: case studies of natural and agricultural ecosytems in a Western Australian setting. . In 'Physiological Plant Ecology.' (Eds. M. C. Press, J. D. Scholes and M. G. Baker) pp 53-173. (Blackwell Science Ltd: Oxford.)

Pate, J. S., Unkovich, M. J., Armstrong, E. L., and Sanford, P. (1994). Selection of reference plants for ^{15}N natural abundance assessment of N₂ fixation by crop and pasture legumes in southwest Australia. *Australian Journal of Agricultural Research* 45, 133-147.

Peoples, M. B., Faizah, A. W., Rerkasem, B., and Herridge, D. F. (1989). Methods for evaluating biological nitrogen fixation by nodulated legumes in the field. Vol. Monograph No. 11. Australian Centre for International Agricultural Research, Canberra.

Peoples, M. B., Lilley, D. M., Burnett, V. F., Ridley, A. M., and Garden, D. L. (1995). Effects of surface applications of lime and superphosphate to acid soils on the growth and N₂ fixation by subterranean clover in mixed pasture swards. *Soil Biology and Biochemistry* 27, 663-672.

Sanford, P., Pate, J. S., and Unkovich, M. J. (1994). A survey of proportional dependence of subterranean clover and other pasture legumes on N₂ fixation in south-west Australia utilizing ^{15}N natural abundance. *Australian Journal of Agricultural Research* 45, 165-181.

Sanford, P., Pate, J. S., Unkovich, M. J., and Thompson, A. N. (1995). Nitrogen fixation in grazed and ungrazed subterranean clover pasture in south-west Australia assessed by the ^{15}N natural abundance technique. *Australian Journal of Agricultural Research* 46, 1427-1443.

Turner, G. L., Bergersen, F., and Tantala, H. (1983). Natural enrichment of ^{15}N during decomposition of plant material in soil. *Soil Biology and Biochemistry* 15, 495-497

Turner, G. L., Gault, R. R., Morthorpe, L., Chase, D. L., and Bergersen, G. L. (1987). Differences in the natural abundance of ^{15}N in the extractable mineral N of cropped and fallowed surface soils. *Australian Journal of Agricultural Research* 38, 15-26.

Unkovich, M.J. and Pate, J.S. (2000). An appraisal of recent field measurements of symbiotic N_2 fixation by annual legumes. *Field Crops Research* 65, 211-228.

Unkovich, M. J., Pate, J. S., and Hamblin, M. J. (1994a). The nitrogen economy of broadacre lupin in southwest Australia. *Australian Journal of Agricultural Research* 45, 149-164.

Unkovich, M.J., Pate, J.S., and Sanford, P. (1993). Preparation of plant samples for high precision nitrogen isotope ratio analysis. *Communications in Soil Science and Plant Analysis* 24, 2093-2106.

Unkovich, M. J., Pate, J. S., Sanford, P., and Armstrong, E. L. (1994b). Potential precision of the $\delta^{15}N$ natural abundance method in field estimates of nitrogen fixation by crop and pasture legumes in S.W. Australia. *Australian Journal of Agricultural Research* 45, 119-132.

Unkovich, M. J., Pate, J. S. and Sanford, P., and Armstrong, E. L. (1997). Nitrogen fixation by annual legumes in Australian Mediterranean agriculture. *Australian Journal of Agricultural Research* 48, 267-293.

Unkovich, M. J., Sanford, P., Pate, J. S., and Hyder, M. (1998). Effects of grazing on plant and soil nitrogen relations of pasture-crop rotations. *Australian Journal of Agricultural Research* 49, 475-485.

Warembourg, F. R. (1993). Nitrogen fixation in soil and plant systems. In 'Nitrogen Isotope Techniques.' (Ed. R Knowles and T H Blackburn.) pp.127-156. (Academic Press Inc.: San Diego.)

Wetselaar, R. and Farquhar, G. D. (1980). Nitrogen losses from tops of plants. *Advances in Agronomy* 33, 263-302

Yoneyama, T., Fujita, K., Toshida, T., Matsumoto, T., Kambayashi, I., and Yazaki, J. (1986). Variation in natural abundance of ^{15}N among plant parts and in ^{15}N/^{14}N fractionation during N_2 fixation in the legume-Rhiozobia symbiotic system. *Plant Cell Physiology* 27, 791-799.

Chapter 7

The Use of ^{15}N to study Biological Nitrogen Fixation by Perennial Legumes

Mark B. Peoples[1], Brian Palmer[2] and Robert M. Boddey[3]
[1]CSIRO Plant Industry, GPO Box 1600, Canberra, ACT 2601 Australia E-mail: mark.peoples@pi.csiro.au: [2]CSIRO Tropical Agriculture, Davies Lab, PMB PO Aitkenvale, Queensland 4814 Australia: [3]Embrapa-Agrobiologia, Km 47, Seropédica, 23851-970, Rio de Janeiro, Brazil

Key words: *Cajanus cajan, Medicago sativa, Trifolium repens*, tree legumes, ^{15}N enrichment, ^{15}N natural abundance

1.   INTRODUCTION

In virtually all terrestrial ecosystems, and especially in those under management by man, the most significant contributions from biological nitrogen fixation are likely to be those associated with nodulating leguminous plants. While many of the species concerned are annuals, there are also a number of important perennials. These would include temperate pasture legumes such as lucerne (alfalfa, *Medicago sativa)* and white clover (*Trifolium repens*), covercrops and forages (*Calopogonium* and *Pueraria* spp.) in rubber (*Hevea brasiliensis*) and oil palm (*Elaeis guineensis*) plantations or pastures in the tropics, pigeon pea (*Cajanus cajan*) in the cropping systems of Asia, and woody perennials in agroforestry systems, plantations, and forests. Included in the latter category are not only leguminous trees and shrubs nodulated by soil bacteria of the *Rhizobium* group (*Acacia, Leucaena, Gliricidia, Calliandra* spp.), but also species which form nodules when infected by actinomycetes of the genus *Frankia* (*Alnus, Myrica, Casuarina* spp.). This paper will primarily review the various ^{15}N-based approaches which have been used to study symbiotic N$_2$

119

M. Unkovich et al. (eds.),
Stable Isotope Techniques in the Study of Biological Processes and Functioning of Ecosystems, 119–144.
© 2001 *Kluwer Academic Publishers. Printed in the Netherlands.*

fixation by tree and shrub legumes in agroforestry systems, and also include examples of the application of ^{15}N methodologies to quantify inputs of fixed nitrogen (N) by pigeon pea and perennial pasture species.

2.   ASSESSING BIOLOGICAL NITROGEN FIXATION BY PERENNIALS

It must be stressed in the first place that a number of legume species are not capable of forming N_2-fixing nodules with rhizobia, and this applies especially to members of the Caesalpinioideae (e.g. *Senna* spp.). Furthermore, even those legume species that can form effective symbioses may derive little N from N_2 fixation if N is well supplied and conserved in the ecosystem. As a further complication, nodules can be difficult to find in species such as lucerne (Gault *et al.* 1995), pigeon pea (Kumar Rao and Dart 1987) and woody perennials (Boddey *et al.* 2000), simply because of the extensive nature of their roots systems. Finally there can be marked seasonality in formation and longevity of nodules and appreciable changes in degree of nodulation in response to management or climatic and soil variables (Sanginga *et al.* 1995, Boddey *et al.* 2000). Consequently it can be both difficult and time consuming to ascertain whether a host plant has the potential to fix N based purely on visual assessments for presence or absence of nodulation.

Unfortunately the application of any technique to quantify contributions of fixed N by perennial plant species poses special problems, any one or more of which may limit the accuracy with which estimates of N_2 fixation are obtained. These include:

1. The long-term, perennial nature of growth and the seasonal or year-to-year changes in patterns of N assimilation,
2. The large plant-to-plant variation in growth and N demand which one typically encounters in perennial pasture legumes or within a single landrace or provenance of a tree species,
3. Experimental restrictions to the study of the growth and N_2 fixation in agroforestry systems due to small numbers of trees available, or, during the establishment phase of saplings, simply because of logistical limitations of working with large rapidly growing trees,
4. Difficulties in accurately quantifying the often large amounts of standing biomass and N produced by the N_2-fixing component of a system, or the progressive losses of material from the plant canopy caused by fall of foliage and branches from trees and shrubs.

When dealing with natural ecosystems one may encounter further difficulties associated with the taxonomic diversity of occurrence and the possibly high degrees of variability in distribution of N_2-fixing species over gradients in availability of soil water, nutrients and other resources. These aspects are discussed in greater depth in the chapter in this volume by Unkovich and Pate, and in a recent review by Boddey *et al.* (2000).

In addition to the general constraints mentioned above, specific problems arise which limit the use of particular procedures to quantify N_2 fixation by perennial legumes (Kumar Rao and Dart 1987, Peoples *et al.* 1988, Danso *et al.* 1992, Ledgard and Steele 1992, Sanginga *et al.* 1995, Boddey *et al.* 2000). However, under the right circumstances, those methodologies which rely upon the analysis of the heavy isotope of nitrogen, ^{15}N, are generally considered as potentially able to provide the most accurate measures of N_2 fixation (Chalk 1985, Danso *et al.* 1993).

The use of ^{15}N isotope techniques provides time-integrated estimates of the proportion of legume N derived from atmospheric N_2 (%Ndfa, sometimes also abbreviated as %Pfix). This is calculated by comparing the isotopic compositions of the legume with that of plant-available soil N as:

$$\%Ndfa = 100(x - y)/(x - B) \tag{1}$$

where: $x = {}^{15}N$ content of plant-available soil N (usually provided by a non-N_2-fixing reference plant).

$y = {}^{15}N$ content of the N_2-fixing species.

B = a measure of isotopic discrimination which occurs during N_2 fixation.

When the ^{15}N concentration of soil mineral-N is higher than that present in atmospheric N_2, symbiotic N_2 fixation will result in a gradual decline in the ^{15}N composition of the N-fixing species (y) from the level measured in the reference plant (x), as N is progressively assimilated from the air but should approach B as plant reliance upon N_2 fixation for growth increases.

The amounts of N_2 fixed can be subsequently calculated as the product of %Ndfa and the amount of plant N accumulated during growth:

$$\text{Amount of N}_2 \text{ fixed} = (\%Ndfa/100) \text{ x (legume N accumulated)} \tag{2}$$

A key requirement for the success of ^{15}N methodologies is that the ^{15}N concentration in plant-available soil N differs significantly from atmospheric N_2 (0.3663 atoms% ^{15}N). This condition is usually met by introducing ^{15}N-enriched compounds (often as N-fertiliser) to the soil. In most cases the reference plant receives the same fertiliser regime as the accompanying legume. However, in some situations it is considered prudent to supply the reference plant with higher rates of N-fertiliser (the 'A'-value method, Chalk 1985, Danso *et al.* 1993) to ensure comparable growth between reference plant and legume.

With attention to analytical procedures and a suitably precise mass spectrometer, it should also be possible to utilise the slight enrichment in ^{15}N that occurs naturally in many soils (commonly falling in the range of 0.368 to 0.373 atoms% [15]N) to estimate the levels of N_2 fixation (Shearer and Kohl 1986). These small differences in ^{15}N between soil N and atmospheric N_2 (usually described in terms of $\delta^{15}N$ or parts per thousand, ‰, relative to atmospheric N_2) result from the net effect of fractionation between the lighter and heavier isotopes of N, ^{14}N and ^{15}N, which occurs during almost all soil N transformations. For instance chemical and biochemical reactions such as ammonia volatilisation, nitrification, and denitrification are believed to favour the retention of ^{15}N in soil (Shearer and Kohl 1986, Högberg 1997).

Procedures based on both the applications of ^{15}N enriched materials and the use of ^{15}N natural abundance have been applied to a number of studies on perennial legumes. Particular problems are associated with either procedure:

1. Representative sampling of whole plant ^{15}N composition,
2. The choice of appropriate reference plants,
3. Potential transfer of fixed N from N_2-fixing trees to associated reference species,
4. The implications of seasonal relocation and recycling of N within the N_2-fixing legume.

The remainder of the paper reviews the assumptions, advantages and interpretative difficulties associated with the application of both ^{15}N enrichment and ^{15}N natural abundance methods to N_2-fixing perennials.

3.   [15]N ENRICHMENT METHODOLOGIES

The following discussion will summarise the strategies used by various ^{15}N enrichment studies with lucerne (Heichel *et al.* 1984, Danso *et al.* 1988, Hardarson *et al.* 1988), or white clover in temperate pastures (Steele and Littler 1987, Labandera *et al.* 1988, McNeill and Wood 1990, Jørgensen *et al.* 1999), perennial legumes in tropical pastures (Vallis *et al.* 1977, Cadisch *et al.* 1989), pigeon pea (Kumar Rao *et al.* 1987), and tree legumes (Danso *et al.* 1992, Liyanage *et al.* 1994, Sanginga *et al.* 1994, Danso *et al.* 1995, Sanginga *et al.* 1995). The reader is referred to these publications and several comprehensive reviews on ^{15}N enrichment techniques (Chalk 1985, Danso *et al.* 1993) for further details.

3.1 Labelling the soil

The objective of all ^{15}N enrichment studies is to artificially widen the difference in ^{15}N content between the N source of the soil and atmospheric N_2 by applying ^{15}N-enriched materials. The greater the ^{15}N enrichment of the plant-available soil N pool that can be achieved, the greater the accuracy of subsequent calculations (see Fig. 1). Ideally the soil N should be labelled uniformly throughout the rooting zone, in order that the resultant ^{15}N enrichment of soil mineral N remains approximately stable with time over the period of the study (line 'A' in Fig. 2). This proves to be extremely difficult to achieve in field experiments, particularly where long-lived perennials are involved. Indeed, one usually finds a highly enriched N pool close to the soil surface, whose enrichment declines rapidly with time as N is assimilated by plant roots and mineralised from unenriched native soil organic matter (line 'C', Fig. 2). These changes over time in ^{15}N enrichment down the soil profile can result in the introduction of substantial errors into the enrichment technique, especially where different temporal and spatial patterns of N uptake exist between the reference plant and legume (Chalk 1985, Danso *et al.* 1993).

Various forms of ^{15}N enriched fertiliser have been applied in studies with perennial legumes. There is some evidence from pasture systems that the type of fertiliser used may influence subsequent calculations of %Ndfa if differences exist between legume and reference plant in respective preferences for nitrate *versus* ammonium (Steele and Littler 1987). Urea, ammonium sulphate, ammonium nitrate, and potassium nitrate are most commonly utilised at enrichments usually in the range of $3 - 10$ atom% [15]N, although occasionally enrichments as high as $30 - 95$ atom% [15]N have been used. Labelled fertiliser may be broadcast mixed with sand (Cadisch *et al.* 1989), or applied as solution uniformly onto the soil surface (eg Hardarson *et al.* 1988, Peoples *et al.* 1996), but some studies have attempted to introduce ^{15}N further down the soil profile by either incorporating labelled fertiliser to 15cm (Heichel *et al.* 1984, Jørgensen *et al.* 1999), or injecting ^{15}N solution into the soil (Steele and Littler 1987).

Because most labelling techniques initially result in a localised highly enriched zone at the soil surface there could be some justification in using nitrate rather than an ammonium-based fertiliser since the former, as an anion, should be progressively leached and therefore distributed more deeply down the soil profile. For example in an agroforestry experiment in northern Queensland, Australia, involving applications of ^{15}N-enriched nitrate (Peoples *et al.* 1996), the soil was found to be enriched to levels measurably above natural abundance to at least $30 - 60$ cm within 12 months following repeated surface applications.

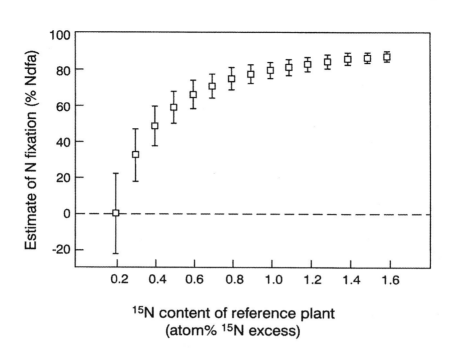

Figure 1. Estimated values of %Ndfa in a hypothetical legume containing 0.2 atom% [15]N
±0.02 SD and a range of assumed atom% [15]N excess ± 10% SD in a non-N₂-fixing reference
plant. The bars represent the upper and lower limits of standard deviation of the estimates.
After Hardarson *et al.* (1988).

Commonly investigators have applied a single application of 20 − 45 kg
of ^{15}N-enriched N/ha to perennial legumes. These relatively high application
rates attempt to ensure sufficient enrichment of the biomass accumulated
over the period of study. However, not only does soil enrichment change
rapidly with time ('C', Fig. 2) using this method, but the fertiliser treatment
in itself may impact on the subsequent measurements of N₂ fixation by
delaying nodulation and possibly also suppressing %Ndfa. This may be less
of a problem for legumes growing in competition with aggressively N-
accumulating grasses in mixed pasture swards (McNeill and Wood 1990),
but rates as low as 20 mg N/kg soil have been found to depress %Ndfa by
N₂-fixing trees (Sanginga *et al.* 1995). To try and overcome these potential
problems a number of researchers have used multiple application of small
amounts (<1 − 5 kg N/ha) of labelled fertiliser in studies of sub-tropical
(Vallis *et al.* 1977) and temperate pastures (Steele and Littler 1987, Danso *et
al.* 1988, Labandera *et al.* 1988) and tree legumes (Peoples *et al.* 1996). As
indicated in the agroforestry example shown in Table 1 this procedure may

be expected to maintain reasonably consistent reference plant enrichments over prolonged periods.

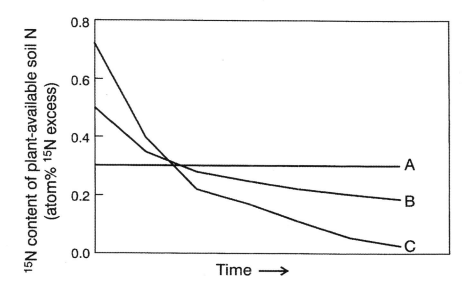

Figure 2. Hypothetical decline in soil atom% ^{15}N excess with time. 'A' indicates the theoretical ideal situation, 'B' represents changes in enrichment with a slow-release form, and 'C' shows a typical decline following application of ^{15}N enriched fertiliser to the soil surface. After Danso *et al.* (1993).

Alternative approaches have also been used. One strategy involves the application of a soluble inorganic ^{15}N source combined with sucrose and straw to stimulate microbial growth within the soil (Jørgensen *et al.* 1999). The ^{15}N label is then rapidly taken up by microbes and incorporated into relatively stable pools of ^{15}N-labelled organic N, while the straw should immobilise ^{15}N enriched mineral N. The expectation then is that the label will be released slowly over time and that the rate of decline in ^{15}N enrichment of the plant-available soil pool will also be slower (line 'B', Fig. 2). Another possible means of achieving a 'slow-release' of ^{15}N label is to incorporate ^{15}N-enriched organic matter well in advance of experimentation or to take advantage of relatively homogeneously distributed 'residual ^{15}N' remaining in soil from a previous experiment (Chalk 1985, Danso *et al.* 1988, Labandera *et al.* 1988).

Table 1. Examples of levels of atom% ^{15}N excess achieved in regrowth of N_2-fixing *Gliricidia* and non-N_2-fixing reference plants in a tree legume study using multiple applications of ^{15}N enriched fertiliser. Estimates of the proportion of *Gliricidia* N derived from N_2 fixation (%Ndfa) are presented using ^{15}N data from both reference species.[a]

Months since ^{15}N last applied	Reference plants Guinea grass	*Senna*	N-fixing species *Gliricidia*	% Ndfa (using grass)	% Ndfa (using *Senna*)
	(^{15}N excess)	(^{15}N excess)	(^{15}N excess)		
3	0.299	0.203	0.103	65	49
6 *(9 after initial)*	0.338	0.227	0.028	92	88
3 *(12 after initial)*	nd	0.466	0.065	nd	86
3 *(15 after initial)*	nd	0.345	0.113	nd	67

[a] Derived from Peoples *et al.* (1996) and unpublished data.

A range of protocols has been used when applying labelled fertiliser to agroforestry studies. In some instances the ^{15}N has been applied within a 1m circle around the base of each tree, while in other situations the ^{15}N has been watered in uniformly over a large area ($10 - 15$ m^2) while restricting sampling to trees at the centre of the treated area (Peoples *et al.* 1996). In other cases the ^{15}N has been applied to 'microplots' analogous to those used in annual pasture and crop systems. In such cases the trees may be isolated from their neighbours by deep trenches ($45 - 150$ cm deep) lined with either galvanised iron sheets or plastic, or may be grown in large concrete cylinders each containing a reconstituted soil profile (Danso *et al.* 1992, Liyanage *et al.* 1994). The latter approach has also been applied to pigeon pea (Kumar Rao *et al.* 1987). These precautions have been employed to avoid lateral migration of ^{15}N while also restricting root movement beyond the enriched zone and intermingling of roots from different trees.

3.2   Choice of reference used

Estimates of N_2 fixation could theoretically be calculated on the basis of direct measures of the ^{15}N content of soil mineral-N (Chalk 1985). However, it is more usual and convenient to use a non-N_2-fixing reference plant totally reliant upon soil N for growth to provide an integrated measure of the ^{15}N composition of plant-available soil N. It is assumed that the ratio of $^{15}N/^{14}N$ of the soil N assimilated by the reference plant is identical to that of the legume. In studies involving pastures, companion grasses or herbaceous weeds are normally utilised as a reference species. The range of reference plants used for ^{15}N-enrichment studies with trees include uninoculated legume trees, grass, rice, or maize intercrops in alley cropping systems, and non-N_2-fixing trees such as *Eucalyptus*, or non-nodulating legumes such as *Cassia / Senna* species (Chalk 1985, Danso *et al.* 1992, Liyanage *et al.*

1994). Some of these possible reference species are likely to be more suitable than others. Apart from the obvious requirement that the reference plant does not fix N_2, it should ideally:

1. explore and exploit the same soil N pool as the legume because of similarities in root morphology and physiology, and
2. have the same duration of growth and pattern of N uptake as the legume.

It would also be preferable if legume and reference species exhibited similar patterns of root breakdown and turnover of below-ground biomass, but whether this has been the case is extremely difficult to determine experimentally.

Given these requirements, choice of reference plant emerges as the critically important factor likely to affect the accuracy of estimates of N_2 fixation by perennials (Danso *et al.* 1992, Sanginga *et al.* 1995). For instance, if the reference plant assimilates soil N predominantly from the highly enriched zone at the soil surface, any unenriched soil N taken up by a deeper rooted legume below the enriched zone would be indistinguishable from fixed N, and therefore lead to an overestimation of %Ndfa. In the reverse situation where the roots of the reference were recovering soil N deeper in the soil profile than the legume, N_2 fixation would be considerably underestimated.

An example of the effect that the choice of reference species can have on estimates of %Ndfa may be seen in Table 1 for data obtained three months after ^{15}N application. The majority of the guinea grass (*Panicum maximum*) roots were located in the top 20 cm, while both the N_2-fixing gliricidia (*Gliricidia sepium*) and non-fixing tree legume senna (*Senna spectabilis*) had roots down to 150 – 200 cm (Peoples *et al.* 1996). As might be expected, the resultant enrichment in the guinea grass was substantially higher than the senna. Consequently, where ^{15}N data for the former were used to estimate the dependence of gliricidia upon N_2 fixation, higher results were obtained than when using senna as the reference (Table 1).

Given the above difficulties in selecting appropriate reference species, it appears to be a sensible strategy to sample a range of non-N_2-fixing species and then derive a %Ndfa value simply as an average of those obtained from all reference sources (Boddey *et al.* 1995). The potential impact of any variations between legume and reference root distribution and/or pattern of soil N uptake become particularly important when attempting to quantify low levels of N_2 fixation. However, the absolute ^{15}N enrichment of the reference plant becomes of lesser impact when the legume is deriving a high proportion of its N requirements from atmospheric N_2 - see chapter in this volume by Unkovich and Pate, and the reference comparison presented in Table 1 for the sampling 9 months after the initial ^{15}N application.

3.3   **Sampling protocols**

A major source of error in N_2 fixation studies with perennials revolves around difficulties in obtaining truly representative samples of biomass for ^{15}N and total N analysis.   However, further potentially more important constraints relate to estimating annual dry matter production (net primary productivity, NPP) when assessing inputs of fixed N on an annual per hectare basis. With perennial pasture legumes this is usually undertaken using combinations of animal exclusion cages and sequential harvests (Peoples *et al.* 1998), or by timing harvests to coincide with hay cuts (Heichel *et al.* 1984), but it is obviously more difficult in agroforestry systems, as shown below.

In some cases it may be possible to sequentially harvest representative trees or shrubs in agroforestry systems and thereby calculate aerial NPP (and total N accumulation) from incremental changes in biomass, while in other situations the cumulative regrowth of coppiced or hedge-rowed trees might be used (Ladha *et al.* 1993, Liyanage *et al.* 1994, Peoples *et al.* 1996, Unkovich *et al.* 2000). However, this is rarely possible in mature plantations and natural forests or woodlands where one may only have recourse to assessing annual growth increments by applying allometric regression equations based on trunk diameter (Brown 1997). The other problematic component when assessing aerial NPP is quantifying amounts of dry matter and N added to the system annually in litter fall (Stocker *et al.* 1995, Unkovich *et al.* 2000). These measurements ideally require intensive and repeated sampling. The methodologies involved are beyond the scope of this review.

Further logistical complications become apparent whenever individual trees are harvested, separated into component parts and the ^{15}N enrichment determined for each part. Differences in ^{15}N enrichment of different components can lead to possible errors in calculating %Ndfa (Table 2, see also Danso *et al.* 1995). Use of a weighted ^{15}N composition based on the ^{15}N enrichment and proportional amounts of N in each organ of a whole plant has then been suggested as a strategy to improve the accuracy of assessments of plant ^{15}N and total N content (Liyanage *et al.* 1994).

Cutting down and sampling large individual trees inevitably generates very large amounts of materials which then result in practical limitations on the number of replicate trees that can be effectively transported for processing back in the laboratory. There can also be problems in preparing subsamples of material on site. Thus while it is relatively easy to grind large samples of leaves, the preparation of woody trunks, branches and roots for analysis poses greater difficulties. One useful approach is to 'subsample' woody components by using a saw and collecting saw-dust which is then ground for N and ^{15}N analysis (Peoples *et al.* 1996).

Table 2. Examples of measures of atom% ^{15}N excess in different tree organs, and estimates of the proportion of plant N derived from N_2 fixation (%Ndfa).

Species parameter	Plant component					Whole tree total
	Root	Trunk	Branches	Leaves	Shoot regrowth	
Uncut trees[a]						
Senna						
^{15}N excess	0.363	0.346	0.358	0.313	nd	0.337
Gliricidia						
(1)^{15}N excess	0.167	0.166	0.219	0.150	nd	0.163
% Ndfa	52	48	36	52		52
(2)^{15}N excess	0.151	0.135	0.137	0.094	nd	0.120
% Ndfa	57	61	61	70		64
Trees pruned as hedge-rows[b]						
Senna						
^{15}N excess	0.158	0.227	nd	nd	0.345	0.313
Gliricidia						
^{15}N excess	0.037	0.039	nd	nd	0.113	0.097
% Ndfa	76	82			67	70
Calliandra						
^{15}N excess	0.052	0.065	nd	nd	0.143	0.123
% Ndfa	67	71			58	60

[a] Comparsion of two provenances of *Gliricidia* (Liyanage *et al.* 1994)
[b] Comparison of two different tree legume species (Peoples *et al.* 1996)

Inspection of the data in Table 2 suggests that where it is not possible to sample whole trees, a satisfactory alternative could be to subsample leaves or regrowth. Collections of 20 – 60 leaves at prescribed intervals, sampling of most recently formed leaves, or the use of regrowth from trees that are regularly pruned and maintained as hedgerows are all approaches that have been used to estimate %Ndfa in studies with trees and shrubs (Peoples *et al.* 1996, Unkovich *et al.* 2000). However, such strategies may not always be appropriate (see Danso *et al.* 1995), particularly when attempting to quantify amounts of N_2 fixed. For example, the omission of roots from the calculations may lead to underestimation of amounts of N_2 fixed by trees by up to 30 – 40% (Danso *et al.* 1995). Even where roots are harvested from soil and included in the analysis it should be recognised that the resulting measures of plant N and estimates of amounts of N_2 fixed are still likely to be conservative since it is virtually impossible to recover all live roots and nodules, let alone materials which had senesced during the study period

(Sanginga *et al.* 1995). Studies using ^{15}N shoot-labelling, as described in the chapter in this volume by McNeill, suggest that the below-ground components of perennial pasture legumes or pigeon pea may represent as much as 40 – 70% of plant total N (Zebarth *et al.* 1991, Jørgensen and Ledgard 1997, Peoples unpubl.).

Measurements undertaken in stands of mature leucaena (*Leucaena leucocephala*) by van Kessel *et al.* (1994), have suggested that estimates of inputs of fixed N based solely on foliar analysis might provide unreliable estimates of N_2 fixation, since foliage of this species represents only 4 – 27 % of the total N of the above-ground biomass. Other field investigations with different provenances of gliricidia have indicated that leaves can variously represent from as little as 9% to up to 51% of total above-ground N (Liyanage *et al.* 1994). Nevertheless, it is likely that leaves will represent the largest single pool of above-ground N in many agroforestry systems, particularly where trees are maintained as pruned or coppiced hedge-rows or where foliage and shoots are periodically harvested for green manure or forage. For example, up to 70 – 80 % of all above-ground N has been reported to be harvested in prunings where species such as gliricidia, calliandra (*Calliandra calothyrus*), codariocalyx (*Codariocalyx gyroides*), or the non-N_2-fixing shrub legume senna are regularly hedge-rowed (Ladha *et al.* 1993, Sanginga et al. 1994, Peoples *et al.* 1996).

4.   [15]N NATURAL ABUNDANCE APPROACH

Use of this approach is considered in detail in the chapter in this volume by Unkovich and Pate. To summarise, the ^{15}N natural abundance ($\delta^{15}N$) technique should work most effectively if the following situations apply:
1. There are just two pools of N for legume growth - a plant-available form of soil N and N_2 from the air,
2. The two pools are sufficiently different in natural ^{15}N abundance to allow accurate measurement with a mass spectrometer, and
3. The biological variability of these abundances is small compared to the difference between them.

The ^{15}N natural abundance method has been used with apparent success in field trials and on-farm studies with lucerne (Brockwell *et al.* 1995, Gault *et al.* 1995, Peoples *et al.* 1998, Dear *et al.* 1999, McCallum *et al.* 2000), white clover (Riffkin *et al.* 1999a), pigeon pea (Brockwell *et al.* 1991, Ladha *et al.* 1996), and in N_2-fixing trees for both legume (Hamilton *et al.* 1993, Ladha *et al.* 1993, Yoneyama *et al.* 1993, Peoples *et al.* 1996, Unkovich *et al.* 2000) and actinorhizal associations (Domenach *et al.* 1989, Kurdali *et al.* 1990, Mariotti *et al.* 1992). Although some of these studies were undertaken in native forests other investigations on the distribution of $\delta^{15}N$ signals in natural ecosystems have raised concerns about the interpretation of ^{15}N data and the use of the technique to quantify inputs of fixed N (Hansen and Pate, 1987, van Kessel *et al.* 1994, Boddey *et al.* 2000). Specific difficulties are related to:

1. the levels and spatial and temporal variability of the ^{15}N natural abundance of plant-available soil N,
2. possible selective uptake from amongst several inorganic and organic N sources with different ^{15}N signatures and potential isotope discrimination effects of mycorrhizae (see chapter in this volume by Stewart),
3. the potential for the $\delta^{15}N$ of soil N to differ under N_2-fixing and reference trees over time as a result of the fall of leaf litter of lower ^{15}N abundance from the N_2-fixing species.

However, virtually all agricultural ecosystems are established after removal of original native vegetation followed by burning or mechanical incorporation of litter. The mechanical disturbance caused by ploughing the soil is then likely to stimulate the mineralisation of soil organic N so that several of the potential sources of error listed above may not be as important in agroforestry, pasture or cropping systems as in natural ecosystems. These aspects are discussed in Boddey *et al.* (2000).

4.1   Isotopic fractionation during N_2 fixation

Potential isotopic discrimination during N_2 fixation can be ignored in enrichment studies as effects will be very small relative to the concentrations of ^{15}N being measured in plant tissues. However, it can be an important factor to consider when analysing low levels of ^{15}N using the ^{15}N natural abundance method. The extent to which such discrimination occurs is usually determined by growing nodulated plants in an inert growing medium such as sand and providing a culture solution containing N-free nutrients. Analyses of the $\delta^{15}N$ of plants grown in this way can provide a measure of 'B' to be used in Equation 1 above.

Ideally a *B* value should be obtained for each new species studied, since different legume symbioses characteristically exhibit slight enrichment or depletion of $\delta^{15}N$ amongst their component plant parts. The shoots are commonly depleted in $\delta^{15}N$ relative to atmospheric N_2, while nodules are correspondingly enriched in $\delta^{15}N$ (see Table 3, and comparable data for annual legumes in the chapter in this volume by Unkovich and Pate). Although there is generally little effect due to host cultivar on shoot $\delta^{15}N$, there can be large differences in effects of rhizobial strains in symbiosis within the same species. For example in the data for white clover presented in Table 3, shoot $\delta^{15}N$ values were similar across several varieties inoculated with a single strain (-1.58 to -1.48 ‰), but ranged from -3.93 to -1.48‰ (unrelated to the amount of N accumulated) across nine effective strains (Riffkin *et al.* 1999b). This raises problems as to what *B* value to use for field-grown legume material collected where a range of unknown rhizobial strains might be involved. Application of a correct value for *B* becomes especially important when %Ndfa is >85%. The limitations of experimentally determining *B* and its importance in calculating %Ndfa are discussed in detail in Peoples *et al.* (1997) and elsewhere in this volume.

Table 3. Examples of the ^{15}N natural abundance (δ^{15}N) detected in effectively nodulated perennial legumes grown in the absence of combined N.[a]

Legume	δ^{15}N of plant part (‰)	
	Nodules or nodulated root	Shoot
Crop legume		
Pigeon pea	+10.60	-0.90
Forages/cover-crops		
Lucerne	nd	-0.44
White clover	nd	-3.93 to -1.48
Calopogonium	+6.92	-0.95
Pueraria	+8.33	-1.22
Shrubs and trees		
Calliandra	+10.05	-1.29
Codariocalyx	+4.53	-1.83
Gliricidia	+4.78	-1.45
Leucaena	+10.11	-0.34
Sesbania grandiflora	+12.30	-2.89
Tagasaste	-0.53	-0.47

[a] Derived from unpublished findings of Peoples except for tagasaste (*Chamaecytisus proliferus*, Unkovich *et al.* 2000) and white clover (Riffkin *et al.* 1999b)

4.2   The natural abundance of plant-available soil N

The ^{15}N content of the ideal reference plant is assumed to provide an integrated determination of the δ^{15}N of the soil N available for legume growth over the duration of the study. Just as in enrichment studies, the higher the ^{15}N content of plant-available soil N, the greater the potential accuracy of the subsequent estimates of %Ndfa. In any case, δ^{15}N levels in the reference plant will have greatest impact on the accuracy of %Ndfa determinations when N_2 fixation is low. Ideally the δ^{15}N of reference material should be at least 2‰ greater than *B* to provide reliable estimates of %Ndfa (see Peoples *et al.* 1997 and discussion elsewhere in this volume). Although the levels of δ^{15}N detected in reference plants show a large range across different farming systems and different regions of the world, values are mostly sufficiently high for quantitative studies of N_2 fixation in many cropping, perennial pasture and agroforestry systems (Table 4).

Table 4. Examples of levels of ^{15}N natural abundance ($\delta^{15}N$) detected in non-N_2-fixing reference plants growing under different patterns of land use.

Land use	Number of observations	Level of $\delta^{15}N$ detected (‰)		Source[a]
		Range	Mean	
Australia				
Cropping	148	+0.4 to +17.5	+7.8	1
Lucerne pastures	50	-0.4 to +10.5	+4.6	2,3,4,5
Agroforestry	27	+0.9 to +8.6	+4.3	6,7,8
SE Asia				
Cropping	60	+0.1 to +16.6	+5.4	1,9
Agroforestry/trees	58	+1.0 to +9.5	+4.6	6,10,11,12
S America				
Agroforestry/trees	24	+1.5 to +8.5	+4.5	11

[a] 1: Peoples *et al.* (1997); 2: Brockwell *et al.* (1995); 3: Gault *et al.* (1995); 4: Peoples *et al.* (1998); 5: McCallum *et al.* (2000); 6: Peoples *et al.* (1991); 7: Peoples *et al.* (1996); 8: Unkovich *et al.* (2000); 9: Ladha *et al.* (1996); 10: Ladha *et al.* (1993); 11: Yoneyama *et al.* (1993); 12: Rowe *et al.* (1999).

Table 5. Examples of levels of ^{15}N natural abundance detected in reference plants sampled from the same field site over a series of growing seasons under different land use systems.[a]

Year	Land use								
	Forest	Pasture				Cropping		Agroforestry	
	Site locations[b]								
	(1)	(2)	(3)	(4)	(5)	(6)	(7)	(8)	(9)
	Level of $\delta^{15}N$ detected (‰)								
1	+15.3	+3.30	+5.55	+5.30	+6.68	+10.80	+6.50	+4.32	+6.47
2	+15.8	+3.00	+4.66	+4.90	+5.24	nd	+5.90	+4.37	+6.82
3		+3.21	+5.14	+3.92	+5.24	+9.90		+4.10	
4				+4.01		+8.90		nd	
5						+7.50		+4.23	

[a] Derived from Hamilton *et al.* (1993), Ladha *et al.* (1993), Peoples *et al.* (1996, 1998), Dear *et al.* (1999), and unpublished data.
[b] (1) Wombat State Forest, Musk Creek, and (2) Rutherglen, Victoria, (3) Wagga Wagga, (4) Junee Reefs, (5) Trangie, and (6) North Star, New South Wales, (7) Kingaroy, and (8) Townsville, Queensland, Australia; (9) Claveria, North Mindanao, Philippines.
[c] Standard errors of the mean $\delta^{15}N$ at various sites ranged from ±0.12 to ±1.19 ‰.

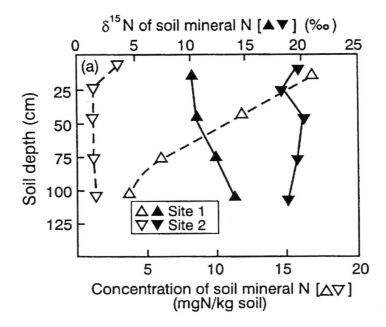

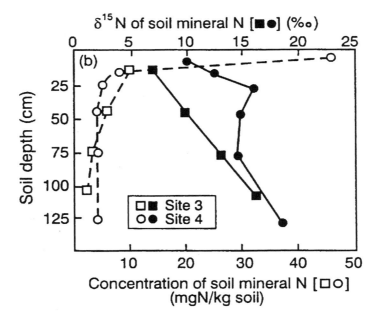

Figure 3. Levels of ^{15}N natural abundance (‰) and concentration (mg/kg soil) of mineral N extracted from various agricultural soils. Examples indicating where δ^{15}N may (a) remain relatively constant, or (b) vary with depth (after Boddey *et al.* 2000).

Changes in the ^{15}N natural abundance of soil N with depth and time have been examined by a number of investigators. Apart from the general finding that the δ^{15}N of plant-available soil N is usually unrelated to concentrations of mineral-N present (Fig. 3), there is little consensus between data of different researchers in observed trends across soil types or environments. In some situations δ^{15}N values can be relatively uniform down a soil profile (Fig. 3a). Data collected from a range of different land use systems and environments (from cool temperate to tropical climates) also suggest that δ^{15}N values for reference plants remain relatively constant over prolonged periods, or may change only gradually within or between seasons (Table 5, see also Ladha *et al.* 1993, Dear *et al.* 1999). However, there is also evidence that the δ^{15}N of plant-available soil N can change rapidly with depth (Fig. 3b), and reference δ^{15}N values exhibit fluctuations within, or between growing seasons (Unkovich *et al.* 2000), or show considerable localised variability (Hansen and Pate 1987). Some studies have even detected a substantial decline in the δ^{15}N of reference plant material over a number of years in lucerne pastures (Gault *et al.* 1995) or tree legume plantations (van Kessel *et al.* 1994).

Table 6. Comparison of levels of ^{15}N natural abundance (‰) detected in potential reference species growing within Australian lucerne pastures.[a]

Reference material	Site location[b]						
	(1)	(2)	(3)	(4)	(5)	(6)	(7)
Perennial grass				3.70		5.72	9.12
Annual grass	3.00	1.99	2.97	3.23	3.74		9.25
Dicot weeds	3.70	2.16	3.08	3.92	3.80	4.67	

[a] Derived from Brockwell *et al.* (1995); Gault *et al.* (1995); Peoples *et al.* (1998); McCallum *et al.* (2000), and unpublished data.
[b] (1) Horsham, Victoria, Australia; (2) Lockhart, (3) Wagga Wagga, and (4) Junee, New South Wales; (5) – (7) Canberra, Australian Capital Territory.

Under the latter conditions significant errors in determinations of N_2 fixation are to be expected, especially if the legume and various reference sources exhibit different rooting depths and/or N uptake characteristics. Certainly there are cases where appreciably different δ^{15}N values have been detected between alternative reference plants (see agroforestry sites 11 and 12 in Table 7), but it is comforting to find that at many locations different reference plants have closely similar δ^{15}N values (e.g. Tables 6 and 7). Collectively the data of Tables 6 and 7 suggest that reasonable estimates of N_2 fixation for perennials should often be possible with the ^{15}N natural abundance procedure even across a range of species with radically differing rooting patterns and growth habits (see also van Kessel *et al.* 1994,

Unkovich *et al.* 2000). However, it is important not to become complacent when choosing a reference plant, particularly in agroforestry systems. Table 7 provides examples of situations where the $\delta^{15}N$ of reference material from outside tree legume stands differs from samples collected from amongst the trees themselves. Crops growing in cultivated fields adjacent to agroforestry sites also typically show elevated $\delta^{15}N$ values and would therefore be inappropriate references for N_2-fixing trees. However, this difficulty may well not apply to alley crop hedgerows (Table 7).

4.3   Sampling protocols

Since field-grown crop legumes can show considerable differences in the $\delta^{15}N$ of their component parts (Peoples *et al.* 1997), it is recommended that wherever possible entire plants, or at least whole shoots, should be sampled to assess overall ^{15}N abundance for estimation of N_2 fixation with the ^{15}N natural abundance technique. Sampling of entire trees may of course raise difficulties but, from the limited evidence available, variation between parts of woody species is often less than observed in annual species. For example, young and old leaflets and branches and leaves taken from different heights from 4 – 5 m high *Desmodium rensonii* trees growing in Indonesia ranged from only +0.3 to +0.6 ‰ in their ^{15}N abundance (Peoples *et al.* 1991). Similarly there may be only minor differences in $\delta^{15}N$ of trunk, branch and leaf material in prunings (Table 8), indicating that subsampling may be a realistic option for many studies based on ^{15}N natural abundance. Under these situations the seasonal relocation and recycling of N within N_2-fixing trees, considered by some to be a potential source of error, may turn out to be relatively unimportant in the calculation of N_2 fixation.

Nevertheless, differences in ^{15}N natural abundance between different tissues of trees and shrubs are not always small as suggested above. For example the levels of $\delta^{15}N$ detected in the 3 – 10 cm apical region of coppiced shoots collected during a study of the N-dynamics of the perennial fodder shrub legume tagasaste were found to be >1‰ higher than measured in the whole coppice dry matter (Unkovich *et al.* 2000). Differences of up to 3‰ between leaves and branches have been reported for leucaena (van Kessel *et al.* 1994), while a range of 5.2 ‰ has been observed between different above-ground tissue of *Prosopis grandis* in a desert ecosystem (Shearer *et al.* 1983). Therefore, one would recommend that at least a preliminary survey be made of $\delta^{15}N$ values for different plant parts (leaves, twigs, branches, trunks and roots) of both legumes and reference plants and these used to assess qualitatively errors likely to be involved in using only foliar samples.

Table 7. Comparison of levels of ^{15}N natural abundance (‰) detected in potential reference species either growing within or adjacent to tree and shrub legume stands.[a]

Reference material	Site location[b]					
	(1)	(2)	(3)	(4)	(5)	(6)
Trees						
Non-legume	1.67	4.15	2.59			
Non-nod legume					3.83	2.80
Grass						
Under trees	1.87			4.45		2.84
Outside plots	1.80	4.32	2.50		1.30	
Alley crop						
Maize						
Rice						
Nearby crops						
Cassava	7.09		4.50	7.48		
Corn				7.39		

Reference material	Site location[b]						
	(7)	(8)	(9)	(10)	(11)	(12)	(13)
Trees							
Non-legume			2.88		7.64		2.94
Non-nod legume	7.00	8.39		4.35		5.5	
Grass							
Under trees		9.22	2.98	4.37	5.98		2.68
Outside plots	5.94	3.07					4.18
Alley crop							
Maize	7.34				4.92	6.0	
Rice						7.9	

[a] Derived from Peoples *et al.* (1991), Ladha *et al.* (1993), Peoples *et al.* (1996), Gathumbi (1999), Rowe *et al.* (1999), and unpublished data.
[b] (1) Ciawi, east Java, (2) Sembawa, south Sumatra, (3) Sei Putih, north Sumatra, (4) Tembuku Bangli, Bali, (5) and (6) Matalom, Leyte, (7) Claveria, North Mindanao, Philippines; (8) Redland Bay, south-east Queensland, Australia; (9) Silkwood, and (10) Townsville, north Queensland, Australia; (11) western Kenya, Africa; (12) north Lampung, Indonesia; (13) Seropedica, Rio de Janeiro, Brazil.

Table 8. Levels of ^{15}N natural abundance (‰) in plant parts of various tree legumes from plantation and agroforestry systems.[a]

Location and species	Uncut trees		Pruned hedgerows			
	Old growth	New growth	Re-growth	Leaf	Stem	Trunk
Philippines						
Acacia mangium	+0.55	+0.65				
A. auriculiformis	+1.41	+1.20				
Albizia falcataria	-0.88	-0.96				
Cassia spectabilis			+6.51	+6.77	+6.22	+5.06
Gliricidia			+3.72	+4.04	+3.14	+3.39
Gliricidia	+0.74	+0.68				
Enterolobium spp.	+1.30	+1.48				
Indonesia						
Desmodium rensonii	+0.60	+0.50				
Calliandra	+1.09	+1.37	+1.13[b]			
Leucaena	+5.26	+6.25	+5.80[b]			
Australia						
Calliandra				+1.06	+1.16	
Calliandra	+2.52	+2.32				
Cassia spectabilis				+4.81	+3.78	
Codariocalyx				+2.14	+1.44	
Gliricidia				+2.10	+2.03	
Gliricidia	+1.80	+1.39				
Sesbania sesban	+3.22	+3.03				
Sesbania sesban				+5.83	+6.38	

[a] Data adapted from Boddey *et al.* (2000)
[b] One third of the branches of individual trees were cut and after 3 months the $\delta^{15}N$ of the regrowth was compared with the old and young growth on the remaining uncut portion of the same trees.

4.4   Effect of litter deposition

In certain cases the litter (senescent leaves and fruits etc.) from actively N$_2$-fixing perennial species can exhibit a lower $\delta^{15}N$ signature than that of non-fixing plants reliant on soil N (Table 9). After a year or more of litter deposition this could have important consequences. A large accumulation of N-rich residues under the trees would be likely to increase N availability to the trees to such an extent that subsequent N$_2$ fixation might be inhibited. Indeed it has even been suggested that once canopy closure has occurred there may no longer be large requirement for N$_2$ fixation in the system. However, where trees are pruned for animal feed (Peoples *et al.* 1996) or green manure mulch for arable crops (Ladha *et al.* 1993, Sanginga *et al.* 1995) with most canopy N not returned as litter, one would expect contributions of fixed N to continue unabated.

Table 9. Comparisons of the levels of ^{15}N natural abundance (‰) in the fallen leaf litter of different tree legumes growing at the same location in north Queensland, Australia.[a]

Species	Fallen leaves
Non-fixing	
Senna spectabilis	+2.51
N₂-fixing	
Calliandra	-0.75
Gliricidia	-0.95
Codariocalyx	-1.55

[a] Derived from unpublished data.

Another possible effect of deposition and decomposition of litter is that the ^{15}N signatures of pools of available N beneath the N_2-fixing trees might, over a period of time, decline below those values needed to unambiguously quantify N_2 fixation (van Kessel *et al.* 1994). Possible transfer of fixed N from trees to understorey plants, or from legume to neighbouring grass in pasture swards has also been raised as a source of error in ^{15}N-enrichment studies (McNeill and Wood 1990, Danso *et al.* 1992, Sanginga *et al.* 1995). Although such transfers will undoubtably occur over time in perennial systems (eg Jørgensen *et al.* 1999), their significance in estimating %Ndfa would become important only if the N_2-fixing legume is denied direct access to its own released N (for example if transfer is mediated directly to non-fixing species by mycorrhizal hyphae).

The impact of growing N_2-fixing and non-fixing species in close association has been specifically examined as a potential source of error for N_2 fixation methods in an agroforestry system in north Queensland and in a pasture study in southern New South Wales, Australia. For example in the agroforestry trial the non-nodulating legume *Senna spectabilis* was grown in plots immediately beside three other species of effectively nodulating trees. Substantial and significant 'edge-effects' were measured in regrowth of senna (Table 10) indicating greater competitive ability by senna roots in assimilation of soil mineral from beneath its nodulated neighbours, and/or the uptake of N released from fallen leaf litter or exuded from the roots of the fixing trees. However, at no time over the 12 month period of observations did the $\delta^{15}N$ of the senna prunings from trees growing immediately beside (within 0.5 m) a fixing species differ significantly from material collected from the trees in the centre of the 24 m^2 plot (Table 10). Similarly in the pasture study, no effect of pasture composition could be detected at any time over three years in the levels of $\delta^{15}N$ of the companion perennial grass grown at a range of different densities (5 – 40 plants/m^2) in association with legumes (Dear *et al.* 1999). These data suggest that errors induced by the physical location of the reference plant in relation to the N_2-

fixing component are not necessarily a major concern in studies involving ^{15}N natural abundance.

Table 10. Comparisons of the levels of dry matter (DM) production (t/ha) and the ^{15}N natural abundance (‰) of shoot prunings collected from the non-nodulating tree legume *Senna spectabilis* either growing within the centre of 24 m[2] experimental plots, or immediately alongside (within 0.5 m) to a N_2-fixing tree.[a]

Sampling area	Time after planting (weeks)							
	52		65		91		104	
	DM	‰	DM	‰	DM	‰	DM	‰
Mid-plot	1.61	3.61	3.28	4.28	4.51	4.05	5.34	4.17
Growing next to:								
Calliandra	3.22	3.90	7.38	4.57	7.11	3.75	8.62	4.08
Gliricidia	4.05	4.21	5.18	3.84	10.73	4.63	7.21	3.91
Codariocalyx	5.08	4.12	8.66	4.11	9.29	4.08	11.17	3.72

[a] Derived from Boddey *et al.* (2000)

5. CONCLUDING REMARKS

The ^{15}N-based methods reviewed in this chapter comprise one subset of several techniques available for studying N_2 fixation. Unfortunately most other alternative techniques are generally unsuitable for perennial legumes. A large number of different ^{15}N enrichment procedures have been applied or evaluated, but there is still no one technique to label soil for long-term investigations with perennials that will be reliable in all situations. The ^{15}N natural abundance method shows considerable promise for use in cropping, pasture and agroforestry systems, but it is clear that if the $\delta^{15}N$ of plant-available soil N (as determined from the non-N_2-fixing reference plant) is little different to zero or very variable, and/or B values differ greatly with rhizobial strain of the particular legume species under study, very significant errors might well be experienced when estimating %Ndfa. Both the enrichment and ^{15}N natural abundance procedures have distinct limitations. There will always be uncertainty associated with the choice of reference plants and representative sampling strategies. However, with careful use either technique should provide valuable information. This review aims to help researchers working with N_2-fixing perennials to determine where, when and how ^{15}N-based methodologies might best be applied and to indicate possible constraints to interpretation of the data obtained.

ACKNOWLEDGMENTS

Special thanks go to the Australian Centre for International Agriculture (ACIAR) and the Grains Research & Development Corporation (GRDC) whose support allowed the collection of much of the published and unpublished data presented in this review. Critical input from John Pate at the management stage is gratefully acknowledged.

REFERENCES

Boddey, R.M., Oliveira, O.C. de, Alves, B.J.R., and Urquiaga, S. (1995). Field application of the ^{15}N isotope diluiton technique for the reliable quantification of plant-associated biological nitrogen fixation. *Fertility Research* 42, 77-87.

Boddey, R.M., Peoples, M.B., Palmer, B., and Dart, P.J. (2000). Use of the ^{15}N natural abundance technique to quantify biological nitrogen fixation by woody perennials. *Nutrient Cycling in Agroecosystems* 57, 235-270.

Brockwell, J., Andrews, J.A., Gault, R.R., Gemell, L.G., Griffith, G.W., Herridge, D.F., Holland, J.F., Karsono, S., Peoples, M.B., Roughley, R.J., Thompson, J.A., Thompson, J.A., and Troedson, R.J. (1991). Erratic nodulation and nitrogen fixation in field-grown pigeon pea [*Cajanus cajan* (L.) Millsp.]. *Australian Journal of Experimental Agriculture* 31, 653-661.

Brockwell, J., Gault, R.R., Peoples, M.B., Turner, G.L., Lilley, D.M., and Bergersen, F.J. (1995). Nitrogen fixation in irrigated lucerne grown for hay. *Soil Biology & Biochemistry* 27, 589-594.

Brown, S. (1997). Estimating biomass and biomass change of tropical forests: a primer. *Forestry Paper* 134, FAO, Rome.

Cadisch, G., Sylvester-Bradley, R., and Nösberger, J. (1989). ^{15}N-based estimation of nitrogen fixation by eight tropical forage-legumes at two levels of P:K supply. *Field Crops Research* 22, 181-194.

Chalk, P.M. (1985). Estimation of N_2 fixation by isotope dilution: an appraisal of techniques involving ^{15}N enrichment and their application. *Soil Biology & Biochemistry* 17, 389-410.

Danso, S.K.A., Bowen, G.D., and Sanginga, N. (1992). Biological nitrogen fixation in trees in agro-ecosystems. *Plant and Soil* 141, 177-196.

Danso, S.K.A., Hardarson, G., and Zapata, F. (1988). Dinitrogen fixation estimates in alfalfa-ryegrass swards using different nitrogen-15 labelling methods. *Crop Science* 28, 106-110.

Danso, S.K.A., Hardarson, G. and Zapata, F. (1993). Misconceptions and practical problems in the use of ^{15}N soil enrichment techniques for estimating N_2 fixation. *Plant and Soil* 152, 25-52.

Danso, S.K.A., Zapata, F., and Awonaike, K.O. (1995). Measurement of biological N_2 fixation in field-grown *Robinia pseudoacacia* L. *Soil Biology & Biochemistry* 27, 415-419.

Dear, B.S., Cocks, P.S., Peoples, M.B., Swan, A.D., and Smith, A.B. (1999). Nitrogen fixation by subterranean clover (*Trifolium subterranean* L.) growing in pure culture and in mixtures with varying densities of lucerne (*Medicago sativa* L.) or phalaris (*Phalaris aquatica* L.). *Australian Journal of Agricultural Research* 50, 1047-1058.

Domenach, A.M., Kurdali, F., and Bardin, R. (1989). Estimation of symbiotic dinitrogen fixation in alder forest by the method based on natural ^{15}N abundance. *Plant and Soil* 118, 51-59.

Gault, R.R., Peoples, M.B., Turner, G.L., Lilley, D.M., Brockwell, J., and Bergersen, F.J. (1995). Nitrogen fixation by irrigated lucerne during the first three years after establishment. *Australian Journal of Agricultural Research* 46, 1401-1425.

Gathumbi, S. (1999). Nitrogen sourcing by mixed species fallows of fast growing legumes in western Kenya. PhD thesis, Wye College, University of London, United Kingdom.

Hamilton, S.D., Hopmans, P., Chalk, P.M., and Smith, C.J. (1993). Field estimation of N_2 fixation by *Acacia* spp. using ^{15}N isotope dilution and labelling with ^{35}S. *Forest Ecology and Management* 56, 297-313.

Hansen, A.P. and Pate, J.S. (1987). Evaluation of the ^{15}N natural abundance method and xylem sap analysis for assessing N_2 fixation of understorey legumes in jarrah (*Eucalyptus marginata*) Donn ex Sm.) forest in S.W. Australia. *Journal of Experimental Botany* 38, 1446-1458.

Hardarson, G., Danso, S.K.A., and Zapata, F. (1988). Dinitrogen fixation measurements in alfalfa-ryegrass swards using nitrogen-15 and the influence of the reference crop. *Crop Science* 28, 101-105.

Heichel, G.H., Barnes, D.K., and Henjum, K.I. (1984). N_2 fixation, and N and dry matter partitioning during a 4-year alfalfa stand. *Crop Science* 24, 811-815.

Högberg, P. (1997). ^{15}N natural abundance in soil-plant systems. *New Phytologist* 137, 179-203.

Jørgensen, F.V. and Ledgard, S.T. (1997). Contribution from stolons and roots to estimates of the total amount of N_2 fixed by white clover (*Trifolium repens* L.). *Annals of Botany* 80, 641-648.

Jørgensen, F.V., Jensen, E.S. and Schjoerring, J.K. (1999). Dinitrogen fixation in white clover grown in pure stand and mixture with ryegrass estimated by the immobilized ^{15}N isotope dilution method. *Plant and Soil* 208, 293-305.

Kumar Rao, J.V.D.K. and Dart, P.J. (1987). Nodulation, nitrogen fixation and nitrogen uptake in pigeon pea (*Cajanus cajan* (L.) Millsp.) of different maturity groups. *Plant and Soil* 99, 255-266.

Kumar Rao, J.V.D.K., Thompson, J.A., Sastry, P.V.S.S., Giller, K.E., and Day, J.M. (1987). Measurement of N_2-fixation in field-grown pigeon pea (*Cajanus cajan* (L.) Millsp.) using ^{15}N-labelled fertilizer. *Plant and Soil* 101, 107-113.

Kurdali, F., Domenach, A.M., and Bardin, R. (1990). Alder-poplar associations: Determination of plant nitrogen sources by isotope techniques. *Biology and Fertility of Soils* 9, 321-329.

Labandera, C., Danso, S.K.A., Pastorini, D., Curbelo, S., and Martin, V. (1988). Nitrogen fixation in a white clover-fescue pasture using three methods of nitrogen-15 application and residual nitrogen-15 uptake. *Agronomy Journal* 80, 265-268.

Ladha, J.K., Peoples, M.B., Garrity, D.P., Capuno, V.T., and Dart, P.J. (1993). Estimating dinitrogen fixation of hedgerow vegetation using the nitrogen-15 natural abundance method. *Soil Science Society of America Journal* 57, 732-737.

Ladha, J.K., Kundu, D.K., Angelo-Van Coppenolle, M.G., Peoples, M.B., Carangal, V.R., and Dart, P.J. (1996). Legume productivity and soil nitrogen dynamics in lowland rice-based cropping systems. *Soil Science Society of America Journal* 60, 183-192.

Ledgard, S.F. and Steele, K.W. (1992). Biological nitrogen fixation in mixed legume/grass pastures. *Plant and Soil* 141, 137-153.

Liyanage, M. deS, Danso, S.K.A., and Jayasundara, H.P.S. (1994). Biological nitrogen fixation in four *Gliricidia sepium* genotypes. *Plant and Soil* 161, 267-274.

Mariotti, A., Sougoufara, B., and Dommergues, Y.R. (1992). Estimation de la fixation d'azote atmospherique par le tracage isotopique naturel dans une plantation de Casuarina equisetifolia (forst). *Soil Biology & Biochemistry* 24, 647-653.

McCallum, M.H., Peoples, M.B., and Connor, D.J. (2000). Contributions of nitrogen by field pea (*Pisum sativum* L.) in a continuous cropping sequence compared with a lucerne (*Medicago sativa* L.)-based pasture ley in the Victorian Wimmera. *Australian Journal of Agricultural Research* 51, 13-22.

McNeill, A.M. and Wood, M. (1990). ^{15}N estimates of nitrogen fixation by white clover (*Trifolium repens* L.) growing in a mixture with ryegrass (*Lolium perenne* L.). *Plant and Soil* 128, 265-273.

Peoples, M.B., Herridge, D.F., and Bergersen, F.J. (1988). Measurement of nitrogen fixation in crop and shrub legumes. In 'Sustainable Agriculture: Green Manure in Rice Farming.' pp. 223-237. (IRRI: Philippines.)

Peoples, M.B., Bergersen, F.J., Turner, G.L., Sampat, C., Rerkasem, B., Bhromsiri, A., Nurhayati, D.P., Faizah, A.W. Sudin, M.N., Norhayati, M., and Herridge, D.F. (1991). Use of the natural enrichment of ^{15}N in plant available soil N for the measurement of symbiotic N_2 fixation. In 'Stable Isotopes in Plant Nutrition, Soil Fertility and Environmental Studies.' pp. 117-129. (IAEA: Vienna, Austria.)

Peoples, M.B., Palmer, B., Lilley, D.M., Duc, L.M. and Herridge, D.F. (1996). Application of ^{15}N and xylem ureide methods for assessing N_2 fixation of three shrub legumes periodically pruned for forage. *Plant and Soil* 182, 125-137.

Peoples, M.B., Turner, G.L., Shah, Z., Shah, S., Aslam, M., Ali, S., Maskey, S., Bhattari, S., Afandi, F., Schwenke, G.D., and Herridge, D.F. (1997). Evaluation of the ^{15}N natural abundance technique for measuring N_2 fixation in experimental plots and farmers' fields. In 'Extending Nitrogen Fixation Research to Farmers' Fields.' *Proc. International Workshop on Managing Legume Nitrogen Fixation in Cropping Systems of Asia.* (Eds O.P. Rupela, C. Johansen, D.F. Herridge,) pp 57-75. (ICRISAT, Hyderabad, India.)

Peoples, M.B., Gault, R.R., Scammell, G.J., Dear, B.S., Virgona, J., Sandral, G.A., Paul, J., Wolfe, E.C., and Angus, J.F. (1998). The effect of pasture management on the contribution of fixed N to the N-economy of ley-framing systems. *Australian Journal of Agricultural Research* 49, 459-474.

Riffkin, P.A., Quigley, P.E., Cameron, F.J., Peoples, M.B., and Thies, J.E. (1999a), Annual nitrogen fixation in grazed dairy pastures in south-western Victoria. *Australian Journal of Agricultural Research* 50, 273-281.

Riffkin, P.A., Quigley, P.E., Kearney, G.A., Cameron, F.J., Gault, R.R., Peoples, M.B., and Thies, J.E. (1999b). Factors associated with biological nitrogen fixation in dairy pastures in south-western Victoria. *Australian Journal of Agricultural Research* 50, 261-272.

Rowe, E.C., Hairiah, K., Giller, K.E., Van Noordwijk, M. and Cadisch, G. (1999). Testing the safety-net role of hedgerow tree roots by ^{15}N placement at different soil depths. *Agroforestry Systems* 43, 81-93.

Sanginga, N., Danso, S.K.A., Zapata, F., and Bowen, G. (1994). Influence of pruning management on P and N distribution and use efficiency by N_2 fixing and non-N_2 fixing trees used in alley cropping systems. *Plant and Soil* 167, 219-226,

Sanginga, N., Vanlauwe, B., and Danso, S.K.A. (1995). Management of biological N_2 fixation in alley cropping systems: estimation and contribution to N balance. *Plant and Soil* 174, 119-141.

Shearer, G. and Kohl, D.H. (1986). N_2-fixation in field settings: estimations based on natural ^{15}N abundance. *Australian Journal of Plant Physiology* 13, 699-756.

Shearer, G., Kohl, D.H., Virginia, R.A., Bryan, B.A., Skeeters, J.L., Nilsen, E.T., Sharifi, M.R., and Rundel, P.W. (1983). Estimates of N_2 fixation from variation in the natural abundance of N-15 in Sonoran desert ecosystems. *Oecologia* 56, 365-373.

Steele, K.W. and Littler, R.A. (1987). Field evaluation of some factors affecting nitrogen fixation in pastures by ^{15}N isotope dilution. *Australian Journal of Agricultural Research* 38, 153-161.

Stocker, G.C., Thompson, W.A., Irvine, A.K., Fitzsimon, J.D., and Thomas, P.R. (1995). Annual patterns of litterfall in a lowland and tableland rainforest in tropical Australia. *Biotropica* 27, 412-420.

Turner, G.L., Gault, R.R., Morthorpe, L., Chase, D.L. and Bergersen, F.J. (1987). Differences in the natural abundance of ^{15}N in the extractable mineral nitrogen of cropped and fallowed surface soils. *Australian Journal of Agricultural Research* 38, 15-25.

Unkovich, M.J., Pate, J.P., Lefroy, E.C., and Arthur, D.J. (2000). Inputs of fixed N by the fodder tree legume tagasaste (*Chamaecytisus proliferus*) in deep sands of Western Australia assessed using the ^{15}N natural abundance technique. *Australian Journal of Plant Physiology* 27, 921-929.

van Kessel, C., Farrell, R.E., Roskoski, J.P., and Keane, K.M. (1994). Recycling of the naturally-occurring ^{15}N in an established stand of *Leucaena leucocephala*. *Soil Biology & Biochemistry* 26, 757-762.

Vallis, I., Henzell, E.F., and Evans, T.R. (1977). Uptake of soil nitrogen by legumes in mixed swards. *Australian Journal of Agricultural Research* 28, 413-225.

Yoneyama, T., Muraoka, T., Murakami, T., and Boonkerd, N. (1993). Natural abundance of ^{15}N in tropical plants with emphasis on tree legumes. *Plant and Soil* 153, 295-304.

Zebarth, B.J., Alder, V., and Sheard, R.W. (1991). *In situ* labelling of legume residues with a foliar application of a ^{15}N-enriched urea solution. *Communications in Soil Science and Plant Analysis* 22, 437-447.

Chapter 8

Source/Sink Interactions in Crop Plants
Application of $^{13}CO_2$ and Urea-^{15}N Techniques in Quantitative Analysis

Jairo A. Palta
CSIRO Plant Industry, Centre for Mediterranean Agricultural Research, P.O. Private Bag 5, Wembley, WA 6913 Australia, and Centre for Legumes in Mediterranean Agriculture (CLIMA), The University of Western Australia, Crawley, WA 6009 Australia. Email: j.palta@ccmar.csiro.au

Key words: stable isotope techniques, carbon, nitrogen, accumulation, allocation, remobilisation

1. INTRODUCTION

This chapter will describe two innovative experimental approaches for assessing how resources generated in the autotrophic assimilatory processes of photosynthesis and N_2 fixation, and the uptake of nitrogen from soil, are allocated and mobilised between competing sinks within a typical crop plant. It will demonstrate how such information might be useful for formulating and predicting the source/sink interactions of crop plants when grown under potentially stressful conditions in dryland agricultural systems. The two experimental approaches presented involve feeding of the stable isotopes ^{13}C and ^{15}N in the form of $^{13}CO_2$ and urea-^{15}N, followed by sequential sampling for measurement of dry matter accumulation and carbon and nitrogen content. The topics that are covered include protocols for labelling plant material by leaf feeding of ^{15}N-urea or atmospheric feeding of $^{13}CO_2$, and description of procedures for evaluating parameters of nitrogen and carbon distribution and mobilisation within the labelled plants. Crop plants to be considered are wheat (*Triticum aestivum* L.), narrow-leafed lupin (*Lupinus angustifolius* L.) and chickpea (*Cicer arietinum* L.).

145

M. Unkovich et al. (eds.),
Stable Isotope Techniques in the Study of Biological Processes and Functioning of Ecosystems, 145–165.
© 2001 *Kluwer Academic Publishers. Printed in the Netherlands.*

A full quantitative analysis of source/sink interaction in crop plants requires accurate measurement of the amounts of carbon and nitrogen available in the various parts of the plant, particularly vegetative organs, coupled with similarly detailed information on subsequent patterns of mobilisation to reproductive parts. Much of our earlier knowledge on the availability, export and import of carbon stemmed from studies in which the assimilation, loss and redistribution of carbon were measured concurrently in plants fed with the radioactive isotope ^{14}C (Cock and Yoshida 1972, Bidinger *et al.* 1977, Austin *et al.* 1977, 1980). In recent years however, problems associated with the use of radioactive material in field studies, particularly in relation to the handling and disposal of wastes and possible contamination of soils and ground water, have limited the extent to which ^{14}C can be used in these studies. The use of the stable isotopes ^{13}C and ^{15}N have accordingly become adopted as an alternative means for studying carbon and nitrogen flow in crop plants. The isotopes concerned are not radioactive, enabling field experiments to be conducted without risk of health or environmental hazards. In addition, photosynthetic and metabolic discrimination against ^{13}C *versus* ^{12}C is less than against the heavier ^{14}C (see Van Norman and Brown 1952, Svejcar *et al.* 1990).

As tracers, ^{13}C and ^{15}N offer great potential for generally defining supplier-consumer relationships in terms of which plant sources are net exporters of carbon and nitrogen and how such products are mobilized to reproductive parts. Special advantages from use of ^{13}C and ^{15}N come from the ability to apply either as a series of pulses to label specific organs such as specific leaves or to a whole plant canopy at a range of critical times in plant growth. One can then follow temporary accumulation in the plant followed by later mobilization to the grain. Several studies of this nature have already used ^{13}C and ^{15}N as tracer of plant metabolism and allocation (Warembourg *et al.* 1982, Blacklow 1982, Yoneyama 1980, Kouchi and Yoneyama 1984). However, relatively few of such investigations have yielded quantitative information on the contribution to grain of carbon and nitrogen from reserves previously accumulated in vegetative parts (Cliquet *et al.* 1990, Palta *et al.* 1991a, 1994). Such studies become particularly constructive when integrated with agronomic or environmental aspects such as time of sowing, timing of fertiliser applications and effects of periodic waterlogging or drought (Palta *et al.* 1994, Palta and Fillery 1995).

To provide elements of continuity and authenticity in this chapter, it is proposed to concentrate on techniques used in experimental work of the author on wheat, narrow-leafed lupin, rice and chickpea. Strategies for effectively labelling plant material will first be considered, focusing especially on the development and use of $^{13}CO_2$ and urea-^{15}N feeding techniques. The chapter will then concentrate on effective assessment of parameters of nitrogen and carbon allocation and mobilisation followed by a series of specific examples illustrating how resulting data can be used to assess source/sink interactions in crop plants under drought conditions.

2.   LABELLING PLANT MATERIAL

Highly enriched sources of ^{13}C and ^{15}N, coupled with feeding techniques which uniformly incorporate the ^{13}C and ^{15}N into plant tissues and metabolic compartments, are clearly essential for effective labelling of plant material (Fried 1978). Highly enriched sources of ^{13}C and ^{15}N in the form of $^{13}CO_2$ (~99 atom %) and ^{15}N-urea (~99 atom %) are the most commonly used stable isotopes for labelling carbon and nitrogen pools in plant material. Both isotopes are naturally present at low concentrations in the atmosphere, 1.1% for ^{13}C and 0.36% for ^{15}N, so background enrichments are likely to confuse labelling patterns only where the added stable isotopes have been poorly absorbed by the plant. Labelling with highly enriched $^{13}CO_2$ and ^{15}N-urea and (~99 atom %) usually results in detectable levels of the isotope within a few hours. Full isotope equilibration with the circulatory pathways and pools can be expected within a few days (Cliquet *et al.* 1990, Palta *et al.* 1991a, 1994). Until relatively recently the use of ^{13}C and ^{15}N has been limited because of the high cost of mass spectrometric analysis and the time and costs involved in sample processing and preparation. However, with the advent of systems combining automatic gas-sample preparation and assays on a mass spectrometer, high analytical precision (0.1 – 0.5%) is now possible for coupled analyses of ^{13}C and ^{15}N within a single sample of combusted dry matter. In addition, the recent use of CO_2-controlled assimilation chambers in CO_2-enriched experiments have provided a fully effective means for steady state ^{13}C labelling where plants are feeding on an atmosphere enriched with $^{13}CO_2$.

As mentioned above isotopic analyses are now relatively easy for both ^{13}C and ^{15}N and can be performed simultaneously on the same sample. However, it is much more difficult to accomplish dual labelling of plants with ^{13}C and ^{15}N using the same feeding technique (Palta *et al.* 1991a). Dual labelling procedures using gaseous $^{15}NH_3$ and $^{13}CO_2$ simultaneously are unpractical since sophisticated assimilation chambers and CO_2 environmental control equipment is required when undertaking the prolonged periods of feeding required (Warembourg *et al.* 1982). A possible technique for dual labelling wheat plants with ^{15}N and ^{13}C, trialled by Palta *et al.* (1991a) would be to use double-labelled urea as a source of $^{15}NH_3$ and $^{13}CO_2$. The double labelled urea (99 atom % ^{15}N and 99 atom. % ^{13}C) is supplied to plants by immersing their cut leaf tips into a 0.5% solution at different times during the plant development. However, since high rates of plant transpiration need to be maintained to promote sufficient uptake of urea through the cut tips, substantial losses of ^{13}C can occur from respiration of the urea following export from the source leaf. This results in considerably lower recoveries of ^{13}C compared to ^{15}N in plant parts at each time of feeding (Table 1). Thus, while providing an adequate level of ^{15}N this urea-feeding strategy is unlikely to provide sufficient ^{13}C for long-term remobilisation studies (Palta *et al.* 1991a). The more suitable alternative, therefore, is to feed highly enriched $^{13}CO_2$ to the plant atmospherically, while synchronously engaging in leaf feeding of ^{15}N urea.

Table 1. Excess of ^{15}N and ^{13}C in the main stem (MS), individual tillers (T1, T2, T3), roots and ears of wheat 48 h after leaves were fed with double labelled urea at the stages of tillering, stem elongation and anthesis. (From Palta *et al.* 1991).

Plant part	Tillering		Stem elongation		Anthesis	
	^{15}N	^{13}C	^{15}N	^{13}C	^{15}N	^{13}C
Main stem	832.8	199.7	1512.2	190.0	1867.9	240.8
Tiller 1	35.7	2.2	33.7	25.5	32.9	22.5
Tiller 2	33.1	3.9	77.0	20.9	31.8	19.7
Tiller 3	13.6	1.7	27.4	10.4	13.3	16.2
Roots	20.9	7.5	93.1	51.5	35.2	21.1
Ear MS					71.8	26.7
Ear T1					5.3	12.5
Ear T2					14.1	10.3
Ear T3					2.3	8.0
Whole plant	926.1	215.0	1744.0	299.2	2074.6	377.0
Recovery (%)	63.7	34.1	90.0	35.6	85.7	35.9
48 h respiratory losses[a] (µg)		88.9		128.6		110.8
Non-respiratory losses[a] (µg)		304.8		411.8		561.8

[a] Respiratory losses of ^{13}C after leaves were fed and non-respiratory losses of ^{13}C during feeding were estimated from respiration measurements.

3.   LEAF FEEDING WITH ^{15}N-LABELLED SUBSTRATES

Foliage feeding with gaseous $^{15}NH_3$ has often been used for labelling plant material with ^{15}N. However, feeding of plant canopies with $^{15}NH_3$ is a slow process and therefore best suited to targeted short-term exposures. Moreover, with such feeding one is unlikely to achieve the level of enrichment necessary when monitoring the long-term translocation and remobilisation of nitrogen in field-grown plants (Warembourg *et al.* 1982, Palta *et al.* 1991b). Long-term gaseous exposures to $^{15}NH_3$ ideally require expensive assimilation chambers incorporating high levels of precision for application of CO_2 and controlling of environmental variables to maintain steady rates of photoassimilation (Warembourg *et al.* 1982, Lockyer and Whitehead 1987, Janzen and Bruinsma 1989). Feeding individual leaves with $K^{15}NO_3$ solutions is an alternative way of labelling with ^{15}N (Pate 1973, Blacklow 1982) to avoid the limitations associated with the use of gaseous $^{15}NH_3$ (Palta *et al.* 1991b). While this procedure has provided the necessary uniformity in ^{15}N enrichment of various organs of a wheat plant and thereby estimate short-term movement of nitrogen from senescing flag leaves (Blacklow 1982), it is considered unlikely to provide sufficient ^{15}N enrichment for monitoring long-term mobilization of nitrogen. This problem has been overcome by using 99 atom % ^{15}N-urea, a compound which has a higher nitrogen content per unit weight than $K^{15}NO_3$ (Palta *et al.* 1991b). With the important *proviso* that the fed plant has the capacity to assimilate urea and that the applied concentrations are not toxic, this form of labelling should be more effective in most circumstances than feeding studies using $^{15}NH_4$ or $^{15}NO_3$.

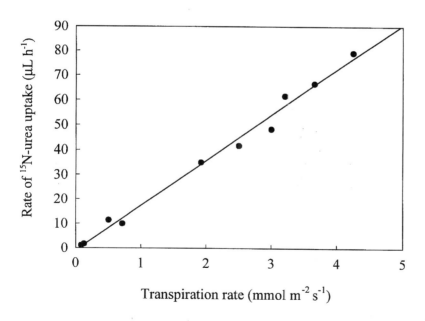

Figure 1. Relationship between the rate of ^{15}N-urea uptake and the transpiration rate of the leaves of wheat fed at the stage of stem elongation. (Adapted from Palta *et al.* 1991a)

As indicated above leaf feeding with ^{15}N-urea is the preferred option for labelling plant material for long term studies of storage and remobilisation of nitrogen (Palta *et al.* 1991b, Palta *et al.* 1994, Palta and Fillery 1993, Palta and Fillery 1995). This approach has been successfully used for wheat, rice, grasses, lupin and chickpea regardless of whether plants are raised in growth cabinets, glasshouse or field conditions. All of the above-named plant species possess active urease systems in their foliage, allowing them to assimilate the fed compound effectively even when relatively high concentrations are applied. In a typical feeding study leaves are fed continuously for intervals up to 4 days by dipping their cut tips into ^{15}N-urea solutions (99 atom %). The uptake of ^{15}N-urea closely follows transpiration loss by the fed leaf, so that higher uptake rates occur at mid-day when transpiration rates are high, lower rates at early morning or late afternoon when transpiration is less. The data in Figure 1 illustrate the close correlation between transpiration rate and uptake of label fed. However, in studies involving uptake by wheat, ^{15}N-urea per unit of dry weight has been found to be 1.5 – 4 times greater for main shoot leaves than for companion leaves of subtending tillers, so care has to be taken when interpreting data for plants fed through only one class of leaf. Simultaneous feeding of more than one leaf per shoot is of course likely to increase the overall uptake per plant of specifically enriched source, and this procedure should also lead to more uniform labelling than when feeding through single leaves.

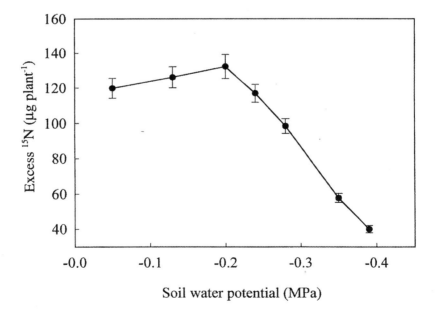

Figure 2. Influence of changes in soil water potential in the root zone on the total uptake of ^{15}N by a main stem leaf of wheat plants at the stage of stem elongation. Bars indicate ± s.e.m for four plants. (From Palta *et al.* 1991b).

As to be expected, our initial experiments on wheat showed that the uptake of ^{15}N-urea by leaf feeding was greatest for well-watered plants in which the soil water potential was above -0.2 MPa. When soil water potentials fell below -0.28 MPa, uptake declined progressively until at -0.4 MPa it had been reduced by 55% (Fig. 2). Fast and effective labelling would therefore be generally expected of wheat growing in soils of water content close to field capacity. From the well-known deleterious effects of foliar fertilisation with high levels of urea and the variations in susceptibility of species in this respect (Krogmeier *et al.* 1989, Bremner 1990), one would expect leaf burning to occur. For example, we have found that the necrotic area of fed wheat leaves increased to 50% as the concentration of the urea solution increased from 0.4 to 3% (Fig. 3). As a general rule therefore, one would recommend that the concentrations of ^{15}N-urea solutions for leaf feeding of wheat should not exceed 2.5%.

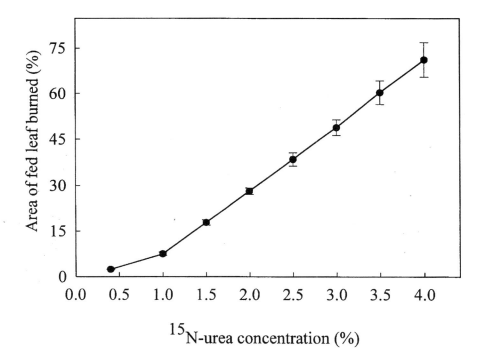

Figure 3. Effect of the concentration of ^{15}N-urea on the damage of the fed leaf of wheat at the early stage of tillering. Bars indicate ± s.e.m for four plants. (From Palta *et al.* 1991b).

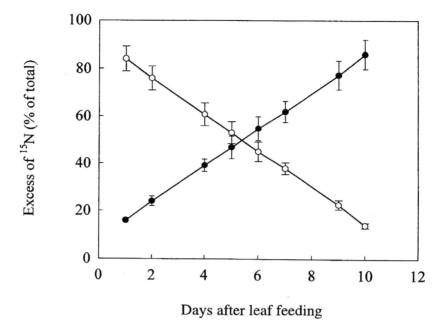

Figure 4. Increase with time after leaf feeding in the excess of ^{15}N in the rest of the plant (•) and decrease in fed leaf (o). Bars indicate ± s.e.m for four plants. (Adapted from Palta *et al.* 1991b).

Detectable amounts of ^{15}N can be recovered from wheat plants 12 h after the commencement of leaf feeding of a 1.5% solution of ^{15}N-urea. Then, 24 h after commencement of feeding, 84% of the total ^{15}N taken up remains in the fed leaf, after which export to the rest of the plant continues at a slow rate (Palta *et al.* 1991b). Altogether it requires up to 10 days for ^{15}N to be fully distributed throughout all regions of a plant (Fig. 4), suggesting that at least this period of time should elapse before sampling of plants for assessing initial distribution of ^{15}N. As a guide to the levels of labelling likely to be achieved, feeding of a single leaf of a wheat plant for 4 days at the early stage of tillering, using a 1% ^{15}N-urea (99 atom %) solution, would be expected to label plant parts by about 2 atom % ^{15}N excess. A similar level of enrichment would be achieved when labelling is undertaken at the stages of stem elongation and flowering, using 2 and 2.5% ^{15}N-urea solutions respectively (Fig. 5).

The experience of the author is that enrichments of this order of magnitude are more than appropriate for studying long-term distribution and mobilisation of nitrogen in wheat. Before the commencement of leaf feeding, consideration should be given to the amounts of ^{15}N and ^{15}N-urea required to achieve a target labelling of 2 atom % above natural abundance.

As a general rule one can estimate, the quantity of ^{15}N required ($^{15}N_R$) to achieve a particular target enrichment by using the percentage of nitrogen in dry matter (N%), the total dry weight of the plant (DW), the amount of ^{15}N natural abundance in the plant ($^{15}N_{NA}$), and the enrichment of the ^{15}N-urea supplied, according to:

$$^{15}N_R = \frac{[(N\% \times DW/100)(2A\%^{15}N)] - {}^{15}N_{NA}}{A\%^{15}N_s} \tag{1}$$

The amount of ^{15}N-urea (~99 atom %) required for leaf feeding is then obtained simply by dividing the quantity of ^{15}N required ($^{15}N_R$) by the proportion of N in urea (N%$_C$).

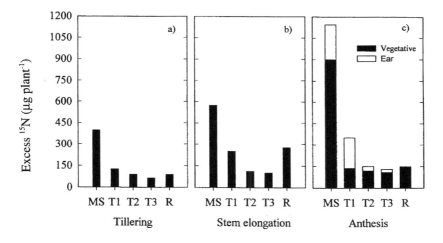

Figure 5. Excess ^{15}N in the main stem, tiller one, two and three, and roots in the wheat plant 10 days after feeding with ^{15}N-urea. Plants were fed at the stages of (a) tillering, (b) stem elongation and (c) flowering. Shadow bars in (c) indicate the excess ^{15}N in the ears of the main stem and individual tillers. (From Palta *et al.* 1991b).

4. ATMOSPHERIC FEEDING OF ^{13}C

Atmospheric feeding with $^{13}CO_2$ was first introduced in 1980 as a means of overcoming the already mentioned limitations relating to the use of $^{14}CO_2$. Although information of a semi-quantitative nature on sink:source identities can be obtained from cuvette-based spot feeding, fully effective use of atmospheric feeding with $^{13}CO_2$ requires the availability of an automatically controlled $^{13}CO_2$ feeding system (Yoneyama *et al.* 1980, Kouchi and Yoneyama 1984). The apparatus involved is both sophisticated and expensive, and, because of its complexity, has been mostly used under growth cabinet and glasshouse conditions. However to extend studies to the field, the CSIRO Plant Industry Laboratory in Perth has designed a portable, inexpensive, manually-operated system for atmospheric feeding of $^{13}CO_2$. This is reliable, relatively simple to use and has now been successfully used in wheat (Palta *et al.* 1994), rice (Ros *et al.* 2001), narrow-leafed lupin (J. A. Palta and M. Dracup, unpubl.) and chickpea (Davies *et al.* 2000) under both controlled environment and field conditions.

In the above approach, entire plant canopies are fed with $^{13}CO_2$. Prior to feeding, two semicircles of polyvinyl chloride (PVC) are placed on top of the soil in the pots and covered with plastic beads (5 mm diameter) so as to isolate the shoot atmosphere from that of the soil and prevent loss of fed $^{13}CO_2$ to the latter. Previously built Mylar film (~200 μm, Dow Co., Melbourne, Australia) enclosures supported on aluminum frames are then placed over the plants enclosing them completely. A vial containing $NaH^{13}CO_3$ (99 atom %), previously attached solidly to one of the internal walls of the chamber, is used as source of the label. The chambers are then carefully sealed onto the polyethylene sheet on the ground using masking tape. Air within the chamber is stirred with a 20-cm diameter fan, the transparent chamber is then covered with black polyethylene and lactic acid introduced into the vial to release $^{13}CO_2$. The chamber is retained in darkness for 2 – 5 min to provide uniform distribution of the $^{13}CO_2$ within the atmosphere of the chamber. After this the chamber is exposed to full light so that photosynthesis can proceed. Total CO_2 concentration in the chamber prior to and following the injection of lactic acid is monitored using a LI-6251 CO_2 infrared gas analyser (LI-COR, Inc. Lincoln, NE). A relationship previously established between the analyser reading and the actual $^{13}CO_2$ concentration (Fig. 6) can then be used to assess changes in $^{13}CO_2$ concentration in the chamber throughout the labelling period. It is possible to examine this relationship because a proportion of the gas analyser signal arises from the absorbance by $^{13}CO_2$ in the bands between 4.31 and 4.33 μm a region where there is overlapping infrared gas absorbance by $^{13}CO_2$ and $^{12}CO_2$ (D. McDermitt, LI-COR, Inc., personal communication). Once the total CO_2 concentration has decreased to about 120 μl L^{-1}, air enriched in CO_2 to 1000 μl L^{-1} CO_2 is fed directly via an entry port into the chamber until the CO_2 concentration regains a near atmospheric value (`360 μl L^{-1}). This operation is repeated at least three times before the chamber is removed, so that CO_2 concentrations are maintained close to atmospheric throughout the feeding period.

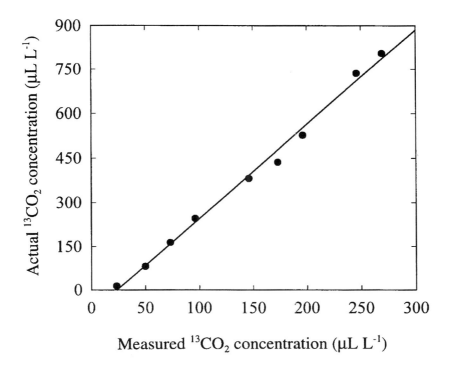

Figure 6. Relationship between actual $^{13}CO_2$ concentration (calculated from chamber volume with air free of CO_2) and measured $^{13}CO_2$ concentration using a LI-625 portable infrared gas-analyser. The points are calculated and measured values after each of nine injections of 20 mL of $^{13}CO_2$ (99.6 % Atom) into the chamber. Y= -61+3.1 X, r^2=0.989 (P=0.0001). (From J. A. Palta, unpubl.).

A target enrichment within the plant can easily be achieved in this manner, for example in studies when a feeding procedure is carried out between 1000 – 1500 h on a clear day, with photosynthetically active radiation (PAR, 400 – 700 nm) in the plane of the leaves maintained within the range 800 – 1200 µmol $m^{-2}s^{-1}$ and ambient temperatures within the range 22 – 28°C. Temperatures inside the chamber during feeding are ± 3°C of ambient. Note that the technique prevents $^{13}CO_2$ loss to the soil during feeding while also ensuring that the atmosphere inside the chamber is uniformly enriched at all times. As a general rule, after 48 h of feeding the main shoot and individual tillers or branches, and roots are found to be more or less uniformly labelled to levels approximating the target enrichment. A target enrichment of about 5 atom % ^{13}C has shown to be more than sufficient for measuring the long-term storage and mobilisation of carbon in wheat, lupin and chickpea (Palta *et al.* 1994, Davies *et al.* 2000, J. A. Palta and M. Dracup, unpubl.).

Based on experiences using the above-mentioned apparatus and protocols of isotope application, effective estimates of distribution and proportional fluxes of carbon and nitrogen can be obtained using only one feeding of $^{13}CO_2$ at an appropriate stage of development (Palta and Gregory 1997). However, definitive estimates of pre-anthesis storage of carbon and nitrogen and eventual contributions of carbon and nitrogen to grain or seed filling would ideally require at least three feedings during the pre-anthesis stage to ensure that the ^{13}C content is homogeneously distributed between different plant organs prior to the mobilisation phase of the plant growth cycle (Palta *et al.* 1994, Davies *et al.* 2000).

5.   PARAMETERS OF NITROGEN AND CARBON DISTRIBUTION AND REMOBILISATION AFTER LABELLING

In a typical feeding study, above and below ground biomass of fed and non-fed plants (control plants for measuring natural abundance) are sampled 48 or 72 h after feeding. Plants are divided into principal components (main stem, individual tillers or branches, grains or seeds and roots) and resulting parts then freeze-dried and weighed. All plant parts are then ground to less than 1 mm for analysis and quantification of ^{15}N and ^{13}C. Analysis for percentage of nitrogen and carbon, and the stable isotope enrichments of ^{15}N and ^{13}C are determined by mass spectrometery. In the analysis the most common standard used is a solution of ^{15}N-glycine and ^{13}C-glucose. The ^{15}N and ^{13}C enrichments and nitrogen and carbon content of these standards are originally determined against standards from the International Atomic Energy Agency. The results of the analysis (atom % excess) of ^{15}N and ^{13}C in plant parts are converted into percentages and milligrams of nitrogen and carbon. It is then possible to derive values for Relative Specific Allocation (RSA) and the amount of carbon (or nitrogen) derived from the fed ^{13}C or ^{15}N. We define RSA as the proportion of newly incorporated carbon (or nitrogen) relative to total carbon (or nitrogen) in a given sample. RSA is expressed as percentage of the total carbon (or nitrogen) of a plant part and is estimated using the atom % ^{13}C or ^{15}N measured in the enriched part (A% ^{13}Ce or ^{15}Ne), in the control plant (A% ^{13}Ccp or ^{15}Ncp) and in the $^{13}CO_2$ or ^{15}N-urea supplied.

Accordingly, for carbon:

$$RSA_C(\%) = \frac{[(A\%^{13}Ce) - (A\%^{13}Ccp)]}{[(A\%^{13}Cas) - (A\%^{13}Ccp)]} \times 100 \qquad (2)$$

For nitrogen:

$$RSA_N(\%) = \frac{[(A\%^{15}Ne) - (A\%^{15}Ncp)]}{[(A\%^{15}Nas) - (A\%^{15}Ncp)]} \times 100 \qquad (3)$$

The quantity (Q) of new carbon or nitrogen in a plant part (pp) is then calculated using the plant part dry weight (DW_{pp}), the carbon or nitrogen concentration and the RSA, as follows:
For carbon:

$$Q_C = RSA_C \times DW_{pp} \times C\ concentration \qquad (4)$$

For nitrogen:

$$Q_N = RSA_N \times DW_{pp} \times N\ concentration \qquad (5)$$

Partitioning (P) is defined as the proportion of the total input allocated into the different plant parts and is estimated as follows using the carbon and nitrogen content in the plant part (C_{pp} or N_{pp}) and whole plant (C_{plant} or N_{plant}), and the excess of ^{13}C or ^{15}N in the plant part ($^{13}C_{pp\ or}\ ^{15}N_{pp}$) and whole plant ($^{13}C_{plant\ or}\ ^{15}N_{plant}$).
For carbon:

$$\%P_C = \frac{(C_{pp} \times A\%\ excess\ ^{13}C_{pp})}{(C_{plant} \times A\%excess\ ^{13}C_{plant})} \times 100 \qquad (6)$$

For nitrogen:

$$\%P_N = \frac{(N_{pp} \times A\%\ excess\ ^{15}N_{pp})}{(N_{plant} \times A\%excess\ ^{15}N_{plant})} \times 100 \qquad (7)$$

RSA provides an index of the turnover of carbon or nitrogen in a given plant part (Palta and Gregory 1997), whereas P represents the proportion of the labelled ^{13}C or ^{15}N in a given plant part relative to its total content of ^{13}C or ^{15}N.

Total ^{13}C or ^{15}N in the plant is then obtained by summing the ^{13}C or ^{15}N in each individual plant part, thus enabling estimates to be made of the amounts of pre-anthesis stored carbon and nitrogen mobilised to the grain in the period between anthesis and maturity. Basic items of information required when making such assessments are the grain ^{13}C or ^{15}N content at maturity ($^{13}C_{gm}$ or $^{15}N_{gm}$), the vegetative carbon or nitrogen content at anthesis (C_{va} or N_{va}) and the vegetative ^{13}C or ^{15}N content at anthesis ($^{13}C_{va}$ or $^{15}N_{va}$). The calculations involved are as shown below:

For carbon:

$$Q_{C_{grain}} = {}^{13}C_{gm} \times \left(\frac{C_{va}}{{}^{13}C_{va}}\right) \tag{8}$$

For nitrogen:

$$Q_{N_{grain}} = {}^{15}N_{gm} \times \left(\frac{N_{va}}{{}^{15}N_{va}}\right) \tag{9}$$

Note that this calculation assumes that pre-anthesis stored ^{13}C and ^{15}N and unlabelled carbon and nitrogen are mobilised to the grain to the same relative extents as those at which they were in the plant material at anthesis.

6.   EXAMPLES OF HOW DATA MAY BE USED TO ASSESS THE SOURCE/SINK INTERACTIONS IN CROP PLANTS UNDER DROUGHT CONDITIONS.

One of the aims of this chapter is to illustrate how use of the above mentioned $^{13}CO_2$ and urea-^{15}N techniques may be directed at evaluating the effect of drought conditions on assimilation, accumulation and mobilisation of carbon and nitrogen in crop plants. Central to this theme is the desirability of assessing how important pre-anthesis assimilates become as providers of carbon and nitrogen for seed growth when plants experience drought conditions. It is already known, for example that early-assimilated carbon and nitrogen are particularly important in wheat (*Triticum aestivum* L.), narrow-leafed lupin (*Lupinus angustifolius* L.) and chickpea (*Cicer arietinum* L.) in low rainfall mediterranean-type environments. This is of course because post-anthesis assimilation of carbon and nitrogen are inevitably restricted if water deficits occur after flowering or when carbon and nitrogen are being mobilised to the grain. Using the above described $^{13}CO_2$ and urea-^{15}N techniques, Palta *et al.* (1994) have shown that the total amount of carbon in grain of a wheat crop can be reduced by 24% when maturing under conditions of rapidly developing water deficits relative to less stressful conditions. In this case severe stress reduced gain in carbon from post-anthesis assimilation by 57%, while concurrently increasing mobilisation of pre-anthesis stored carbon by 36% (Table 2). Somewhat surprisingly total grain nitrogen was not affected by the severity of onset of water deficits. This was found to be because plants were engaging in more effective mobilisation of pre-anthesis stored nitrogen under conditions of fast as opposed to slow onset of water deficits. Furthermore there were lesser losses of pre-anthesis nitrogen associated with rapid onset water deficits and the grain protein contents were higher.

Table 2. Contribution of pre-anthesis and post-anthesis carbon and nitrogen to the grain in wheat subjected to fast and slow rates of development of water deficits after flowering. (Adapted from Palta *et al.* 1994).

Rate	Grain carbon (mg plant^{-1})			Grain nitrogen (mg plant^{-1})		
	Total	Pre-anthesis remobilised	Post-anthesis assimilated	Total	Pre-anthesis remobilised	Post-anthesis uptake
Fast	703	449	254	43	35	8
Slow	922	329	593	42	25	17

A major feature of labelling studies of this kind has been to highlight the role of vegetative or late-formed tillers in grain filling of wheat under terminal drought (see Palta *et al.* 1994). As shown in Figure 7 the gain in carbon and nitrogen by the grain in the main stem and tiller 1 was supported by mobilisation from tiller 2 and 3, but this applied only when these late-formed tillers failed to produce an ear (Fig. 7). Conversely if an ear had been produced, carbon and nitrogen mobilised from the vegetative biomass of these tillers was transported into the ear, which only developed poorly.

In a further example, provided by Palta and Gregory (1997), the allocation of carbon between roots and shoots and the fluxes of carbon to roots and soil were measured in wheat grown when two contrasting soil moisture regimes were imposed during the vegetative stages of growth. The plants in question were fed with $^{13}CO_2$ and the fate of ^{13}C was assessed for each treatment not only in terms of incorporation into root and shoot carbon, but also as losses of ^{13}C in root and microbial respiration, rhizodeposition and shoot respiration. The study showed that relatively more assimilates were allocated to the roots under conditions of limited water supply, as indicated by higher percentages of ^{13}C recovered in roots (Table 3). Changes during vegetative growth in ^{13}C allocated to the roots occurred at the onset of tillering and booting stage, and these effects were associated with changes in the relative activity of respective sinks. The ^{13}C recovered below ground under limited soil water increased by 21% at leaf formation and by 43% at booting, relative to that under adequate soil water (Fig. 8). The increase resulted mainly from an increase in $^{13}CO_2$ respired from roots and microorganisms, as opposed to bound ^{13}C in organic fractions of the soil.

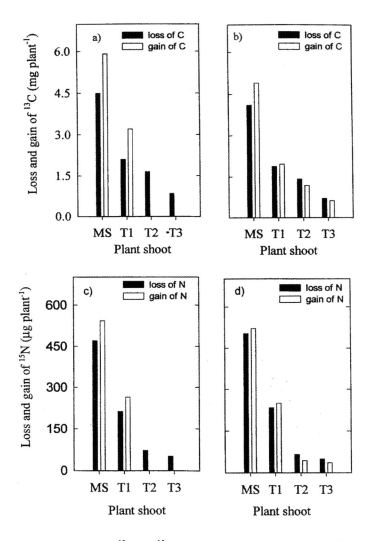

Figure 7. Post-anthesis losses of ^{13}C and ^{15}N from the straw (closed histograms) and the gain by the grain (open histograms) in the main stem (MS) and individual tillers (T1, T2, T3) of wheat under drought. In plots (a) and (c) the late formed tillers (T2, T3) did not produced ears and in (b) and (d) they produced ears. (From J. A. Palta, unpubl.)

Table 3. Allocation of photosynthetically fixed ^{13}C in the plant-soil system a different stages during the vegetative growth of wheat. Values are the absolute quantities of ^{13}C measured in each component of the plant-soil system 48 h after the plants were fed with $^{13}CO_2$. (From Palta and Gregory 1997)

DAS[a]	Shoot	Root	Shoot respiration	Root and microbial respiration	Soil	Total
			(mg ^{13}C per plant)			
Adequate soil watered						
28	0.7	0.4	0.03	0.07	0.01	1.2
36	1.3	1.5	1.20	0.21	0.03	2.2
43	3.4	1.5	0.33	1.06	0.09	6.1
51	8.1	3.9	0.87	2.36	0.22	15.2
64	30.0	7.7	3.08	3.45	0.36	44.7
Limited soil water						
28	0.2	0.2	0.01	0.04	0.004	0.4
36	1.0	0.6	0.05	0.19	0.027	1.9
43	2.0	1.2	0.13	0.63	0.063	4.1
51	4.1	2.4	0.24	1.28	0.117	8.1
64	15.4	6.6	1.15	2.76	0.280	26.3

[a] DAS: Days after sowing.

Glasshouse studies on $^{13}CO_2$ and urea-^{15}N fed narrow-leafed lupin have shown that a rapid onset of water deficit results in accelerated transfer of pre-anthesis carbon and nitrogen to pods. However, high rates of transfer were transient and failed to increase the overall contribution of pre-anthesis assimilates to seeds. In comparable plants under favourable soil-water conditions, pre-anthesis assimilates were transferred preferentially to further growth and reproduction of apical branches, and, in consequence, to a lesser extent to pods and seeds on the main stem. However, this pattern could be reversed following a short period of water deficit induced at the end of flowering on the main stem. Since redirection of assimilates was not reversed following relief of water deficit, possible adjustments to source/sink relations mediated through changes in hormonal balance were indicated (J. A. Palta and M. Dracup, unpubl.). The data from such studies also show that the primary contribution of pre-anthesis assimilated carbon and nitrogen is to support early stages of pod growth (< 21 mm), whereas later pod growth and seed filling rely mainly on the carbon and nitrogen assimilated in branches after flowering.

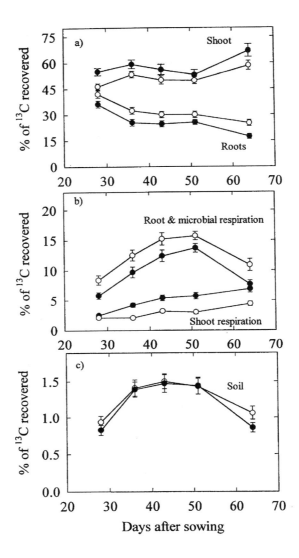

Figure 8. Changes with days after sowing in the allocation of ^{13}C in (a) shoots and roots, (b) respiration of roots and shoots and (c) soil for wheat plants grown under adequate (•) and water-limited conditions (o). Data are the means of four replicates (20 plants) per soil water treatment. Bars indicate ± s.e.m. (From Palta and Gregory 1997.)

The double labelling system has also been used to study the relative significance of pre- and post- anthesis assimilates to pod filling in narrow-leafed lupin (J. A. Palta unpubl.). Under conditions fostering a fast rate of onset of water deficit (0.2 MPa day^{-1}) the use of pre-anthesis carbon and nitrogen for pod set and pod filling was found to increase by 12 and 17%, respectively. However, these increases failed to make a significant contribution to seed yield because, with increasingly severe water deficit, abortion of pods and seeds took place. In comparable plants exposed to slowly progressing water deficit (0.1 MPa day^{-1}) deployment of pre-anthesis carbon and nitrogen for pod set and pod filling increased by 10 and 15%, respectively. Again there was no evidence of a significant contribution to seed yield.

Another potential avenue for use of $^{13}CO_2$ feeding has been in assessing the role of apical branches in pod filling of narrow-leafed lupin exposed to terminal drought (J. A. Palta and C. Ludwig, unpubl.). In a study of this kind, all leaves on each apical branch were fed continuously with $^{13}CO_2$ between 1000 – 1500 h on days with clear skies and the fate of ^{13}C followed through to plant maturity. The data then indicated that each apical branch contributed most of its exportable carbon (80 – 96%) to feeding of seeds of that branch with little carbon transfer to other branches, irrespective of the number of pods being filled on these branches.

Studies of chickpea grown in Mediterranean-type environments are particularly relevant to the labelling approach considered in this chapter since severe effects of drought are frequently encountered in this species, especially in terms of markedly reduced photosynthesis of foliage during seed feeding (Leport *et al.* 1998, 1999). In a recent study by Davies *et al.* (2000) the contribution of carbon and nitrogen fixed during vegetative growth to seed growth has been measured in two chickpea genotypes, in each case comparing adequately watered plants to those subjected to terminal drought. In the adequately watered genotypes 7 – 9% of the carbon in the seed came from pre-podding carbon while this increased to 9 – 16% when the same pair of genotypes was subjected to terminal drought. The quantity of pre-podding nitrogen mobilised to the seed was 62 – 85% in the adequately watered chickpea versus 91 – 97% in those exposed to terminal drought. Interestingly, genotypic variation in mobilisation of both carbon and nitrogen has been observed in chickpea and such variation is shown to be correlated with differences in yield under terminal drought (Davies *et al.* 2000).

In summary, this chapter illustrates the potential value of using experiments involving combined feeding of $^{13}CO_2$ and urea-^{15}N to assess source:sink interactions in crop plants. The use of this stable isotope instead of potentially hazardous radioactive labelling with ^{14}C has turned out to be fully quantitative and accurate when analysing the mobilizable resources of carbon and nitrogen in the vegetative parts of a plant and the subsequent patterns of transfer of these to the reproductive parts. The instrumentation required to measure the ^{13}C and ^{15}N in the labelled and unlabelled plant material exist in most geology, chemistry, plant biology and soil science departments of many Universities as well as in appropriated oriented government laboratories, so it is hoped that the substance of this chapter encourages future use of such resources in studies of assimilate partitioning.

ACKNOWLEDGMENTS

The published and unpublished research cited in this paper was supported by the Grains Research and Development Corporation funds (GRDC). Mr D. McDermitt from LI-COR Inc. provided detailed information on the infrared absorbance by $^{12}CO_2$ and $^{13}CO_2$ of the LI-6251 CO_2 analyser. The technical inputs of Ms Elaine Smith and Christiane Ludwig are appreciated and continue to be invaluable. Productive interaction with Drs. Ian R. P. Fillery, Paul C. Pheloung, Craig Russell and Mark B. Peoples is acknowledged. The assistance of Drs Annie McNeill, Murray J. Unkovich and Professor John S. Pate in the final editing of the manuscript is gratefully acknowledged.

REFERENCES

Austin, R.B., Edrich, J.A., Ford, M.A., and Blackwell, R.D. (1977). The fate of dry matter, carbohydrates and ^{14}C lost from the leaves and stems of wheat during grain filling. *Annals of Botany* 41, 1309-1321.

Austin, R.B, Edrich, J.A, Ford, M.A, and Blackwell, R.D (1980). Contribution to grain yield from pre-anthesis assimilation in tall and dwarf barley phenotypes in two contrasting seasons. *Annals of Botany* 45, 309-319.

Bidinger, F., Musgrave, R.B., and Fischer, R.A. (1977). Contribution of stored pre-anthesis assimilate to grain yield in wheat and barley. *Nature* 270, 431-433.

Blacklow, W.M. (1982). ^{15}N moved to the grain of winter wheat when applied as nitrate to senescing flag leaves. *Australian Journal of Plant Physiology* 9, 641-646.

Bremner, J.M. (1990). Problems in the use of urea as a nitrogen fertilizer. *Soil Use and Management* 6, 70-72.

Cliquet, J.B., Deleens, E., Bousser, A., Martin, M., Lescure, J.Ch., Prioul, J.L., Mariotti, A., and Morot-Caundry, J.F. (1990). Estimation of carbon and nitrogen allocation during stalk elongation by ^{13}C and ^{15}N tracing in *Zea mays* L. *Plant Physiology* 92, 79-87.

Cock, J.H., and Yoshida, S. (1972). Accumulation of ^{14}C-labelled carbohydrate before flowering and its subsequent redistribution and respiration in rice plant. *Proceedings Crop Science Society of Japan* 41, 226-234.

Davies, S.L., Turner, N.C., Palta, J.A., Siddique, K.H.M., and Plummer, J.A. (2000). Remobilisation of carbon and nitrogen supports seed filling in desi and kabuli chickpea subject to water deficit. *Australian Journal of Agricultural Research* 51, 855-866.

Fried, M. (1978). Critique of field trials with isotopically labelled N fertilizer. In 'Nitrogen in the Environment'. (Eds D.R. Nielsen and J.G. MacDonald) Vol. 1 pp. 43-62. (Academic Press: New York.)

Janzen, H.H., and Bruinsma, Y. (1989). Methodology for the quantification of root and rhizosphere nitrogen dynamics by exposure of shoots of ^{15}N-labelled ammonia. *Soil Biology and Biochemistry* 21, 189-196.

Kouchi, H., and Yoneyama, T. (1984). Dynamics of carbon photosynthetically assimilated in nodulated soya bean plants under steady-state conditions. 1. Development and application of $^{13}CO_2$ assimilation system of a constant ^{13}C abundance. *Annals of Botany* 53, 875-882.

Krogmeier, M.J., McCarty, G.W., and Bremner, J.M. (1989). Phytotoxicity of foliar-applied urea. *Proceedings of the National Academy of Sciences of the USA* 86, 8189-8191.

Leport, L., Turner, N.C., French, R.J., Barr, M.D., Tennant, D., Thomson, B.D., and Siddique, K.H.M. (1998). Water relations, gas exchange and growth of cool-season grain legumes in a Mediterranean-type environment. *European Journal of Agronomy* 9, 295-303.

Leport, L., Turner, N.C., French, R.J., Barr, M.D., Duda, R., Davies, S.L., Tennant, D., and Siddique, K.H.M. (1999). Physiological responses of chickpea genotypes to terminal drought in a Mediterranean-type environment. *European Journal of Agronomy* 11, 279-291.

Lockyer, D.R., and Whitehead, D.C. (1987). Gaseous ammonia as a source of nitrogen for grass *Journal of Science of Food and Agriculture*. 38, 329-330.

Palta, J.A., Fillery, I.R., Mathews, E.L., and Turner, N.C (1991a). Assessing the use of ^{15}N-^{13}C-urea for studying the long-term storage and remobilisation of carbon and nitrogen in wheat. In 'The Use of Stable Isotopes in Plant Nutrition, Soil Fertility and Environmental Studies.' pp 441-452. (International Atomic Energy Agency: Vienna.)

Palta, J.A., Fillery, I.R., Mathews, E.L., and Turner, N.C. (1991b). Leaf Feeding of ^{15}N-urea for Labelling Wheat with Nitrogen. *Australian Journal of Plant Physiology* 18: 627-36.

Palta, J.A., and Fillery, I.R.P. (1993). Post-anthesis remobilisation and losses of N in wheat in relation to applied nitrogen. *Plant and Soil* 156, 179-181.

Palta, J.A., Kobata, T., Fillery, I.R., and Turner, N.C. (1994). Remobilisation of Carbon and nitrogen in wheat as influenced by post-anthesis water deficits. *Crop Science* 34, 118-124.

Palta, J.A., and Fillery, I.R.P. (1995). N application enhances remobilisation and reduces losses of pre-anthesis N in wheat grown on a duplex soil. *Australian Journal of Agricultural Research* 46, 519-531.

Palta, J.A., and Gregory, P.J. (1997). Drought affects the fluxes of carbon to roots and soil in ^{13}C, pulse-labelled plants of wheat. *Soil Biology and Biochemistry* 29, 1395-1403

Pate, J.S. (1973). Uptake assimilation and transport of nitrogen compounds by plants. *Soil Biology and Biochemistry* 5, 109-119.

Ros, C., Palta, J.A., Bell, R.W., and White, P.F. (2001). Retranslocation of carbon and nitrogen in rice plants during the transplanting recovery period. *Crop Science* (in press).

Svejcar, T.J., and Boutton, T.W. (1985). The use of stable carbon isotope analysis in rooting studies. *Oecologia* 67, 205-208.

Van Norman, R.W., and Brown, A.H. (1952). The relative rates of photosynthetic assimilation of isotopic forms of carbon dioxide. *Plant Physiology* 27, 691-709.

Warembourg, F.R., Montange, D., and Bardin, R. (1982). The simultaneous use of $^{14}CO_2$ and $^{15}N_2$ labelling techniques to study the carbon and nitrogen economy of legumes grown under natural conditions. *Physiologia Plantarum* 56, 46-55.

Yoneyama, T., Arai, K., and Totsuka, T. (1980). Transfer of nitrogen and carbon from a mature sunflower leaf- $^{15}NO_2$ and ^{13}CO feeding studies. *Plant Cell Physiology* 21, 1367-1381.

Chapter 9

Use of Enriched ^{15}N Sources to study Soil N Transformations

Ian R. P. Fillery[1] and Sylvie Recous[2]

[1]CSIRO Plant Industry, P.O. Private Bag 5, Wembley, WA 6913 Australia. Email: i.fillery@ccmar.csiro.au: [2]INRA, Unité d'agronomie, rue Fernand Christ 02007, Laon, France. Email: recous@laon.inra.fr

Key words: ^{15}N recovery, N balance, N loss, ^{15}N dilution techniques, ^{15}N analysis

1.   INTRODUCTION

The large size of the native N pool in soils, compared to fertilizer N inputs, and the inherent variability of native soil N, make assessments of the fate of fertiliser N virtually impossible to assess without the use of a tracer to differentiate the fertiliser input from background soil N. The low natural abundance of the stable isotope, ^{15}N, together with the relatively low cost of ^{15}N-enriched N compounds makes this isotope an ideal tracer for soil N transformations. Consequently, ^{15}N-enriched substrates have been widely used in studies following the fate of N in soil-plant systems. However, the use of ^{15}N to study N processes in the field necessitates large resources of labour and time, and access to a mass spectrometer. Attendant costs of ^{15}N analyses also need to be considered before embarking on a program of research.

In what classes of studies is the use of ^{15}N in nitrogen research most advantageous? Firstly, measurements of ^{15}N enrichment in individual soil N pools, and in plant material, can be employed to follow the fate of applied ^{15}N fertilisers. Figure 1 shows the key transformations in soil that determine the availability of inorganic N for plant uptake. Analysis of added ^{15}N in the NH_4^+, NO_3^- and organic N pools and in plant material enables the recovery of

167

M. Unkovich et al. (eds.),
Stable Isotope Techniques in the Study of Biological Processes and Functioning of Ecosystems, 167–194.
© 2001 *Kluwer Academic Publishers. Printed in the Netherlands.*

added ^{15}N to be determined. Secondly, experiments measuring the difference between added ^{15}N and recovered ^{15}N after a particular time interval can be used as a measure of the loss of fertilizer-derived N from soil-plant systems. Techniques employing ^{15}N balance have been especially useful tools for estimating gaseous losses of N where indirect measurements are often the only practical means of determining the extent of gaseous N transformations. However, where losses of N are determined by balance methods there is always the danger of misinterpreting the significance of gaseous N losses if inadequate technique has resulted in underestimation of ^{15}N recovery.

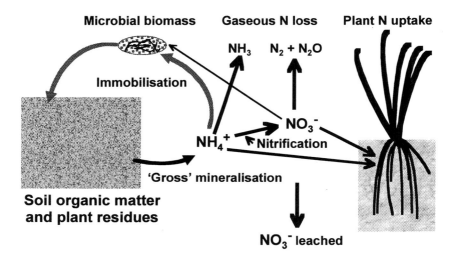

Figure 1. A schematic showing the major transformations of N in soil.

Thirdly, ^{15}N-enriched techniques have been used to determine the efficiency of uptake of N sources by crop plants. However, in this case the processes of mineralisation-immobilisation turnover (MIT) in soil can complicate the issue by diluting the quantity of ^{15}N in the inorganic N pool without changing the overall amount of N available to a crop. As a consequence, different values for the efficiency of use of N sources are often obtained where soil measurements are based on direct ^{15}N assays or on apparent N recoveries assessed by difference between N fertilised and unfertilised treatments (e.g. Hood *et al.* 1999, Schindler and Knighton 1999, Dejoux *et al.* 2000). As a general rule the apparent N recovery approach is the preferred method for establishing the efficiency of use of fertilizer N by plants.

Finally, the dilution of ^{15}N-labelled inorganic pools in soil by natural abundance-derived NH_4^+ and NO_3^- permits studies to be made of the gross or actual rate of soil N processes such as mineralisation and nitrification (Fig. 1). Direct measurements of actual processes rather than data derived from the net change in inorganic N will improve our understanding of N dynamics in soil and thereby help evaluate concepts introduced in simulation models now widely used to predict C and N cycling.

This chapter will describe in detail the range of methods that can now be used to determine the fate of ^{15}N-labelled fertilizers in field studies. In each case experimental detail is accompanied by critical accounts of potential pitfalls, expected outcomes and advice as to how data should be interpreted. Recent developments in the application of ^{15}N dilution techniques to study gross mineralisation, immobilisation and gross nitrification are outlined.

2.   MASS BALANCE ^{15}N RECOVERY STUDIES

There are many publications in the literature that report results from studies using ^{15}N-labelled compounds to determine the extent and timing of losses of N-containing fertilisers which occur following application to soil. The account of Pilbeam (1996) is useful in summarising findings from ^{15}N balance experiments pertinent to wheat while that of Peoples *et al.* (1995) evaluates ^{15}N recovery studies undertaken on other major food and fibre crops. Numerous studies have also been undertaken on the fate of ^{15}N in flooded rice soils (see review by DeDatta 1995).

Saffigna (1987) has described in detail techniques used in field ^{15}N balance experiments. Conceptually, the technique is simple. A known amount of ^{15}N is added, in this case to soil, and soil plus plant material is then sampled to ascertain the quantity of [15]N (derived from the added source) subsequently accumulated in different fractions of soil, and plant material. It is assumed that the ^{15}N unaccounted for in the total soil-plant system may have been lost through any of a number of possible N loss mechanisms. The timing of such losses, knowledge of the distribution of ^{15}N in soil N pools, and information on the soil conditions at the time of loss often may then provide clues as to the mechanism of N loss that eventuated. The 'classical' use of ^{15}N balance involves a single measurement at harvest. The example given here (Fig. 2) allows a comparison of the relative magnitudes of constituent soil processes *viz.* immobilization of N into soil organic N, leaching of NO_3^- and relates these to crop N use efficiency, and ^{15}N unaccounted for. It is notable that in both systems under temperate climatic conditions, the residual soil inorganic ^{15}N at harvest is almost nil and that the main difference between grassland and annual crop is the amount of ^{15}N recovered in organic form in soil either as root residues or microbial biomass.

A

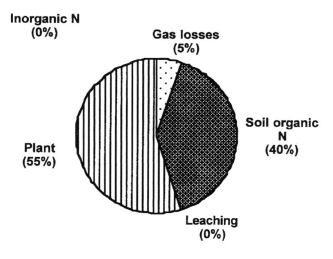

B

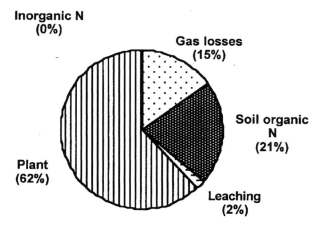

Figure 2. ^{15}N balance at harvest of grassland (A) and corn (B). Adapted from Recous and Loiseau (1997).

Another example of the information that can be obtained from the use of ^{15}N-enriched sources is shown in Table 1, where N applied has been traced dynamically in order to identify the processes involved. In this case, ^{15}N-labelled ammonium sulphate and ^{15}N-labelled urea are applied to the surface of deep sand. The aims of the study are to determine:

a) how much of NH_4^+-based fertiliser was lost to the atmosphere soon after application,

b) the rate of transformation of NH_4^+ to NO_3^-, and

c) to quantify the movement of ^{15}N-labelled NO_3^- through leaching.

By comparing one N source (urea) susceptible to NH_3 volatilisation with another not prone to this loss mechanism on acidic soil (ammonium sulphate), it is thus possible to make a semi-quantitative assessment of the role of NH_3 volatilisation from urea after different application methods. The ability to identify fertiliser-derived NO_3^- in soil profiles enables an assessment to be made of the role of NO_3^- leaching as an N loss mechanism (see Fig. 3). Studies on the fate of fertiliser-N as outlined above have thus demonstrated that, where fertiliser-N is applied correctly in terms of amount and timing, little of the fertiliser N is leached and that soil-derived NO_3^- is the major source of leached N. Similar findings have been found elsewhere for soils in temperate agricultural systems (MacDonald *et al.* 1997). The much larger errors associated with the estimates of total NO_3^- compared to NO_3^- derived from ^{15}N enriched sources (Fig. 3) highlight the increased precision that can be achieved by using ^{15}N enriched substrate in studies of soil N transformations.

The success of the ^{15}N balance technique is of course dependent on it being correct to assume that (a) none of the applied ^{15}N is lost during sampling or subsequent analysis of soil and plant material, (b) samples are not cross-contaminated during sample preparation, and (c) the methods of analysis are fully reliable and reproducible. These attributes need to be tested using trial runs involving known quantities of ^{15}N-labelled soil, thereby confirming the accuracy of each adopted procedure. Sampling should be undertaken immediately after the application of ^{15}N-labelled sources to check that recovery is close to 100% at that time. Unfortunately this is a step that is frequently omitted due to high cost of ^{15}N and isotopic analysis. In any case natural abundance ^{15}N enrichments must be determined using comparable unlabelled materials since the resulting values have to be subtracted from the values of ^{15}N enrichment (obtained from ^{15}N treated soil or plant material) when calculating % recovery of applied ^{15}N.

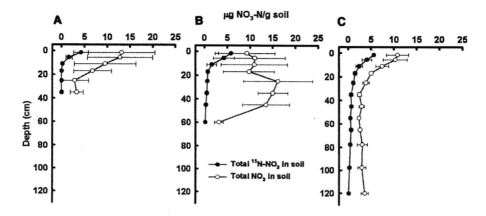

Figure 3. Distribution of total soil NO_3^- and $^{15}NO_3^-$ in deep sand (A) 18, (B) 33 and (C) 45 days after incorporation of the equivalent of 30 kg N/ha of ^{15}N enriched urea (5 atom% ^{15}N) into the surface soil. (I.R.P Fillery unpubl.).

How should studies of the above kind be conducted? The cost of ^{15}N-labelled compounds will usually preclude use at normal field plot scales, e.g. 50 m^2, therefore enriched ^{15}N sources are typically applied to mini-plots (0.1 – 2.0 m^2) positioned within larger plots in broader field studies (see Saffigna 1987). Dimensions of the mini-plots are often determined in terms of units of crop row width (e.g. see Fig. 4). The surface soil in each mini-plot can be surrounded within sheet metal, plate steel or plastic frames, but this additional expense and effort is only to be recommended where runoff is expected, or in lowland rice where floodwater is being used for rice culture. Larger mini-plot dimensions must be used where these are not enclosed to ensure that soil and plant material is sampled from a central area that should be uniformly enriched with ^{15}N. In the case of wheat, two row widths beyond the designated sampling area should provide a sufficient buffer (Fig. 4; Powlson *et al.* 1986). Much smaller microplots, based on PVC or steel pipe (< 24 cm in diameter), have also been used in certain ^{15}N studies; these either being pressed into soil or inserted mechanically using a barrel drill driven by a drilling rig. The insertion of steel or PVC pipe into soil to at least 1 m makes such systems convenient for studying NO_3^- leaching. However, complications can arise in some soils due to the preferential movement of soil water and solutes along the walls of pipes. Cameron *et al.* (1990) describe methods by which such edge-flow effects can be minimised when using undisturbed soil cores.

Table 1. Recovery of ^{15}N (% ^{15}N applied) in soil N fractions after different methods of application of ammonium sulphate (AS) and urea to a deep sand near Regans Ford, WA. Fertilisers were applied at the rate of 50 kg N/ha, 15 June 1990. (I.R.P. Fillery unpubl.).

Application method	Fertiliser	Soil fraction			Total recovered
		NH_4	NO_3	Organic	
18 days after fertiliser application					
Broadcast to soil	AS	78.1	3.0	6.0	88.1
surface	urea	46.2	4.4	8.2	58.8
Incorporated in top	AS	74.4	5.4	10.1	89.9
0.03m	urea	61.2	5.9	8.7	75.8
Band applied at	AS	83.8	8.0	5.7	97.5
0.05m	urea	68.1	7.7	8.7	84.5
33 days after fertiliser application					
Broadcast to soil	AS	45.5	11.3	6.9	73.8
surface	urea	26.0	11.8	9.9	47.7
Incorporated in	AS	51.5	14.5	11.6	77.6
top 0.03m	urea	34.7	19.3	15.1	69.1
Band applied at	AS	61.9	15.6	8.0	85.5
0.05m	urea	52.9	23.9	11.2	88.0
45 days after fertiliser application					
Broadcast to soil	AS	33.2	30.5	nd	-
surface	urea	9.6	23.6	nd	-
Incorporated in top	AS	34.1	33.2	nd	-
0.03m	urea	15.4	33.7	nd	-
Band applied at	AS	51.9	25.3	nd	-
0.05m	urea	28.8	35.0	nd	-
Least Significant	Fertiliser	5.7	ns	1.6	4.0
Differences	application	6.9	3.2	2.0	4.9
	time	6.9	3.2	1.6	4.0

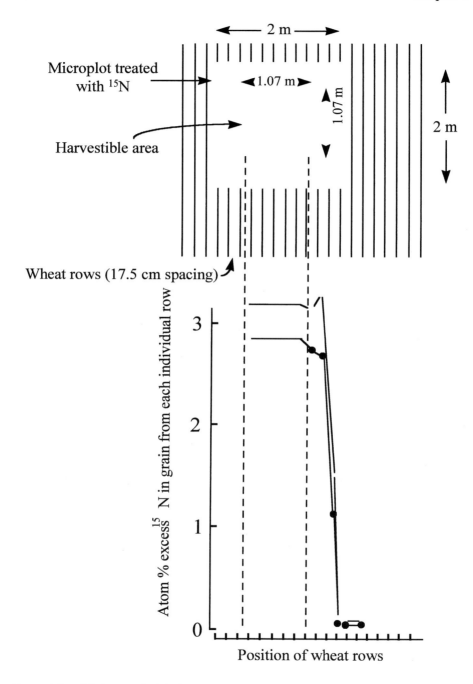

Figure 4. Possible layout of mini-plots, showing the area that can be harvested in relation to wheat rows. Atom % excess ^{15}N in grain was taken from Powlson *et al.* (1986).

A wide range of ^{15}N-labelled inorganic and organic compounds, with enrichments of ^{15}N ranging from 5 to 99 atom % ^{15}N, is available from commercial suppliers (see World Wide Web for suppliers). It is recommended that the ^{15}N-labelled compound used in microplots be the same compound as the bulk fertiliser N source whose fate is being evaluated. ^{15}N-labelled plant residues are not readily available commercially so must be produced by feeding the target plant species with ^{15}N enriched inorganic N. Repeated pulse-labelling, whether applied to an established crop in the field or to plants in pots, is the method most widely used to generate ^{15}N-enriched plant material. The procedure employed may vary according to number of successive ^{15}N applications made, ranging from a single application to weekly applications over a growing season (e.g. Amato *et al.* 1987, Bremer and van Kessel 1992, Thompson and Fillery 1997, Hood *et al.* 1999). Achieving a homogeneous labelling of the different parts of the plants (roots, stems, leaves) or pools of N in each tissue (e.g. water extractable *vs.* non soluble) is not always possible with serial applications of ^{15}N. Since a key aim of using labelled residues is to trace the fate of the whole residue-N in soil, uniform labelling of plant parts is recommended. Nevertheless, reasonable homogeneity in the labelling of tissues has been reported where split application of ^{15}N has been used (Amato *et al.* 1987, Bremer and van Kessel 1992). The best technique for homogeneously enriching plant material with ^{15}N is of course to grow plants in nutrient solutions that can be either easily replaced or adjusted to ensure a uniform feed of ^{15}N (Zaccheo *et al.* 1993, Aita *et al.* 1997, Dejoux *et al.* 2000). However, the high cost of such crop production has often prevented the use of this method. One drawback of growing plants on very artificial and highly enriched N solutions is that significant changes in the biochemical composition of crop tissues can occur, compared to soil- or field-grown crops. Whatever the labelling method with enriched solution, labelling N_2-fixing legumes always results in a lower ^{15}N enrichment of the legume tissues compared to other crops and a very different enrichment of the various components. This is due to the contribution of atmospheric N by N_2 fixation to crop N (Ladd *et al.* 1981, Jensen 1996).

The level of ^{15}N enrichment needed in materials added to soil will depend on a number of factors, including (a) the system being used to measure ^{15}N/^{14}N ratios, (b) the rate of addition of each ^{15}N-labelled compound, (c) the extent of dilution of the added ^{15}N source by native soil N, and (d) the duration of the experiment. Mass spectrometers are 10 to 100 times more sensitive and much more precise than optical ^{15}N analysers (Hauck 1982), so can be routinely used to measure ^{15}N enrichments close to natural abundance as well as variations in ^{15}N natural abundance. As a rule of thumb, compounds enriched with 5 atom % ^{15}N can be applied when mass spectrometers are used to analyse $^{15}N/^{14}N$ ratios in soil samples taken in ^{15}N balance experiments, whereas compounds with at least 20 atom % ^{15}N are needed where an optical analyser is used (see Middleboe 1982). However, it is possible to calculate the ^{15}N atom % needed in applied materials to achieve minimum ^{15}N enrichments in a given pool. The following example is given for the soil organic N pool.

$$\text{atom\% } ^{15}N = \frac{(\text{meq N}_{org} \times \text{NAT}^{15}N_{org}) + (\text{meq N}_{fert} \times \text{atom\% } ^{15}N_{fert})}{(\text{meq N}_{org} + \text{meq N}_{fert})} \quad (1)$$

where meq is the milliequivalents of N (mg N/14.01), N_{org} is soil organic N, N_{fert} is fertiliser N, and NAT is natural abundance.

As a general rule, higher ^{15}N enrichments of added N (10 – 50 %) should be used in studies of actual or gross mineralisation, immobilisation and nitrification, and only small quantities of N must be added to minimise the affect of the N addition on the N transformation under study (Murphy *et al.* 1997, Recous *et al.* 1999). Highly ^{15}N enriched compounds (70 – 99%) are also used in denitrification studies to measure the ^{15}N enrichment in N_2 and N_2O in air since very high dilution of $^{15}N_2$ by atmospheric N_2 is to be expected (e.g. Bronson and Fillery 1998, Mulvaney *et al.* 1997, Weier 1996). Methods for measuring the enrichment of ^{15}N in air are discussed in detail by Smith (1987) and Myrold (1990).

^{15}N-labelled sources are usually sold in powder form. Uniform application of this material within mini-plots is difficult at the best of times, with artificial losses of ^{15}N likely, especially when powdered materials are applied under windy conditions. Pelletising powdered materials (to a mesh commonly used in commercial fertilizers) will overcome this problem. The International Fertilizer Development Center in Muscle Shoals, Alabama, has equipment to pelletise ^{15}N-enriched materials and could be approached. Alternatively, ^{15}N sources can be dissolved in water to give a molar concentration just below saturation for the compound or to provide more dilute solutions. Solutions can be sprayed on the soil surface or if a distribution pattern simulating application of fertiliser granules is desired, applied drop by drop to soil using disposable syringes. The use of concentrated solutions keeps the addition of water to a minimum, which can be advantageous in sandy soils where water content is increased by even small water inputs, thereby modifying N transformations. It is recommended that either a known weight of solid or known volume of ^{15}N solution be added to the area that is to be sampled at a later date.

It is absolutely crucial that rigorous sampling procedures be used. Researchers tend to be fastidious about laboratory techniques and accordingly undertake duplicate or even triplicate analyses of the same soil sample to achieve reproducible assays. However, it should always be remembered that the soil-plant system itself is the likely major source of variability in ^{15}N recovery experiments. Table 2 illustrates this fact by showing the field variability associated with the estimates of the ^{15}N balance components for three experiments. These experiments varied by the type of crop, the microplot design, the method of ^{15}N application, the dates of sampling after ^{15}N applications, the location, and year each study was conducted. The variability in the estimates of ^{15}N content can be seen to depend strongly on the N compartment considered, being least for ^{15}N uptake by the crop and greatest for the ^{15}N in the soil inorganic N.

In the case of ^{15}N uptake by crops, the source of variability largely relates to the determination of dry matter whereas variabilities in % N content and the ^{15}N enrichment of the crop are usually much smaller. The error in analysis of ^{15}N-inorganic N is typically large, particularly where the size of the inorganic N pool varies spatially or temporally, and the variability in the soil inorganic N directly affects the variability of the ^{15}N enrichment. The coefficients of variation for each component of the ^{15}N balance are remarkably constant across experiments, crops and sampling dates. Variabilities in estimates of ^{15}N balance are normally much lower ($4 - 11\%$) than those observed for the individual components of the balance, chiefly because the various compartments (^{15}N in plant, inorganic and organic soil N fractions) are not independent of each other since variability of each reflects overall spatial variabilities of the N processes.

Treatment replication and duplication of soil and plant sampling from individual mini-plots is the preferred strategy to reduce errors in analysis. Good trial design can be undone by incorrectly executed sampling techniques. It is important to note that poor sampling procedures cannot be reversed. Complete recovery of soil from designated sampling areas is recommended for near surface layers of a soil profile where fertiliser has not been applied uniformly or where crops are grown in rows. Soil sampling (either by boxes or replicates of individual cores) should have dimensions and locations that ensure collected soil is representative of row and inter-row areas. This approach overcomes plant-induced variability in the concentration of inorganic N in soil, stemming from uneven distribution of root material in surface soil, and, in the case of row crops, uneven distribution of the immobilized N. The depth to which soil is completely recovered will be dependent on the soil sampled, (deeper in the case of a sand compared to a clay), the depth of placement of fertiliser N and the distribution of roots by depth. Truck or tractor mounted drilling rigs have be used to sample deeper soil, but the difficulty of positioning rigs over small plots together with the damage inflicted on surrounding plots often force one to employ manually-operated augers as the only practical means of obtaining deep soil cores.

Table 2. Error associated with analysis of ^{15}N in different N pools in field ^{15}N balance studies.

Experimental system	Mean coefficient of variation (%) of analyses			
	Plant[15]N	Immobilized [15]N	Inorganic [15]N	[15]N balance
Micro-plot (0.5 m[2]), maize at harvest using 3 replicates, 3 years. Adapted from Normand *et al.* (1997).	9.4	14.5	33.0	7.0
PVC cylinder (0.12 m[2]), eight ^{15}N pulses applied to wheat, sampled 14 days after each application. Adapted from Recous and Machet (1999).	7.0	14.0	28.0	4.0
Micro-plots (0.5 m[2]), canola, 2 times of ^{15}N applications, sampled weekly for 4 weeks after each application. (S Recous, unpubl.)	11.0[a]	14.0	31.0	11.0

[a]Coefficient of variation in estimates of plant dry matter, %N and ^{15}N excess were 11%, 2.5% and 4.4%, respectively.

Accurate recording of soil sample weights and bulk density of sampled layers, and thorough mixing of soil before subsampling are essential. It is preferable to extract soil immediately for inorganic N. However, this is not always possible. Rapid oven drying of soil (40°C) can be used to arrest biologically mediated N transformations, provided samples do not contain N compounds that might volatilise NH_3 under heat and forced-drying. Most researchers freeze soil samples where these cannot be extracted immediately, although freezing has been shown to modify some N constituents in soil (Ross and Bartlett 1990). However, Mattos Junior *et al.* (1995) failed to find any significant changes in NH_4^+ and NO_3^- contents after soil had been frozen at -15°C for 35 or 349 d after sampling, whereas storage of soil at 5°C for as little as one week caused significant changes in inorganic N.

Determination of organic ^{15}N can be undertaken either by difference after completing analyses of total ^{15}N in soil and ^{15}N in inorganic N after KCl extraction or accomplished directly by analysing ^{15}N in soil after extraction, filtration and repeated washing of extracted soil with KCl. Buresh *et al.*

(1982) detail appropriate methods for recovering KCl-extracted soil for organic ^{15}N analysis. Analysis of organic ^{15}N after KCl extraction of soil should always be undertaken where there is a risk of NH_3 loss during the drying of soil and fine grinding required before N analysis using C-N analysers.

3.   USE OF A ^{15}N DILUTION APPROACH TO MEASURE ACTUAL RATES OF AMMONIFICATION, NITRIFICATION AND IMMOBILISATION

Net mineralisation is the outcome of two competing soil processes, ammonification and immobilisation, and knowledge of the actual rates of these processes is pivotal to understanding how soil and crop management might be affecting N turnover. Actual, or gross, N mineralisation (i.e. ammonification) is calculated (Equation 2) from the decrease in ^{15}N enrichment and the change in the size of the soil NH_4^+ pool as microorganisms mineralise the ^{14}N of native soil organic matter to $^{14}NH_4^+$ (Kirkham and Bartholomew 1954). Thus,

$$ m = \frac{(AT_1 - AT_2)}{t} \times \frac{\log((AL_1 \times AT_2)/(AL_2 \times AT_1))}{\log(AT_1/AT_2)} \tag{2} $$

where: m is the gross N mineralisation rate; AT is total NH_4^+-N (unlabelled plus labelled) at time 1 or 2; AL is labelled NH_4^+-N at time 1 or 2, determined from atom % ^{15}N excess; and t is time between initial (time 1) and final (time 2) measurements. Abbreviations used here are those defined by Bjarnason (1988). Units of N are typically µg N g^{-1} dry soil or mg N kg^{-1} dry soil.

Gross nitrification is calculated in a similar manner to gross mineralisation except that NL (labelled NO_3^--N) is used in place of AL and NT (total NO_3^--N) is used in place of AT in Equation 2. Measurements of gross mineralisation and gross nitrification are typically undertaken in paired studies using cores from the same soil, with $^{15}NH_4^+$ added to one core and $^{15}NO_3^-$ added to the other (Barraclough 1991).

Gross immobilisation can be calculated by subtracting the value of gross nitrification from the gross consumption of ^{15}N (see Equation 3) since the consumption component includes any processes that are sinks for NH_4^+ in soil. In the absence of gaseous losses and plant uptake, one should expect the two sinks to be nitrification and immobilisation.

$$c = \frac{(AT_1 - AT_2)}{t} \times \frac{\log(AL_1/AL_2)}{\log(AT_1/AT_2)} \tag{3}$$

where c is the NH_4^+ consumption rate.

It should be emphasized that a number of assumptions are implicit when applying Equations 2 and 3. For example,

1. it is assumed that ^{15}N has been added uniformly to soil and has mixed completely with all pools of indigenous N in soil,
2. where a pool contains labelled N and N carrying natural abundance signal, both such components of the pool are accessed by consuming processes in proportion to their relative amounts in the rooting profile,
3. in the case of N mineralised from soil organic matter, NH_4^+ is the initial product, and
4. the time scale used ensures that there is no remineralisation of immobilised ^{15}N (Barraclough 1991).

3.1   Methods of enriching soil inorganic N pools

Davidson *et al.* (1991) suggested that large errors in the measurement of gross N mineralisation may occur if the distribution of added ^{15}N in soil has not been strictly uniform. Microbial biomass is usually stratified within soil profiles, with the highest rates of activity occurring in the surface few centimetres (Woods 1989, Murphy *et al.* 1998b). Therefore, uniform enrichment of ^{15}N through soil by depth becomes essential to accurate estimation of gross rates of N mineralisation in undisturbed soil cores. The first method (and easiest method) adopted by many researchers for uniformly enriching the soil NH_4^+ pool with ^{15}N consists of mixing soil after the application of solutions containing $(^{15}NH_4)_2SO_4$ or $^{15}NH_4NO_3$ (Nishio *et al.* 1985, Myrold and Tiedje 1986, Bjarnason 1988, Nishio and Fujimoto 1989, Monaghan and Barraclough 1995, Mary *et al.* 1998). However, sieving and mixing of soil might well stimulate microbial activity so that changes in rates of N transformations would be expected (Craswell and Saffigna 1970, Sierra 1992). To avoid problems associated with soil disturbance, Geens *et al.* (1991) have advocated measuring gross N mineralisation by applying a solution of $^{15}NH_4NO_3$ to the soil surface. Water is then used to leach ^{15}N into soil. However, we consider it doubtful that ^{15}N would be uniformly distributed by depth using this procedure. Davidson *et al.* (1991), and Monaghan (1995) describe an apparatus fitted with syringes and needles through which ^{15}N solutions can be injected into intact soil cores. A multiple injection system that ensures even flow of solution from needles as they are withdrawn from the soil, has been developed in the

CSIRO Plant Industry Perth laboratory (Sparling *et al.* 1995). While the injection of small volumes of solutions into loam or fine-textured soils is unlikely to significantly change soil water content, soil water content in sandy soils is increased by as much as 5-fold when solutions are added to ensure uniform distribution of ^{15}N (Sparling *et al.* 1995). Davidson *et al.* (1991) and Barraclough and Puri (1995) also highlight the problem of increased soil water content following the use of ^{15}N solutions to label the soil NH_4^+ pool. Use of smaller volumes of more concentrated NH_4^+ containing solutions would be expected to minimise changes in soil water content (Recous *et al.* 1999), but poorer radial distribution of ^{15}N-NH_4^+ from points of injection might well increase the error in subsequent estimates of gross N mineralisation (Monaghan 1995). If an adequate distribution of label N is achieved through one of the various methods outlined above, a homogeneous labelling of the NH_4^+ soil pool would be expected in those many situations where the initial NH_4^+ pool derives mostly from the label applied. It is often impossible to achieve the same situation when labelling the NO_3^- pool, in which case the initial enrichment of the NO_3^- pool depends on both the relative amounts of the added labelled NO_3^- and on the indigenous unlabelled soil NO_3^-. The latter often changes sharply in concentration with depth (Recous *et al.* 1999).

The use of ^{15}N solutions to label NH_4^+ and NO_3^- as a prerequisite to the measurement of gross mineralisation in dry soil is obviously untenable, based on the discussion above. In this case dry labelling techniques are needed. The addition of ^{15}N-NH_3 to the headspace of chambers containing soil is one method that has been suggested to dry-label the soil NH_4^+ pool, but this method has been found to consistently underestimate gross mineralisation compared to measurements done after the injection of solutions containing ^{15}N (Fig. 5). A decrease in the ^{15}N enrichment with depth in soil as a result of the diffusion of ^{15}NH_3 gas into soil from the chamber atmosphere has been identified as the cause of the underestimation of gross mineralisation (Murphy *et al.* 1999). Another approach is to inject NH_3 into soil. This method requires the construction of an airtight injection system. Murphy *et al.* (1997) have given a detailed description of a multiple needle gas injector that can deliver ^{15}NH_3 contained in air. Displacement of gas from syringes is achieved by moving the syringe barrels and attached needles upwards against a stationary plate that supports syringe pistons. With this arrangement, gas flow from the needles is constant irrespective of the rate at which needles are withdrawn from soil, provided that the injection depth and length of syringe barrels are the same. Gas flow into, and out of syringes, is controlled by a manually operated valve positioned between the syringe barrels and needle assemblies. O-ring seals ensure the manifold valve and syringe barrels are gas-tight (Murphy *et al.* 1997). The volume of gas (or solution) injected can be adjusted by using smaller or larger internal diameter syringe barrels and pistons. Insertion of the needles beyond the intended sampling depth ensures that air trapped inside needles and in the valve manifold is flushed from the injector system before the syringe needle tips (with side port orifices) passes through the zone of soil that is to be sampled. Holes formed in surface soil after the withdrawal of needles should be closed immediately to reduce loss of NH_3 gas.

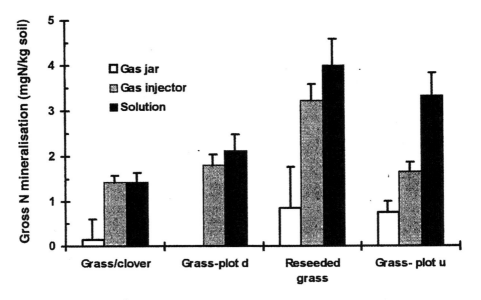

Figure 5. Gross mineralisation in soil cores measured after solution injection of $^{15}NH_4^+$, gas injection of $^{15}NH_3$ and addition of $^{15}NH_3$ to the headspace of jars. Data are adapted from Murphy *et al.* (1999).

3.2   Application of ^{15}N dilution techniques

The short time scales used in ^{15}N dilution measurements, together with the sensitivity of ^{15}N analyses, are attributes that can help studies on the effect of factors such as temperature or substrate addition on the rate of mineralization-immobilization turnover (MIT) of N (e.g. Watkins and Barraclough 1996, Mary *et al.* 1998) or to investigate the processes of the MIT itself (e.g. Barraclough and Puri 1995). Nitrogen-15 dilution techniques are ideally suited to field studies of gross mineralisation, gross nitrification and immobilisation following times when sharp changes in these processes are to be expected (e.g., tillage and summer rainfall). The application of ^{15}N dilution techniques in field studies has yielded useful information on estimates of gross fluxes under grasslands (Schimel 1986, Jamieson *et al.* 1998), forests (Hart *et al.* 1994, Pulleman and Tietema 1999), the stratification of gross mineralisation in soil (Murphy *et al.* 1998b), the effect of crop history on gross mineralisation (Murphy *et al.* 1998a), and the changes in gross mineralisation and immobilisation during decomposition of crop residues (Recous *et al.* 1999). An alternative application of the *in situ* dilution technique uses intact soil cores collected in the field, followed by incubation for 1 to 2 days under controlled conditions after enrichment with ^{15}N. This procedure both standardizes the temperature of incubation, and enables measurements to be undertaken in close proximity to laboratory facilities (Osler *et al.* 2000).

The care needed to obtain uniform ^{15}N enrichment of soil, together with the numerous analytical steps that are inherent to ^{15}N measurements, will inevitably restrict the number of treatments that can be studied concurrently and the overall number of samples that can be collected. As a consequence, gross mineralisation in soil in field studies is often determined between long time sampling intervals. It is questionable whether gross rates measured over 24 or 48 h on a monthly basis for example, can ever adequately characterize a cropping situation and cover the complex patterns of change likely to occur over time (growing season, decomposition kinetics, etc.).

3.3 Use of numerical models to calculate gross N fluxes

Concurrent measurements of gross mineralisation and nitrification, using ^{15}N dilution, typically use paired soil samples, with $^{15}NH_4^+$ added to one pair and $^{15}NO_3^-$ to the other (Barraclough 1991). Alternative numerical methods have been proposed to calculate several N fluxes simultaneously (Myrold and Tiedge 1986, Wessel and Tietema 1992, Smith *et al.* 1994, Mary *et al.* 1998). These methods combine a numerical resolution of the differential equations given by the N and ^{15}N mass equations between four pools (NH_4^+, NO_3^-, organic N, biomass N) over time, using a nonlinear fitting program. This approach is essential to solve the differential system which results when multiple fluxes are simultaneously diluting or enriching the ^{15}N composition of a given pool. It is particularly the case with NH_4^+ or NO_3^- immobilization, either of which can enrich the organic pool, or with mineralisation and remineralisation which are both likely to dilute NH_4^+ pools. The model FLUAZ (Mary *et al.* 1998, Recous *et al.* 1999) is an example of a method that simultaneously calculates the gross mineralisation, gross immobilization, nitrification, remineralization and ^{15}N gas losses after the enrichment of only the NH_4^+ pool.

4.   ANALYSIS OF ^{15}N IN PLANT, SOIL AND EXTRACTS

Many of the wet chemistry techniques used in the past to digest plant and soil for total N and ^{15}N/^{14}N ratio analysis (see reviews by Buresh *et al.* 1982, Hauck 1982) have been supplanted by more convenient dry combustion techniques involving fully automated C:N analysers (Barrie and Workman 1984, see chapter in this volume by Dawson and Brooks). The main function of these analysers is to combust organic N to dinitrogen gas and C to CO_2, which are then fed into the analyser unit of a mass spectrometer. Combustion is performed at high temperature in the presence of oxygen. Computer software controls the amount and time of oxygen release for combustion, and the time when each sample is dropped into the combustion chamber. Successful combustion is dependent on the arrival of oxygen at the point when the sample enters the combustion column. Reduction of nitrogen oxides to dinitrogen gas is achieved by passing combustion gases through a column of reduced copper held at 600°C. Surplus oxygen is also removed in the reduced copper column. Water and CO_2 are removed from the gas stream (helium is used as the carrier gas) before this enters a gas chromatographic column to concentrate dinitrogen gas. A small fraction of the helium plus sample is released into the analyser unit on a mass spectrometer.

Although the introduction of dry combustion analysers has greatly sped ^{15}N analysis, this application is not without pitfalls. Because of the greatly improved sensitivity of linked automated C/N combustion-mass spectrometer systems, only small amounts of soil (20 – 60 mg) and even smaller amounts of plant material (5 – 10 mg) are combusted. Soil and plant material must therefore be ground to the consistency of talcum powder to avoid large sampling errors. A number of ball or ring mills are available that produce finely ground material but each should be researched for effectiveness and grinding efficiency. Care must be taken to clean grinding equipment between samples. A hood outfitted with sand blasting equipment is a very useful facility to clean grinding bowls between samples. Tin capsules are used to deliver the plant and soil samples into the combustion chamber of automated C/N analysers. Good quality electronic balances that weigh to two decimal places beyond the mg range are needed for accurate determination of sample weight.

4.1   Diffusion method to collect N and ^{15}N

The introduction of automated dry combustion techniques has resulted in a radical overhaul of methods typically used to measure ^{15}N in NH_4^+ and

NO_3^- in soil extracts. Steam distillation was widely used before the early 1990s to separate NH_4^+ and NO_3^-, and for concentrating NH_4^+ in boric acid in preparation for back titration with standard acids to determine total N (see Buresh *et al.* 1982). The ammonium sulphate in these solutions was further concentrated using drying-down procedures and the concentrate transferred to vials that were subsequently connected to an apparatus that facilitated the generation of dinitrogen gas by reacting $(NH_4)_2SO_4$ with lithium hypobromide (Buresh *et al.* 1982, Hauck 1982). Manual introduction of dinitrogen gas into inlet bellows on mass spectrometers typically required ~1 mg of N in samples for mass spectrometry. Standard additions of N were often used when NH_4^+ and NO_3^- were measured to ensure adequate N for mass spectrometry (Buresh *et al.* 1982). Use of standard additions of N complicated the application of ^{15}N dilution techniques to determine gross rates of soil N transformations.

The use of dry combustion analysers to prepare N for mass spectrometry necessitated the introduction of techniques that were able to quantitatively transfer relatively small amounts of NH_4^+ and NO_3^- from solution extracts to a medium that could be dried, and then placed in a tin capsule in preparation for combustion. Brooks *et al.* (1989) were the first to describe a microdiffusion technique to collect ^{15}N from NH_4^+ and NO_3^- in KCl extracts. In this technique, the NH_4^+ is displaced from alkaline solution by adding MgO, and NH_3 in the headspace is subsequently trapped on an acidified medium, typically a glass fibre disk impregnated with $KHSO_4$. Other trapping mediums that have been used include vials of concentrated H_2SO_4, HCl or H_3PO_4 (O'Deen and Porter 1979, MacKown *et al.* 1987) and capsules of acid-washed zeolite (Burke *et al.* 1990). Nitrate in the extracts is then reduced to NH_4^+ using Devarda's alloy, and the resulting NH_4^+ also displaced to an acidified glass-fibre disk. The microdiffusion technique acts as a concentration step, a decided advantage for situations where concentrations of NH_4^+ and NO_3^- are low.

Glass food storage jars (120 mL) with gas-tight metal screw top lids are used in the CSIRO Perth laboratory for the diffusion process. A stainless steel wire hook is glued to the under side of the lid to hang a glass-fibre disk in the air space of the jar. Other systems used include plastic specimen containers (Brooks *et al.* 1989) and Mason jars (Saghir *et al.* 1993, Stark and Hart 1996). It is important that glassware be acid-washed to reduce carry-over of [15]N. Where glassware is heated or autoclaved it is recommended that this be etched using 3% HF for 30 min, and thoroughly washed before use. These steps can be avoided by using disposable plastic containers but it is important to ensure that vessels used in the diffusion step are gas-tight.

An aliquot (range from 5 – 40 mL) of extract, calculated to contain 20 – 150 µg N, is pipetted into a clean, dry jar containing sufficient finely-ground

KCl to bring the final concentration to about $2M$. The purpose of the KCl addition is to increase the molality of the solution and thereby decrease the vapour pressure inside the jar. This prevents excessive moisture accumulating on the trapping disk. Experience has shown that NH_3 trapped on acidified disks can be leached back to KCl extract should the disk become saturated. Reagents should be added slowly and carefully to avoid excessively vigorous reaction, contamination of the wall of the jar and the contamination of the acidified trapping disk by the alkaline reagents. Only finely ground MgO (0.7 g) is added when NH_4^+ is to be diffused. The jar is then immediately sealed with a screw-top lid modified to hold an acidified disk to trap the NH_3 released. Nitrate is reduced to NH_4^+ (after the diffusion of the initial NH_4^+) using Devarda's alloy (0.5 g), and a further addition of finely ground MgO ensures NH_3 transfer. In the CSIRO Perth laboratory, two drops of 0.3% Brij35 surfactant are added to moderate the release of hydrogen gas after the addition of the Devarda's alloy. Minor hydrogen gas explosions can result in the contamination of the acidified disk with alkali, reducing trapping efficiency. As is the case for the analysis of NH_4^+, each jar is then immediately sealed with a screw-top lid modified to hold an acidified disk to trap the NH_3 released.

Disks for trapping NH_3 can be punched from glass-fibre filter paper with a 5 mm cork borer. A single disk is mounted on a hook and is acidified by pipetting 10 µL of 2.5M $KHSO_4$ onto the rim of the disk (Brooks *et al.* 1989). After the lids are replaced, the jars should be gently swirled by hand to mix the reagents. Jars are incubated statically, preferably at 20 to 25°C for $7 - 10$ days. Recovery of inorganic N and ^{15}N on the acidified disk in diffusion jars should be tested by adding known amounts of $(^{15}NH_4)_2SO_4$ to KCl solutions. Complete recoveries of NH_4^+ and NO_3^- -N are routinely achieved in the CSIRO Perth laboratories (see Table 3). Each disk is removed from the wire hook with fine-tipped stainless-steel tweezers, and is inserted into a tin foil capsule. Capsules should be dried (~60°C for $1 - 2$ h) before being sealed, and analysis should proceed within 24 h to avoid excessive corrosion of the tin capsule that can arise with lengthy contact with acid in the glass wool disk, and the loss of sample.

The enclosure of an acidified disk in a Teflon envelope (PTFE sealing tape) that is placed into the alkaline solution is an alternative technique for trapping NH_3 (Sorensen and Jensen 1991, Stark and Hart 1996, Sigman *et al.* 1997, Holmes *et al.* 1998). Because PTFE is permeable to gases, but not liquids, the trap can absorb NH_3 without the risk of neutralisation by the alkaline solution. However, the recovery of $^{15}NH_3$ using this procedure is lower than obtained using an acidified glass fibre disk hung above the solution and the reproducibility of recovery of N is much poorer than with the standard technique (Table 3). It is important to ensure that untrapped

NH_3 is removed from the container before the diffusion of NO_3 is undertaken. Stark and Hart (1996) comment that this can be achieved by leaving containers open for at least 5 days. The incomplete recovery of NH_3 also necessitates an independent analysis of the N content in the NH_4^+ and NO_3^- in the sample. This step is not necessary with the acidified glass disk technique used in the Perth CSIRO laboratory where total recovery of NH_3 is routinely obtained (Table 3).

Table 3. Recovery of ^{15}N enriched NH_4^+ and NO_3^- from 2M KCl extracts using the diffusion technique and two trapping methods for NH_3 generated after the addition of MgO (NH_4^+) or Devarda's alloy and MgO (NO_3^-). 75 µg NH_4^+ N and 75 µg NO_3^- N were added. Figures in parentheses are standard deviation of the mean.

Method	NH_4^+ recovered (µg N)	Atom % excess ^{15}N in NH_4^+	NO_3^- recovered (µg N)	Atom % excess ^{15}N in NO_3^-
Acidified glass fibre disk hung from lid of vessel	75.68 (1.06)	1.084 (0.009)	74.71 (1.42)	1.118 (0.012)
Acidified glass fibre disk wrapped in Teflon, immersed in solution	61.85 (15.78)	1.115 (0.036)	58.3 (19.76)	1.160 (0.059)

There are advantages in measuring the N content and $^{15}N/^{14}N$ ratio concurrently by mass spectrometry. The total ^{15}N content in the sample is calculated by multiplying ^{15}N atom % excess by the N content determined by mass spectrometry which take into account natural abundance N in blanks. Where total N is measured separately using an alternative procedure, an isotope dilution equation should be applied to correct for the mass of N in blanks in the diffused sample (Stark and Hart 1996). Stark and Hart (1996) argue that it is not necessary to have complete recovery of ^{15}N where micro-diffusion techniques are used. However, $^{29}N_2$ diffuses at a slower rate than the lighter $^{28}N_2$ molecule, and discrimination against ^{15}N will occur where the transfer of ^{15}N to acid traps is incomplete. The size of the error introduced will depend on the level of ^{15}N enrichment in the samples and is significant in the case of samples close to natural abundance.

Diffusion methods are not entirely devoid of problems. Drawbacks include the use of the same volume to measure N derived from NH_4^+ and N from NO_3^-. In some circumstances the diffusion should be duplicated to set up the appropriate volume for N-NH_4^+ and N-NO_3^-. Colorimetric analysis prior to diffusion is also needed to set up the appropriate aliquot volume when the concentration of N is unknown.

4.2   Analysis of nitrite and nitrate after production of nitrous oxide

A different approach has been described by Stevens and Laughlin (1994) for the measurement of $^{15}NO_2^-$ and $^{15}NO_3^-$ in KCl extracts. This technique involves the analysis of $^{15}N_2O$ by automated continuous flow ratio mass spectrometry after $^{15}NO_2^-$ in KCl extracts is reacted with NH_2OH at pH 1.7 to yield N_2O. Nitrate is then reduced to a mix of NO_2^- and NH_2OH using copperized cadium at pH 4.7. The NO_2^- and NH_2OH react to produce $^{15}N_2O$. The rate of $^{15}NO_3^-$ analysis (100 in 24 h) is comparable to rates achieved using micro-diffusion techniques. However, a gas-phase auto-sampler is required for the mass spectrometer, which would require additional capital expenditure.

4.3   Total Soluble N

There are a number of situations where the analysis of ^{15}N in soluble organic material is required or if available would provide additional information on the fate of applied ^{15}N. For example, the content of soluble N in K_2SO_4 extractions after chloroform fumigation is needed when estimating ^{15}N in microbial biomass. Measurements of soluble organic N in KCl extracts also enable analysis of the movement of soluble organic N in soils amended with ^{15}N-labelled fertilizers.

The standard diffusion technique does not include organic N in extracts. Ross (1992) and Cabrera and Beare (1993) have described persulphate oxidation techniques to measure total (organic plus inorganic) N in KCl or K_2SO_4 extracts of soil. Persulphate oxidation gives quantitative recoveries of N which do not differ significantly from those obtained by Kjeldahl digestion, and have the advantage of simplicity and higher throughput. The persulphate method has also been used to determine the N extracted from chloroform-fumigated soils and hence to estimate the microbial biomass N (Ross 1992, Cabrera and Beare 1993, Sparling and Zhu 1993).

In our laboratory, organic N in the extracts is oxidised using persulphate as described by Sparling *et al.* (1996). An aliquot (usually 40 – 50 mL) of K_2SO_4 extract is pipetted into a diffusion jar of the type described earlier and sufficient $K_2S_2O_8$ added as a finely-ground powder to give a final persulphate concentration of *ca.* 125 mM. The jar is then loosely capped and autoclaved at 121°C for 40 min to oxidise organic N to NH_4^+ and NO_3^--N. It is important that $K_2S_2O_8$-treated extracts be autoclaved for 40 min. Incomplete hydrolysis of $K_2S_2O_8$ causes prolonged effervescence when MgO and Devarda's alloy are added to the mixture. This effervescence causes alkaline reagents to contaminate the acidified disk, and reduces the efficiency of trapping. A slight loss of weight can occur after autoclaving, presumably because of loss of liquids during heating. Sparling *et al.* (1996) found the loss to be equivalent to 5% of the liquid volume, irrespective of the volume autoclaved, and this can easily be corrected for in subsequent calculations.

An aliquot of the autoclaved solution is then transferred to a second, clean jar, taking care that the acidic solution does not contaminate the upper walls of the jar. The transfer of the autoclaved solution removes any chance that droplets of acidic solution on glassware, not in contact with alkaline reagents, will compete with the acidified disk, thereby reducing the efficiency of N recovery. Standard amounts of ^{15}N-labelled KNO_3 can be added to the K_2SO_4-$K_2S_2O_8$ mixture, before and after autoclaving to check the efficacy of the method. Recovery of organic N can be checked using leucine and a range of soluble organic compounds.

An alternative approach is to freeze-dry extracts containing soluble ^{15}N-labelled organic compounds. The resulting salt mixture is mixed and 50 to 70 mg of sample weighed into a tin capsule which is combusted using a C-N analyser linked in series with a mass spectrometer. This method has been successfully used to simultaneously determine C, N, ^{13}C and ^{15}N content of soil extracts in experiments involving ^{13}C and ^{15}N tracing (Aita 1996, Trinsoutrot *et al.* 2000). Again the main strength of this method is the concurrent measurement of total ^{13}C and total ^{15}N in a single sample.

5. SUMMARY

The linkage of C/N combustion analysers with mass spectrometers has resulted in a progressive revolution in the methods used to analyse ^{15}N in soil N fractions. While the semi-automation of procedures used to analyse ^{15}N in soil N fractions has certainly decreased the cost of ^{15}N analysis, and therefore made the application of enriched ^{15}N sources in studies of soil N transformations much more affordable, the smaller quantities of N needed for ^{15}N analysis can lead to increases in analytical error without adequate attention to detail in sample preparation. The use of a C/N analyser inline with a mass spectrometer has resulted in sweeping changes to methods used in the analysis ^{15}N in the NH_4^+, NO_3^-, and soluble N fractions in soil. Although these methods are now routinely used worldwide, it is recommended that each method be carefully evaluated before adoption, and that standards with known N content and ^{15}N enrichment be routinely used to check the accuracy of the diffusion procedure.

^{15}N enriched sources have been traditionally used in field studies that examine the recovery of N in soil fractions and plant material as part of assessments of N loss from agricultural and natural ecosystems. The recent application of ^{15}N techniques to the determination of gross or actual rates of N mineralisation, immobilisation and nitrification in soils is an example of the efficacy of enriched ^{15}N sources in studies of individual soil N processes. Many other novel applications of enriched ^{15}N sources in studies of soil N transformations now become possible and simply await imaginative experimentation.

REFERENCES

Aita, C. (1996). Couplage des cycles du carbone et de l'azote dans les sols cultivés: Etude , au champ, des processus de décomposition après apport de matière organique fraîche. Ph.D thesis, Université Paris VI, p 209.

Aita, C., Recous, S., and Angers, D. (1997). Short term kinetics of residual wheat straw C and N under field conditions: characterisation by ^{15}N^{13}C tracing and soil particle size fractionation. *European Journal of Soil Science* 48, 283-294.

Amato, M., Ladd, J. N., Ellington, A., Ford, G., Mahoney, J.E., Taylor, A. C., and Walsgott, D. (1987). Decomposition of plant material in Australian soils. IV Decomposition in situ of ^{14}C- and ^{15}N-labelled legume and wheat materials in a range of Southern Australian soils. *Australian Journal of Soil Research*, 25, 95-105.

Barraclough, D. (1991). The use of mean pool abundances to interpret ^{15}N tracer experiments. I. Theory. *Plant and Soil* 131, 89-96.

Barraclough, D. and Puri, G. (1995). The use of ^{15}N pool dilution and enrichment to separate the heterotrophic and autotrophic pathways of nitrification. *Soil Biology & Biochemistry* 27, 17-22.

Barrie, A. and Workman, C. T. (1984). An automated analytical system for nutritional investigations using ^{15}N tracers. *Spectroscopy International Journal* 3, 439-447.

Bjarnason, S. (1988). Calculation of gross nitrogen immobilization and mineralization in soil. *Journal of Soil Science* 39, 393-406.

Bremer, E and van Kessel, C. (1992). Seasonal microbial dynamics after addition of lentil and wheat residues. *Soil Science Society of America Journal* 56, 1141-1146.

Bronson, K. F., and Fillery, I. R. P. (1998). Fate of nitrogen-15-labelled urea applied to wheat on a waterlogged texture-contrast soil. *Nutrient Cycling in Agroecosystems* 51, 175-183.

Brooks, P. D., Stark, J. M., McInteer, B. B., and Preston, T. (1989). Diffusion method to prepare soil extracts for automated nitrogen-15 analysis. *Soil Science Society of America Journal* 53, 1701-1711.

Buresh, R. J., Austin, E. R., and Craswell, E. T. (1982). Analytical methods in ^{15}N research. *Fertilizer Research* 3, 37-62.

Burke, I., Mosier, A. R., Porter, L. K., and O'Deen, L. A. (1990). Diffusion of soil extracts for nitrogen and nitrogen-15 analyses by automated combustion/mass spectrometry. *Soil Science Society of America Journal* 54, 1190-1192.

Cabrera, M. L. and Beare, M. H. (1993). Alkaline persulfate oxidation for determining total nitrogen in microbial biomass extracts. *Soil Science Society of America Journal* 57, 1007-1012.

Cameron, K. C., Harrison, D., Smith, N. P., and McLay, C. D. A. (1990). A method to prevent edge-flow in undisturbed soil cores and lysimeters. *Australian Journal Soil Research* 28, 879-886.

Craswell, E. T. and Saffigna ,P. G. (1970). The mineralization of organic nitrogen in dry soil aggregates of different sizes. *Plant and Soil* 33, 383-392.

Davidson, E. A., Hart, S. C., Shanks, C. A., and Firestone, M. K. (1991). Measuring gross nitrogen mineralization, immobilization, and nitrification by ^{15}N isotopic pool dilution in intact soil cores. *Journal of Soil Science* 42, 335-349.

DeDatta, S. K. (1995). Nitrogen transformations in wetland rice ecosystems. *Fertilizer Research* 42, 193-203.

Dejoux, J. F., Recous, S., Meynard, J. M., Trinsoutrot, I., and Leterme, Ph. (2000). Fate of nitrogen from winter-frozen rapeseed leaves: mineralization and uptake by rapeseed crop in spring. *Plant and Soil* 218, 257-272.

Geens, E. L., Davies, G. P., Maggs, J. M., and Barraclough, D. (1991). The use of mean pool abundances to interpret ^{15}N tracer experiments II. Application. *Plant and Soil* 131, 97-105.

Hart, S. C., Nason, G. E., Myrold, D. D., and Perry, D. A. (1994). Dynamics of gross nitrogen mineralisation in an old-growth forest – the carbon connection. *Ecology* 75, 880-891.

Hauck, R. D. (1982). Nitrogen-isotope-ratio analysis. In 'Methods of Soil Analysis, Part 2. Chemical and Microbiological Properties.' (Eds. A. L.Page, R. H. Miller and D. R. Keeney) pp. 735-779. (American Society of Agronomy: Madison, WI.)

Holmes, R. M., McClelland, J. W., Sigman, D. M., Fry, B., and Peterson, B. J. (1998). Measuring ^{15}N-NH$_4^+$ in marine, estuarine and fresh waters: An adaptation of the ammonia diffusion method for samples with low ammonium concentrations. *Marine Chemistry* 60, 235-243.

Hood, R. C., N'Goran, K., Aigner, M., and Hardarson, G. (1999). A comparison of direct and indirect ^{15}N isotope techniques for estimating crop N uptake from organic residues. *Plant and Soil* 208, 259-270.

Jamieson, N., Barraclough, D., Unkovich, M., and Monaghan, R. (1998). Soil N dynamics in a natural calcareous grassland under a changing climate. *Biology and Fertility of Soils* 27, 267-273.

Jensen, E. S. (1996). Compared cycling in a soil-plant system of pea and barley residue nitrogen. *Plant and Soil* 182, 13-23.

Kirkham, D., and Bartholomew, W. V. (1954). Equations for following nutrient transformation in soil, utilizing tracer data. *Soil Science Society of America Proceedings* 18, 33-34.

Ladd, J. N., Oades, J. M., and Amato, M. (1981). Distribution and recovery of nitrogen from legume residues decomposing in soils sown to wheat in the field. *Soil Biology & Biochemistry* 13, 251-256.

MacKown, C. T., Brookes, P. D., and Smith M. S. (1987). Diffusion of nitrogen-15 Kjeldahl digests for isotope analysis. *Soil Science Society America Journal* 51, 87-90.

Mary, B., Recous, S., and Robin D. (1998). A model for calculating nitrogen fluxes in soil using N-15 tracing. *Soil Biology & Biochemistry* 30, 1963-1979.

Mattos Junior, D., Cantarella, H., Raij, B., and van Raij, B. (1995). Handling and storage of soil samples for preservation of inorganic nitrogen. *Revista Brasileira de Ciencia do Solo* 19, 423-431.

MacDonald, A. J., Poulton, P. R., Powlson, D. S., and Jenkinson, D. S. (1997). Effects of season, soil type, and cropping on recoveries, residues and losses of ^{15}N-labelled fertilizer applied to arable crops in spring. *Journal of Agricultural Science, Cambridge* 129, 125-154.

Middleboe, V. (1982). Analysis of nitrogen isotope ratios by emission spectrometry. In 'Soil Analysis. Instrumental techniques and related procedures', (Ed K. A. Smith) pp. 355-375. (Marcel Dekker Inc.: New York, USA.)

Monaghan, R., and Barraclough, D. (1995). Contributions to gross N mineralization from ^{15}N-labelled soil macroorganic matter fractions during laboratory incubation. *Soil Biology & Biochemistry* 27, 1623-1628.

Monaghan, R. (1995). Errors in estimates of gross rates of nitrogen mineralization due to non-uniform distribution of the ^{15}N label. *Soil Biology & Biochemistry* 27, 855-859.

Mulvaney, R. L., Khan, S. A., and Mulvaney, C. S. (1997). Nitrogen fertilizers promote denitrification. *Biology and Fertility of Soils* 24, 211-220.

Murphy, D. V., Bhogal, A., Shepherd, M., Goulding, K. W. T., Jarvis, S. C., Barraclough, D., and Gaunt, J. L. (1999). Comparison of ^{15}N labelling methods to measure gross nitrogen mineralisation. *Soil Biology & Biochemistry* 31, 2015-2024.

Murphy, D. V., Fillery, I. R. P., and Sparling, G. P. (1997). A method to label soil cores with [15]NH₃ gas as a pre-requisite for ^{15}N isotopic dilution and measurement of gross N mineralisation. *Soil Biology & Biochemistry* 35, 1731-1741.

Murphy, D. V., Fillery, I. R. P., and Sparling, G. P. (1998a). Seasonal fluctuations in gross N mineralisation, N consumption and microbial biomass in a Western Australian soil under different land use. *Australian Journal Agricultural Research* 49, 523-535.

Murphy, D. V., Sparling, G. P., and Fillery, I. R. P. (1998b). Stratification of microbial biomass-C and N and gross N mineralisation with soil depth in two contrasting Western Australian agricultural soils. *Australian Journal of Soil Research* 36, 45-55.

Myrold, D. D. (1990). Measuring denitrification in soils using ^{15}N techniques. In 'Denitrification in Soil and Sediment', (Eds N. Revsbech and J. Sorensen) pp. 181-198. (Plenum Press: New York)

Myrold, D. D., and Tiedje, J. M. (1986). Simultaneous estimation of several nitrogen cycle rates using [15]N: Theory and application. *Soil Biology & Biochemistry* 18, 559-568.

Nishio, T., and Fujimoto, T. (1989). Mineralization of soil organic nitrogen in upland fields as determined by a [15]NH₄⁺ dilution technique, and absorption of nitrogen by maize. *Soil Biology & Biochemistry* 21, 661-665.

Nishio, T., Kanamori, T., and Fujimoto, T. (1985). Nitrogen transformations in an aerobic soil as determined by a [15]NH₄⁺ dilution technique. *Soil Biology & Biochemistry* 17, 149-154.

O'Deen, W. A. and Porter, L. K. (1979). Digestion tube diffusion and collection of ammonia for nitrogen-15 and total nitrogen determination. *Analytical Chima Acta* 51, 586-589.

Osler, H. R., Recous, S., Fillery, I. R. P., Gauci. C. S., Zhu, C., and Abbott L. K. (2000). Relationships between mite community structure and N flux rates in Western Australian agricultural soils. (Submitted)

Peoples, M. B., Freney, J. R., and Mosier, A.R. (1995). Minimising gaseous losses of nitrogen. In 'Nitrogen Fertilization in the Environment', (Ed P. E. Bacon) pp. 565-602. (Marcel Dekker Inc.: New York.)

Pilbeam, C. J. (1996). Effect of climate on the recovery in crop and soil of ^{15}N-labelled fertilizer applied to wheat. *Fertilizer Research* 45, 209-215.

Powlson, D. E., Pruden, G., Johnston, A. E., and Jenkinson, D. S. (1986). The nitrogen cycle in the Broadbalk Wheat Experiment: recovery and losses of ^{15}N-labelled fertilizer applied in spring and inputs of nitrogen from the atmosphere. *Journal Agricultural Science Cambridge* 107, 591-609.

Pulleman, M., and Tietema, A. (1999). Microbial C and N transformations during drying and rewetting of coniferous forest floor material. *Soil Biology & Biochemistry* 31, 275-285.

Recous, S., Aita, C., and Mary, B. (1999). In situ changes in gross N transformations in bare soil after the addition of straw. *Soil Biology & Biochemistry* 31, 119-133.

Recous, S., and Loiseau, P. (1997). Transformations and fate of fertiliser-N applied to annual crops and grasslands. In 'Fertilization for Sustainable Plant Production and Soil Fertility'.

11th International World Fertilizer Congress of CIEC, (Eds O. Van Cleemput, S. Haneklaus, G. Hofman, E. Schnug and A. Vermoesen) pp. 302-310. (University of Gent: Belgium.)

Ross, D. J. (1992). Influence of sieve mesh size on estimates of microbial carbon and nitrogen by fumigation-extraction procedures in soils under pasture. *Soil Biology & Biochemistry* 24, 343-350.

Ross, D. S., and Bartlett, R. J. (1990). Effects of extraction methods and sample storage on properties of solutions obtained from forested Spodosols. *Journal of Environmental Quality* 19 108-113.

Saffigna, P. G. (1987). ^{15}N methodology in the field. In Advances in Nitrogen Cycling in Agricultural Ecosystems. (Ed J. R. Wilson) pp. 433-451. (CAB International, Wallingford, Oxon: United Kingdom.)

Saghir, N. S., Mungwari, F. P., Mulvaney, R. L., and Azam, F. (1993). Determination of nitrogen by microdiffusion in mason jars: II. Inorganic nitrogen-15 in soil extracts. *Communications in Soil Science Plant Analysis* 24, 2747-2763.

Schimel, J. P. (1986). Carbon and nitrogen turnover in adjacent grassland and cropland ecosystems. *Biogeochemistry* 2, 345-357.

Schindler, F. V., and Knighton, R. E. (1999). Fate of fertilizer nitrogen applied to corn as estimated by the isotopic and difference methods. *Soil Science Society America Journal* 63, 1734-1740.

Sierra, J. (1992). Relationship between mineral N content and N mineralization rate in disturbed and undisturbed soil samples incubated under field and laboratory conditions. *Australian Journal of Soil Research* 30, 477-492.

Sigman, D. M., Altabet, M. A., Michener, R., McCorkle, D. C. Fry, B., and Holmes, R. M. (1997). Natural abundance-level measurement of the nitrogen isotopic composition of oceanic nitrate: an adaptation of the ammonia diffusion method. *Marine Chemistry* 57, 227-242.

Smith, C. J. (1987). Denitrification in the field. In 'Advances in Nitrogen Cycling in Agricultural Ecosystems'. (Ed J. R. Wilson) pp. 387-398. (CAB International, Wallingford, Oxon: United Kingdom.)

Smith, C. J., Chalk, P. M., Crawford, D. M., and Wood, J. T. (1994). Estimating gross nitrogen mineralisation and immobilisation rates in anaerobic and aerobic soil suspensions. *Soil Science Society of America Journal* 58, 1652-1660.

Sorensen, P. and Jensen, E. S. (1991). Sequential diffusion of ammonium and nitrate from soil extracts to a polytetrafluoroethylene trap for ^{15}N determination. *Analytical Chimica Acta* 252, 210-213.

Sparling, G. P., Murphy, D. V., Thompson, R. D., and Fillery, I. R. P. (1995). Short-term net N mineralization from plant residues and gross and net N mineralisation from soil organic N after rewetting of a seasonally dry soil. *Australian Journal of Soil Research* 33, 961-973.

Sparling, G.P. and Zhu C. (1993). Evaluation and calibration of methods to measure microbial biomass C and N in soils from Western Australia. *Soil Biology & Biochemistry* 25, 1793-1801.

Sparling, G. P., Zhu C., and Fillery, I. R. P. (1996). Microbial immobilisation of ^{15}N from legume residues in soils of differing textures: Measurement by persulphate oxidation and ammonia diffusion methods. *Soil Biology & Biochemistry* 28, 170-175.

Stark, J. M., and Hart, S. C. (1996). Diffusion technique for preparing salt solutions, Kjeldahl digests, and persulfate digests for nitrogen-15 analysis. *Soil Science Society America Journal* 60, 1846-1855.

Stevens, R. J., and R. J. Laughlin (1994). Determining nitrogen-15 in nitrite or nitrate by producing nitrous oxide. *Soil Science Society of America Journal* 58: 1108-1116.

Thompson, R. B., and Fillery, I. R. P. (1997). Transformation in soil and turnover to wheat of N from components of grazed pasture in south Western Australia. *Australian Journal Agricultural Research* 48, 1033-1047.

Trinsoutrot, I., Nicolardot, B., Justes, E., and Recous, S. (2000). Decomposition in the field of residues of oilseed rape grown at two levels of nitrogen fertilisation. Effects on the dynamics of soil mineral nitrogen between successive crops. *Nutrient Cycling in Agroecosystems* 56, 125-137.

Watkins, N and Barraclough, D. (1996). Gross rates of N mineralisation associated with the decomposition of plant residues. *Soil Biology & Biochemistry* 28, 169-175.

Weier, K. L. (1996). Trace gas emissions from a trash blanketed sugarcane field in tropical Australia. In 'Sugarcane: research towards efficient and sustainable production'. Sugarcane 2000 Symposium, pp. 271-272. (CSIRO Tropical Crops and Pastures: Brisbane, Australia.)

Wessel, W. W., and Tietema, A. (1992). Calculating gross N transformations rates of ^{15}N pool dilution experiments with acid forest litter: Analytical and numerical approaches. *Soil Biology & Biochemistry* 24, 931-942.

Woods L. E. (1989). Active organic matter distribution in the surface 15 cm of undisturbed and cultivated soil. *Biology and Fertility of Soils* 8, 271-278.

Zaccheo, P., Crippa, L., and Genevini, P. L. (1993). Nitrogen transformation in soil treated with ^{15}N labelled dried or composted ryegrass. *Plant and Soil* 148, 193-201.

Chapter 10

Stable Isotope Techniques using Enriched ^{15}N and ^{13}C for Studies of Soil Organic Matter Accumulation and Decomposition in Agricultural Systems

Ann McNeill

Agronomy and Farming Systems, The University of Adelaide, Roseworthy Campus, SA 5371 Australia. Email: ann.mcneill@adelaide.edu.au

Key words: plant residues, roots, organic matter, decomposition, carbon, nitrogen, ^{15}N, ^{13}C, legumes

1.   INTRODUCTION

Crop residues, particularly from N_2 fixing legumes in rotations, are considered to be critically important in building up soil organic matter, thereby improving sustainability and providing N and other nutrients to following non-legume crops.   Studies in this regard have traditionally involved the addition to soil of plant residues labelled with isotopes, and subsequent monitoring of total and labelled fractions through time to assess rates of residue turnover.   In most of these studies residues labelled with ^{15}N and ^{13}C, the stable isotopes of nitrogen and carbon, have been used (e.g. Broadbent and Nakashima 1974, Darwis 1993, Aita *et al.* 1997, Jensen 1997), or alternatively with ^{14}C, the radio-active isotope of carbon (Amato and Ladd 1980, Jawson *et al.* 1989, Voroney *et al.* 1989). However, the use of stable isotopes has been favoured due to increasing safety restrictions governing the use of some radio-tracers coupled with rapid technological advancements in measurement of stable isotopes by continuous flow mass spectrometry (CF-MS).   Within this framework incubation studies have provided considerable information concerning the effects of quantity and quality (structural and chemical composition, ratio of C:N) of fresh residue additions on decomposition rates (Franzluebbers *et al.* 1994, Vanlauwe *et al.*

M. Unkovich et al. (eds.),
Stable Isotope Techniques in the Study of Biological Processes and Functioning of Ecosystems, 195–218.
© 2001 *Kluwer Academic Publishers. Printed in the Netherlands.*

1996). Studies have also included investigations of the effects of variable soil conditions including moisture and temperature, and inherent soil chemical and physical properties such as pH and texture, on residue decomposition rate (Clay and Clapp 1990, Ladd *et al.* 1995, Ladd *et al.* 1996).

Research programs investigating organic matter dynamics generally include decomposition studies conducted under field conditions (Jenkinson 1977, Ladd *et al.* 1981, Ladd *et al.* 1985, Ladd and Amato 1986, Harris and Hesterman 1990, Bremer and van Kessel 1992, Haynes 1997), since these, as opposed to pot studies, should provide a more 'realistic picture' of decomposition rates of crop residue components (eg straw, leaves, roots). There is also a need for information concerning the rate at which organic matter accumulates in the soil, particularly in terms of contributions from plant roots. In this case it is of particular interest to assess the actual quantities which are involved in accumulation and turnover, and the rate at which residues subsequently decompose when remaining relatively undisturbed under minimum or zero-tillage systems.

This chapter has two major themes. The first section describes methodologies for conducting simple incubation experiments to measure and compare the fate of C and N from different plant residues following incorporation into soil. The particular case study used to illustrate these methodologies refers to residues added to pots of soil under controlled conditions, but the same principles could be applied widely to studies in pot, plot or field. The second section of the chapter describes *in situ* techniques for enriching growing root systems with ^{15}N thereby allowing the N dynamics of undisturbed root systems, and their contribution to the soil organic matter, to be studied. In this connection, fractionation of the entire labelled soil-plant system, using a combination of dry sieving and freeze-drying techniques, will be shown to provide a means for quantifying the total amount of plant-derived N accumulating below-ground during a growing season, and estimating rhizodeposition of N. Procedures will also be described relating to longer term decomposition of labelled root systems in terms of N uptake by a following crop.

2.   INCUBATION STUDIES USING LABELLED RESIDUE ADDITIONS

This section will describe an incubation experiment designed to follow the fate of C and N from dual-labelled ($^{13}C, ^{15}N$) plant residues incorporated into soils differing in texture and moisture content. Ultimately the rate constants obtained may be extrapolated to derive representative rates of plant-available nutrient supply (via mineralisation) or long term 'tie-up' (immobilisation) in agricultural or natural systems, thus contributing to general understanding of overall nutrient cycling in such systems. Additionally, it has been demonstrated by several researchers (Jenkinson *et al.* 1987, Paustian *et al.* 1992, Nicolardot *et al.* 1994, Parton and Rasmussen 1994, Whitmore and Groot 1994, Grace and Ladd 1995) that the information from incubation studies can be extremely useful when developing, testing and refining models of organic matter decomposition.

The description of a typical incubation experiment given below, including the references to published material, should provide sufficient specific details to allow the reader to design, operate and interpret experiments of a similar nature. The major steps involved in designing and implementing such an incubation study using dual-labelled plant residues, analyses required, expected data outputs and main outcomes, are summarised as a flow diagram in Figure 1.

2.1   Source of labelled residues

The residues for this study were obtained by growing lupin and subterranean clover plants in solution culture in a controlled environment. However, labelled residues of a similar nature might already be available to a researcher, fortuitously, as the by-product of a labelled fertiliser experiment. Using a simplified version of the methodology and apparatus described by Palta (see chapter in this volume) plants were canopy-fed labelled carbon dioxide ($^{13}CO_2$) at weekly intervals for part of their exponential growth phase (total of 4 – 6 times). Labelled ammonium nitrate ($^{15}NH_4{}^{15}NO_3$), in amounts calculated to meet the N requirements of the growing plants, was also added periodically to the solution in which the plants were grown. Solution culture was used as a conveniently rapid method for producing labelled plant biomass, but soil or other media with addition of ^{15}N-labelled solution could also be used. Alternative methods for labelling plants with ^{15}N are leaf or stem feeding of ^{15}N-labelled urea solution (for details see later in this chapter) or the use of $^{15}N_2$ gas for labelling of legumes via N_2 fixation. The last method is costly and technically the most difficult (Warembourg *et al.* 1982, Peoples *et al.* 1989, McNeill *et al.* 1994).

At peak biomass production plants were removed from solution culture, separated into component parts (leaves, roots and stems), and oven-dried at 60°C. For this study the dried residues were reduced in size to pieces about 2mm square by chopping and coarse grinding, but it should be noted that there is some debate concerning the most appropriate size of residues for decomposition experiments (Palm and Rowland 1997). Under field conditions, organic materials are likely to be of very variable size.

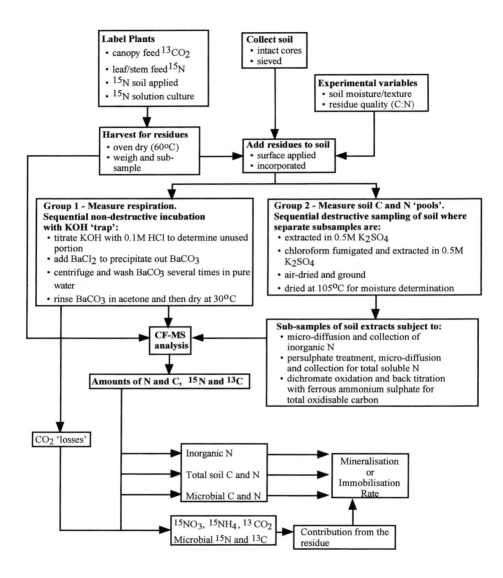

Figure 1. Flow diagram of stages for an incubation experiment where dual-labelled (^{13}C, ^{15}N) plant residues of different quality (C:N ratio) are incorporated into soils of various texture and moisture contents.

2.2 Soil collection

Appropriate samples of top soil were collected from two fields whose N dynamics were being compared. One comprised a loamy sand which had been in pasture for two years, the other, a clay loam, recently cultivated with wheat. It is, of course, imperative to know the cropping/cultivation history of any site being studied and to characterise basic soil properties (texture, pH etc.) which might assist with interpretation of results. It is also highly desirable that stones larger than 5mm in the study area be removed by sieving to reduce variability in results. Soil samples should be stored air dry if they are not to be used immediately. Prior to starting the experiment the moisture content should be adjusted to a required level and the soils then allowed to equilibrate for a minimum of one week. One has also to decide whether it is preferable to use sieved soil mixed with labelled residues or add such residues to the surface of intact soil cores taken from the field (Sparling *et al.* 1995). The latter method would be expected to simulate more closely a situation akin to field conditions, but may be disadvantaged by greater variability between replicates.

2.3 Experimental protocol

In this case study the coarsely ground (>2mm) dried plant residues, dual-labelled with ^{15}N and ^{13}C, were mixed with soil (100g/pot) at a rate equivalent to approximately 2 tonnes dry matter (DM)/ha and incubated at 25°C. Two moisture contents were used for each soil type studied i.e. the clay loam held at 50% and 85% water holding capacity (13.3% and 23.6% w/w), the loamy sand at 30% and 80% (3.8% and 10.0% w/w). These high and low moisture states will be referred to as the 'wet' and 'dry' treatments respectively. Residue materials of C:N ratios, representative of those likely to be involved in decomposition sequences under field conditions, were used: *viz.* lupin stem (C:N=66), lupin root (C:N=25) and subterranean clover shoot (C:N=21). Non-amended soils i.e. with no residues applied, were included as controls. A replicated set of pots representing all treatments was used to monitor soil (microbial) respiration (Anderson 1982), as described in Section 2.5. Other replicated sets of pots were destructively sampled immediately after residue addition and then again at 3, 7, 14, 28, 56 and 112 days following residue addition.

The soil from each destructively sampled replicate pot was split into four fractions (Fig. 1) and treated as follows:

a) A small amount (20g) was dried for 48 hours in an oven at 105°C to assess ambient soil moisture.

b) A similar portion was air-dried and finely ground for total C /^{13}C and total N/^{15}N analysis using continuous flow mass spectrometry (CF-MS, as described in the chapter in this volume by Dawson and Brooks).

c) 25g of soil was shaken in 100mL 0.5M K$_2$SO$_4$ for thirty minutes to extract inorganic N. The filtered extracts were then stored refrigerated (+5°C) or frozen (-15°C).

d) In order to lyse a portion of the microbial biomass a further 25g of soil was fumigated for seven days (Jenkinson and Powlson 1976a,b, Jenkinson 1988) using purified chloroform. Prior to fumigation the moisture content of soils from the 'dry' treatment was adjusted to that of the 'wet' treatment soils to facilitate effective fumigation. The fumigated sample was then shaken in 100mL 0.5M K$_2$SO$_4$ for thirty minutes to extract inorganic N, including that released by lysis of the microbial biomass, and the filtered extract stored either refrigerated or frozen.

2.4   Analysis

A sub-sample of each of the residues used for amending the soils was finely ground and analysed for N and C content (%N$_R$ and %C$_R$), and ^{15}N and ^{13}C content (atom %), using CF-MS. The amounts of carbon (C$_R$), nitrogen (N$_R$), ^{13}C in excess of natural abundance (^{13}C$_R$) and ^{15}N in excess of natural abundance (^{15}N$_R$) added as residue were calculated using the following equations:

$$C_R \text{ (or } N_R) = \frac{DM_R \text{ x } \%C_R \text{ (or } \%N_R)}{100} \tag{1}$$

$$^{13}C_R \text{ (or } ^{15}N_R) = \frac{C_R \text{ (or } N_R) \text{ x } ^{13}C_R \text{ (or } ^{15}N_R) \text{ at.}\%xs}{100} \tag{2}$$

where DM$_R$ is the dry weight of the added residue. The values for enrichment (atom % excess) of residue carbon (^{13}C$_R$ at. %xs) and nitrogen (^{15}N$_R$ at. %xs) were calculated by subtracting the known natural abundance values for unlabelled residue from the measured atom % values for the labelled residues. It is important to note that the enrichment of any labelled C or N pool referred to in this chapter is derived by subtracting the known (or measured) natural abundance value for that pool from the measured atom % value of the labelled pool.

The extracts from unfumigated and fumigated soils were analysed as follows (Fig. 1):

a) The ammonium plus nitrate in sub-samples (15 – 25mL) of the soil extracts was collected using a micro-diffusion technique (see Brooks *et al*. 1989 and chapter in this volume by Fillery and Recous), and the N (μg) and [15]N (atom %) measured using CF-MS.

b) Total soluble N (TSN) was recovered from another sub-sample (10 – 20mL) of each extract, using a combination of the persulphate oxidation and the micro-diffusion techniques described by Ross (1992), Cabrera and Beare (1993) and Sparling *et al*. (1996). Total N (μg) and [15]N (atom %) were then measured using CF-MS.

c) Total oxidisable carbon (TOC) was determined using dichromate oxidation followed by back-titration with ferrous ammonium sulphate (Kalembasa and Jenkinson 1973). In this case study we did not attempt to measure ^{13}C in the extracts, but recent work (Recous - pers.comm., Aita 1996) has demonstrated that this is possible, thus enabling ^{13}C content of microbial biomass to be calculated.

2.5   Respiration - microbial

Respiration was monitored from individual pots (four replicates per treatment) sealed inside one litre jars together with a dish containing 20mL of 0.5M KOH solution. CO_2 evolution per pot was measured as the amount of 'trapped' CO_2 in the KOH solution determined by precipitation as $BaCO_3$ (after addition of $BaCl_2$); followed by back-titration of any unused KOH with 0.1M HCl using phenolphthalein as an indicator. The moisture content of the soil in each pot was determined at the end of the incubation thereby allowing results to be expressed as micrograms of respired C per gram of dry soil (Fig. 2).

Sampling of the CO_2 'traps' was carried out at seven day intervals over a period of four weeks after residue addition, and thereafter at fourteen day intervals until the end of the experiment. The $BaCO_3$ precipitate was repeatedly centrifuged and washed with water, and then finally washed in acetone before being dried at 30°C and analysed for ^{13}C using CF-MS. Harris *et al*. (1997) recommend that strontium carbonate should be used in preference to $BaCO_3$, since the latter does not combust as reliably in automated N/C analysers, thereby giving rise to large quantitative errors. We have not found this to be a problem.

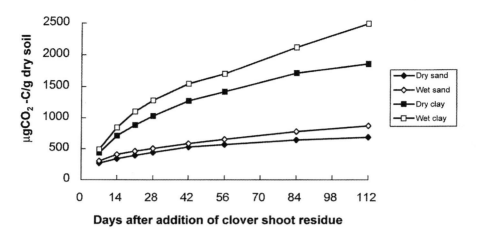

Figure 2. Cumulative soil respiration (μgCO_2-C/g dry soil) of a loamy sand (maintained at 30% (dry) or 80% (wet) of water holding capacity) and a clay loam soil (maintained at 50% (dry) or 85% (wet) of water holding capacity) after amendment with clover shoot residues. (McNeill, unpubl.)

Using values for total C (C_{tot}) and CO_2 evolution (soil CO_2) of the amended soil, combined with the calculated enrichment of the CO_2 (soil $^{13}CO_2$ at. %xs) and the amount of excess ^{13}C added in residue ($^{13}C_R$, see Equation 2), it becomes possible to calculate the proportion of total soil C which has been lost as respiration from the amended soil (C_{resp}), while also specifically determining how much of that carbon has come from decomposition of the residue ($^{13}C_{resp}$).

$$C_{resp} = \frac{\text{soil } CO_2}{C_{tot}} \tag{3}$$

$$^{13}C_{resp} = \frac{\text{soil } CO_2 \times \text{soil } ^{13}CO_2 \text{ at.\%xs}}{100} \tag{4}$$

The proportion of the residue carbon that has been respired can then be calculated as $^{13}C_{resp}/C_{resp}$. Data can thus be obtained on cumulative total losses of residue carbon via respiration (Table 1) over the short term (days) and the longer term (months).

Table 1. Cumulative proportion of residue C respired following incorporation of various legume residues into loamy sand maintained at 30% (dry) or 80% (wet) of water holding capacity or clay loam soil maintained at 50% (dry) or 85% (wet) of water holding capacity. (McNeill, unpubl.)

Soil moisture treatment		Dry			Wet		
Weeks after addition of residue		1	4	16	1	4	16
Soil texture	Residue type						
Loamy	Lupin stem	0.34	0.44	0.47	0.32	0.43	0.56
	Lupin root	0.29	0.39	0.41	0.30	0.38	0.43
	Clover shoot	0.39	0.47	0.49	0.39	0.50	0.55
Clay loam	Lupin stem	0.19	0.28	0.34	0.27	0.44	0.48
	Lupin root	0.21	0.30	0.31	0.26	0.42	0.53
	Clover shoot	0.30	0.42	0.47	0.43	0.53	0.58

2.6 Microbial N and C and microbial ^{15}N

Microbial N and C can be estimated by the fumigation-extraction method outlined by Brookes *et al.* (1985) and Wu *et al.* (1990). In this procedure the amount of nitrogen released following fumigation (i.e N_{flush}) can be calculated as the difference between the total soluble nitrogen (TSN) of the fumigated and unfumigated soil extracts (Sparling *et al.* 1996). The corresponding 'flush' of carbon can then be calculated in a similar fashion as the difference between the total oxidisable carbon (TOC) of the fumigated and unfumigated soil extracts (Vance *et al.* 1987, Tate *et al.* 1988). Amounts of microbial carbon and nitrogen, traditionally expressed as micrograms per gram (μg/g) of dry soil, can then be estimated using the conversion factors (k_{EC} and k_{EN}) and procedures discussed by Jenkinson (1988). The factors used for the soils in this case study were 0.38 for microbial N and 0.30 for microbial carbon (Sparling and Zhu 1993).

The measured amounts of total soluble N in the fumigated and unfumigated soil extracts (TSN_f and TSN_{uf}) together with the ^{15}N enrichments of these total soluble N pools (TSN_f at. %xs and TSN_{uf} at. %xs) are then employed to determine the contribution of the residue to microbial N. Firstly, the ^{15}N enrichment of the nitrogen 'flush' (N_{flush} at. %xs) is calculated as follows:

$$N_{flush} \text{ at.\%xs} = \frac{(TSN_f \times TSN_f \text{ at.\%xs}) - (TSN_{uf} \times TSN_{uf} \text{ at.\%xs})}{N_{flush}} \qquad (5)$$

The amount of nitrogen in the N 'flush' derived from the residue ($^{15}N_{flush}$) is calculated as:

$$^{15}N_{flush} = N_{flush} \, at.\%xs \times N_{flush} \qquad (6)$$

The proportion of N from the residue that has been incorporated into the microbial biomass is determined as $^{15}N_{flush}/^{15}N_R$, where $^{15}N_R$ is the amount of excess ^{15}N added as residue (see Equation 2).

2.7   Inorganic N and ^{15}N

Using data from the measured N and ^{15}N contents of the extracted inorganic N (for details of the micro-diffusion process see chapter in this volume by Fillery and Recous), and a broadly similar approach to that described above for microbial N, it becomes possible to calculate (a) the rate at which mineralisation or immobilisation of N has occurred following addition of residues, and (b) the contribution of the added residues to the mineral N pool. Such data enable the fate of N from the added residues to be determined in both the short and long term (Fig. 3) and permit a number of useful inferences to be made concerning the effects of soil moisture, texture and type of residue on rates of residue decomposition.

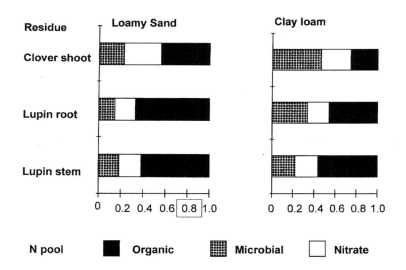

Figure 3. Proportion of legume residue N recovered as nitrate and microbial N, or remaining as organic N, four months after incorporation into a loamy sand maintained at 80% water holding capacity or a clay loam soil maintained at 85% water holding capacity (McNeill, unpubl.).

3. *IN SITU* DETERMINATION OF THE ACCUMULATION AND DECOMPOSITION OF PLANT ROOT N IN SOIL

This section outlines the use of simple techniques for ^{15}N enrichment of growing root systems *in situ* in order to determine total below-ground N accumulation by a plant and also quantify decomposition of below-ground N in terms of uptake by a subsequent plant. As in the previous case study experimental protocols and analytical methods will be explained in detail, and procedures given for calculations required when processing data. The case study is summarised in Figure 4.

3.1 Feeding ^{15}N in solution to plants

As to be expected from the marked capacity of plants to circulate recently acquired N within their system, feeding ^{15}N-labelled urea solution to a plant via the stem (Russell and Fillery 1996a), leaf (McNeill *et al.* 1997, 1998) or petiole (Rochester *et al.* 1998) would normally be expected to achieve a high level of *in situ* isotope labelling of roots. However it should be remembered that, unlike leaf and petiole immersion which apparently involve both active and passive mechanisms of N uptake (Wittwer *et al.* 1963), uptake by the wick technique is driven by the transpiration stream. It is unknown whether this difference affects the relative extent of root labelling obtained by the two techniques.

Urea is most commonly selected as a source for fed ^{15}N because it is a non-polar, undissociated molecule of low salt index. Additionally it carries a high concentration of N relative to its mass and is readily metabolised by plant tissue possessing effective levels of the enzyme urease. Russell and Fillery (1996a) mention that urea, in possessing dual amino groups resembles the amides, asparagine and glutamine, both of which are used as principal N carriers in the xylem of a number of legumes. However, some studies have used other N containing compounds such as KNO_3 (Oghoghorie and Pate 1972), although subsequent transfer to roots of the plant would tisrt have to involve assimilation by nitrate reductase in shoots since nitrate is not mobile in phloem.

3.2 Stem feeding

A schematic diagram of the stem or wick feeding technique is shown (Fig. 5). For further detail the reader is referred to Russell and Fillery (1996a). A small hole, from 0.25 – 0.5mm in diameter is drilled into the basal part (1 – 3cm from soil surface) of the stem of a plant (such as lupin, faba bean, field pea) and, using a fine needle, a thick cotton thread is passed through the hole. The thread either side of the stem is encased in flexible plastic tubing to prevent evaporative loss of solution from the wick. In the initial work by Russell and Fillery (1996a) plasticine or putty was used to seal the tubing to the stem, and a clip used to fix the tubing in place, but we have subsequently found these unnecessary provided a tight contact can be achieved between tubing and stem. The two threads encased in tubing fit tightly through holes in the cap of a clear plastic vial with the 'bare' or non-encased ends of the threads coiled at the base of the vial. A third hole in the cap of the vial is made large enough to allow insertion of the tip of a pipette for addition of ^{15}N-labelled urea solution. Following such addition, this hole is plugged with a rubber bung. The vial should be wrapped with aluminium foil to minimise condensation of solution on the sides of the vial so that all of the solution remains in contact with the wick.

3.3   Leaf or petiole feeding

Feeding of ^{15}N via target leaves of a plant has been used for a number of years in plant physiology experiments (Oghoghorie and Pate 1972, Pate 1973). Usually the youngest fully expanded leaf is chosen for feeding as this is likely to be metabolically the most active in assimilation and subsequent export of N. There are a number of possible approaches to leaf feeding, depending on the type of plant under study. For example, the tip (1 – 2mm) of a single leaf may be fed in the case of monocotyledenous plants such as wheat (Palta *et al.* 1991), the tips of all leaflets of a compound leaf in the case of certain pasture legumes such as clover or serradella (McNeill *et al.* 1997), or a V-notch flap surrounding the cut mid-vein of a leaf in the case of large foliage plants such as cowpea (Pate *et al.* 1984). In each of these cases the appropriate part of the leaf is cut under water, the excess water shaken off and the leaf inserted into ^{15}N-labelled solution contained in a plastic vial. The vial is sealed around the petiole of the leaf, or over the top of the flap, using a flexible non-porous plug of material. During leaf-feeding of several hundred leaves of subterranean clover pasture plants in the field the author found it impractical to cut individual leaves under water, yet despite dispensing with this protocol high rates of solution uptake were still obtained.

Cut petioles of soybean have been used for *in situ* ^{15}N feeding to soybean plants in studies concerned with the estimation of below-ground N (Rochester *et al.* 1998), and the technique uses essentially similar equipment to leaf-feeding. However, we have found that rates of solution uptake by petioles of subterranean clover and serradella can be 1.5 – 2.5 times less than those exhibited by comparable cut leaves (McNeill *et al.* 1997).

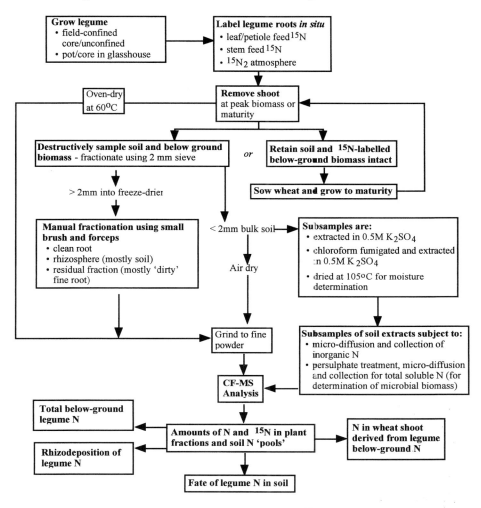

Figure 4. Flow diagram of stages in an experiment using ^{15}N labelling to determine total below-ground N accumulation by a legume and the uptake of this N by a subsequent crop.

3.4 Uptake and concentration of urea

Provided the feeding protocol is carefully set up and conducted and climatic conditions are adequate for plant growth and maintenance of high rates of transpiration, solution uptake from vials should be consistent and reliable. Indeed, uptake rates as high as $0.5 - 1.5$ mL per day from leaf and stem feeds have been recorded for annual pasture and crop legume species. However, uptake of solution can be impeded when cool cloudy conditions prevail. For example, in one field study on subterranean clover, overcast conditions resulted in solution uptake rates less than 0.15mL per day. Evaluation of the extent of intake of ^{15}N from a feed can be made by recording the weight (volume) of the solution in the vial before and after a feeding interval. This is necessary particularly where only small amounts of solution have been taken up. However, when using this approach it may be incorrect to assume that uptake of N in solution has been constant, so a safer strategy is to aim for conditions which ensure complete uptake of solution. The maximum volume of solution ever reported to have been taken up via a single stem feed is of the order of 4.0mL (Russell and Fillery 1996a). For leaf-feeding the corresponding values are typically within the range 1.5 to 2.0mL (McNeill *et al.* 1997, Palta *et al.* 1991). As a general rule highly enriched urea (99.8 atom % ^{15}N) is used in order to promote maximum subsequent enrichment of the root system, but there is no certainty that the label entering the shoot will be effectively partitioned to the root.

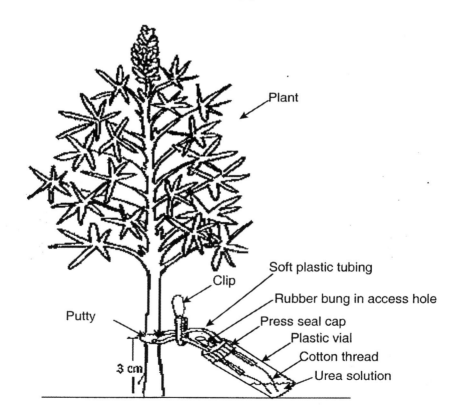

Figure 5. A schematic diagram of the stem or wick technique for *in situ* feeding of ^{15}N-labelled urea to plants. (Source: Russell and Fillery 1996a).

Before embarking on any study it is important to ascertain the tolerance to urea shown by the plant being investigated, since urea can be highly toxic to plant growth of certain species at relatively low concentrations. Our data suggest that urea concentrations ≤ 0.5% (w/w) are suitable for leaf-feeding legume pasture species such as serradella and subterranean clover without causing necrosis, whereas 0.9% (0.15M) urea solution has been used for petiole feeding soybean (Rochester *et al.* 1998) and concentrations as high as 2.7% to 4.0% have been used for stem-feeding of lupins (Russell and Fillery 1996a). The necrotic effects of different concentrations of urea solution leaf-fed to wheat have been described by Palta *et al.* (1991).

3.5 Experimental protocols relating to the 'what' and 'where' of labelling

The major aim of the labelling exercises described here is to obtain uniform and measurable enrichment of the plant root system. Generally, young plants seem to take up solution more rapidly, and to apportion relatively more of the absorbed label to their root systems, than do more mature plants (McNeill *et al.* 1997). For example, a single stem feed of 1mL of 0.5% urea at 98.5 atom % ^{15}N enrichment to young glasshouse grown plants enriched the root systems of lupin, faba bean and field pea within the range 0.2294 – 1.3613 ^{15}N atom % excess. Similarly, a single leaf feed of 2mL of 0.5% urea at 98.5 atom % ^{15}N enrichment to individual pasture plants resulted in reasonably high enrichments of up to 0.7858 ^{15}N atom % excess. Some researchers have employed multiple feeds to a single plant (Russell and Fillery 1996a) with the aim of effecting higher and more uniform labelling, but it is debatable whether there are advantages in doing this. As a further issue, the time and frequency of feeding ^{15}N can be compromised by practical constraints such as numbers of plants to be fed or frequency of visitation where remote field sites are involved. Furthermore, uniform labelling of legume root systems may be more difficult to achieve where a large mass of nodules actively fixing N_2 is concentrated in a small volume of total root, thereby resulting in a localised dilution of the added ^{15}N label.

Where measurement of total ^{15}N recovery is required of the study it would be appropriate to use confined plant culture systems such as PVC cylinders or pots, preferably with the base sealed to prevent leaching losses of mobile ^{15}N. Alternatively, as we have done in field studies, ion exchange resin bags can be attached at the base of otherwise open-ended cylinders to 'trap' leached mineral N, mostly as nitrate. Atmospheric losses of N from the plant, and to a lesser extent from the soil, are of course impossible to exclude. Where total recovery of ^{15}N is not required unconfined individual plants or groups of plants in micro-plots would be the preferred option. Since the shoots of fed plants may be much more highly enriched than the below-ground parts it is imperative to prevent contamination of the soil by falling shoot materials. For obvious reasons, the highly-labelled fed leaf itself should be removed immediately following the feed. Exclusion of shoot material from the soil is relatively easy to carry out by daily monitoring in a glasshouse study, but is more difficult to accomplish at remote field sites experiencing full vagaries of the weather. In an effort to overcome this problem nets may be placed under labelled crop material to prevent the senescent fallen leaves from becoming incorporated into the soil. However, this solution is impossible to apply in most pasture systems due to the prostrate growth habit of the plants which results in most of the soil surface area being covered by plant material. In pastures also, there is a continuous cycle of foliage senescence and relatively rapid incorporation into the soil. It is generally recommended in such cases that samplings for estimation of below-ground N should be conducted prior to major episodes of leaf senescence.

3.6   Soil core fractions

Bulk soil (BS), for the purpose of this technique, is defined as comprising all material capable of passing through the mesh of a sieve when an intact soil-root core is gently 'broken' apart on that sieve. The mesh of the sieve should be 2mm or less to facilitate maximum recovery of root material, although it is recognised that a relatively coarse mesh (5mm) may be required if soil is of a high clay content and particularly where the soil is wet. The total wet weight of the BS should be recorded and soil moisture determined, as described in the first case study. A sub-sample of the BS should be air-dried.

All material remaining on the sieve is placed in a plastic bag and immediately freeze-dried. Where this is not possible samples are frozen to assist subsequent freeze-drying. The freeze dried sample can then be fractionated as far as possible into:
− Rhizosphere (RH) - soil which is easily separated from the root and often accumulates at the bottom of the bag used for freeze drying.
− Clean root (CRT) - which comprises all root material that can be thoroughly cleansed of soil using a dry brush. The outcome from this procedure is less satisfactory where the soil is of high clay content.
− Residual fraction (RF) - comprising very fine root fragments and fine soil which remain after root cleaning

In some experiments we have partitioned only into CRT fraction and RF in an effort to reduce the numbers of samples required for analysis.

However, when calculating rhizodeposition it is imperative that the RH fraction is separated since, in our experience RF generally contains a high proportion of fine root which, if included, would lead to over-estimation of rhizodeposition. All freeze-dried fractions should be weighed and, these together with a sub-sample of the air-dried BS, finely ground prior to analysis for N and ^{15}N content using CF-MS (Fig. 4). The N and ^{15}N data from the soil fractions can be used subsequently to calculate the below-ground N for the plant.

3.7   Calculation of total below-ground N

The excess ^{15}N of the CRT fraction can be related to the N content of the CRT as μg excess ^{15}N per mg N_{CRT}. This is termed here as the specific enrichment of the CRT fraction (Sp.E_{CRT}) and is calculated as follows:

$$\text{Sp.E}_{CRT} = \frac{\text{Total N}_{CRT}}{(\text{Total N}_{CRT} \times at.\%xs^{15}\text{N}_{CRT})} \tag{7}$$

Note that in making these calculations it is assumed that (i) the ^{15}N is uniformly distributed throughout the root system and (ii) all the enriched N in the fractions other than CRT is root-derived N. The root or root-derived N (N′) can then be calculated sequentially for each fraction from the respective ^{15}N contents in excess of natural abundance divided by the specific enrichment of the CRT. That is,

$$N'_{BS} = \frac{(N_{BS} \times at.\%xs^{15}\text{N}_{BS})}{\text{Sp.E}_{CRT}} \tag{8}$$

$$N'_{RH} = \frac{(N_{RH} \times at.\%xs^{15}\text{N}_{RH})}{\text{Sp.E}_{CRT}} \tag{9}$$

$$N'_{RF} = \frac{(N_{RF} \times at.\%xs^{15}\text{N}_{RF})}{\text{Sp.E}_{CRT}} \tag{10}$$

Total below-ground plant-derived N (N_{BG}) can then be estimated using the following equation:

$$N_{BG} = N_{CRT} + N'_{RH} + N'_{RF} + N'_{BS} \qquad (11)$$

where N_{CRT} is measured clean macro-root N, and N'_{RH}, N'_{RF} and N'_{BS} are the estimated amounts of root or root-derived N within the total N of the rhizosphere, residual fraction, and bulk soil fractions respectively. Table 2 provides an example from an experiment on subterranean clover (McNeill *et al.* 1998) showing how data are employed when calculating total below-ground N accumulation.

Table 2. Enrichment (atom %xs), N content (mg) and amount of excess ^{15}N (µg) of the shoot and below-ground soil-plant fractions (CRT, RF, RH and BS) from an intact soil core containing vegetative subterranean clover which had been fed six weeks previously with ^{15}N.

Fraction	^{15}N enrichment (atom %xs)	N content (mg)	Excess ^{15}N content (µg)
Shoot	1.6599	204	3350
CRT	0.4495	56	249
RF	0.1630	149 (N'_{RF} = 55)	248
RH	0.0203	213 (N'_{RH} = 6)	41
BS	0.0064	1193 (N'_{BS} = 16)	73

Estimated total plant-derived BGN = 56 + 55 + 6 + 16

Data taken from McNeill *et al.* 1998.

3.8   Quantifying decomposition of below-ground N in terms of uptake of legume-derived N by a subsequent plant

Pots or cylinders containing soil with ^{15}N-labelled BGN can be used to grow plants which can be expected to derive an appreciable proportion of their total biomass N from the decomposition of this labelled material.

Two approaches have been adopted when calculating this potential N benefit to a following plant. Russell and Fillery (1996b) have employed a weighted average value for ^{15}N enrichment (at. % excess) of total lupin BGN to assess the quantity of N mineralised which is likely to be subsequently available to a following crop. Then, assuming that N released into the soil from the BGN and taken up by the subsequent plant has the same ^{15}N enrichment as that of the BGN the amount of the subsequent crop total N (N_{Plant}) derived from the decomposition of the lupin BGN (df_{BGN}) can be estimated as follows:

$$N_{plant} \, df_{BGN} \; = \; (\frac{^{15}Nat.\%xs \; of \; N_{plant}}{^{15}Nat.\%xs \; BGN}) \; x \; N_{plant} \qquad (12)$$

Alternatively, McNeill *et al.* (1998) have used the mean value for excess ^{15}N content of estimated total legume BGN ($\mu gXS^{15}N_{BGN}$) at peak production combined with the measured total excess ^{15}N content of the subsequent wheat plant ($\mu gXS^{15}N_{PLANT}$) to assess the percentage of legume BGN recovered by the wheat (% BGN_{WHT}) as follows:

$$\%BGN_{WHT} \; = \; (\frac{\mu gXS^{15}N_{plant}}{\mu gXS^{15}N_{BGN}}) \; x \; 100 \qquad (13)$$

Using this approach the N benefit of the legume to the wheat can then be expressed quantitatively by reference to the estimated total BGN, as shown in Figure 6.

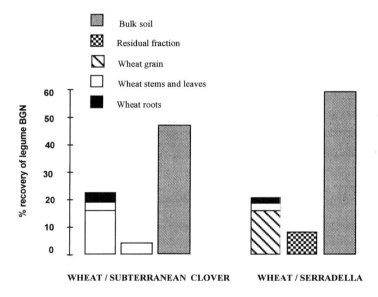

Figure 6. Fate of legume derived below-ground N in wheat-soil system following (a) subterranean clover and (b) serradella (McNeill *et al.* 1998).

4.   CONCLUSIONS

The soil/plant residue incubation experiments and *in situ* labelling methodologies described in this chapter employ relatively simple basic techniques which will not be difficult to perform by suitably qualified persons in any adequately equipped laboratory. However, major limitations in some cases may relate to the high cost of the isotope-labelled compounds required to generate sufficiently large quantities of labelled residues, provision of or access to a mass spectrometer of high precision for stable isotope analyses and the generally high cost of analyses. The major assumption underlying any techniques involving enrichment of plant and soil components with isotopes, is of course that the isotope in question is uniformly distributed within each component of the system under study. Although it is generally conceded that uniform labelling of the root system is paramount to accurate *in situ* estimation of total BGN (McNeill *et al.* 1997), further evaluations are required in specific situations to validate this assumption.

It is hoped that the material in this chapter stimulates you to embark upon some innovative and exciting experiments using stable isotopes to investigate the accumulation and decomposition of soil organic matter. In any event the technique of dual (^{13}C and ^{15}N) labelling of plants *in situ* offers a highly promising approach towards more precise quantitative understandings of soil organic matter inputs by crops and pastures, and, in a broader context, for obtaining a clearer idea of the manner in which management practices might influence the dynamics of soil organic matter decomposition, potential productivity and sustainability of agricultural systems.

ACKNOWLEDGEMENTS

The author acknowledges the contributions made by Ian Fillery, Daniel Murphy, Craig Russell, Graham Sparling and Chunya Zhu to the innovation and development of techniques described in this chapter. Also thanks to John Pate and Murray Unkovich for comments on the manuscript.

REFERENCES

Aita C (1996). Couplage des cycles du carbone et de l'azote dans les sols cultivés : étude au champ des processus de décomposition après apport de matière organique fraîche. pp. 196 PhD Thesis University Pierre et Marie Curie, Paris 6, France.

Aita, C., Recous, S., and Angers, D.A. (1997). Short-term kinetics of residual wheat straw C and N under field conditions: characterization by $^{13}C^{15}N$ tracing and soil particle size fractionation. *European Journal of Soil Science* 48, 283-294.

Amato, M., and Ladd, J.N. (1980). Studies of nitrogen immobilisation and mineralisation in calcareous soils: V. Formation and distribution of isotope-labelled biomass during decomposition of ^{14}C- and ^{15}N-labelled plant material. *Soil Biology and Biochemistry* 12, 405-411.

Anderson, J.P.E. (1982). Soil Respiration In 'Methods of Soil Analysis, Part 2, Chemical and Microbiological Properties' pp. 831-871. (American Society of Agronomy: Madison,WI.)

Bremer, E., and van Kessel, C. (1992). Seasonal microbial biomass dynamics after addition of lentil and wheat residues. *Soil Science Society of America Journal* 56, 1141-1146.

Broadbent, F.E., and Nakashima, T. (1974). Mineralization of carbon and nitrogen in soil amended with carbon-13 and nitrogen-15 labeled plant material. *Soil Science Society of America Proceedings* 38, 313-315.

Brookes, P.C., Landman, A., Pruden, G., and Jenkinson, D.S.(1985). Chloroform fumigation and the release of soil nitrogen: a rapid direct extraction method to measure microbial biomass nitrogen. *Soil Biology and Biochemistry* 17, 837-842.

Brooks, P.D., Stark, J.M., McInteer, B.B., and Preston, T. (1989). Diffusion method to prepare soil extracts for automated nitrogen-15 analysis. *Soil Science Society of America Journal* 53, 1701-1711.

Cabrera, M.L., and Beare, M.H.(1993). Alkaline persulfate oxidation for determining total nitrogen in microbial biomass extracts. *Soil Science Society of America Journal* 57, 1007-1012.

Clay, D.E., and Clapp, C.E. (1990). Mineralization of low C- to N- ratio corn residues in soils fertilised with NH_4^+ fertiliser. *Soil Biology and Biochemistry* 22, 355-360.

Darwis, S (1993). Effet des modalités de gestation de la paille de blé sur l'évolution du carbon et de l'azote au cours de sa décomposition dans le sol. PhD Thesis, Institut National Agronomique Paris-Grignon, France.

Franzluebbers, K., Weaver, R.W., and Juo, A.S.R. (1994). Mineralization of labelled N from cowpea (*Vigna unguiculata* (L.) Walp.) plant parts at two growth stages in sandy soil. *Plant and Soil* 160, 259-266.

Grace, P.R., and Ladd, J.N. (1995). SOCRATES (Soil Organic Carbon Reserves And Transformations in agro-EcoSystems: A Decision Support System for Sustainable Farming Systems in Southern Australia. (Co-operative Research Centre for Soil and Land Management: Adelaide, Australia.)

Harris, G.H., and Hesterman, O.B. (1990). Quantifying the nitrogen contribution from alfalfa to soil and two succeeding crops using nitrogen-15. *Agronomy Journal* 82, 129-134.

Harris, D., Porter, L.K. and Paul, E.A. (1997). Continuous flow isotope ratio mass spectrometry of carbon dioxide trapped as strontium carbonate. *Communications in Soil Science and Plant Analysis* 28, (9&10), 747-757.

Haynes, R.J. (1997). Fate and recovery of ^{15}N derived from grass/clover residues when incorporated into a soil and cropped with spring or winter wheat for two succeeding seasons. *Biology and Fertility of Soils* 25, 130-135.

Jawson, M.D., Elliott, L.F., Papendick, R.I., and Campbell, G.S. (1989). The decomposition of ^{14}C-labelled wheat straw and ^{15}N-labelled microbial material. *Soil Biology and Biochemistry* 21, 417-422.

Jenkinson, D.S. (1977). Studies on the decomposition of plant material in soil: V. The effects of plant cover and soil type on the loss of carbon from ^{14}C labelled ryegrass decomposing under field conditions. *Journal of Soil Science* 28, 424-434.

Jenkinson, D.S. (1988). Determination of microbial biomass carbon and nitrogen in soil. In 'Advances in Nitrogen Cycling in Agricultural Ecosystems' (Ed. J.R. Wilson). pp. 368-386. (CAB International: Wallingford, UK).

Jenkinson, D.S., and Powlson, D.S. (1976a). The effects of biocidal treatments on metabolism in soil - 1. Fumigation with chloroform. *Soil Biology and Biochemistry* 8, 167-177.

Jenkinson, D.S., and Powlson, D.S. (1976b). The effects of biocidal treatments on metabolism in soil - V. A method to measure soil microbial biomass. *Soil Biology and Biochemistry* 8, 209-213.

Jenkinson, D.S., Hart, P.B.S., Rayner, J.H., and Parry, L.C. (1987). Modelling the turnover of organic matter in long-term experiments at Rothamsted. *Intecol Bulletin* 15, 1-8.

Jensen, E.S. (1997). Nitrogen immobilisation and mineralisation during initial decomposition of ^{15}N-labelled pea and barley residues. *Biology and Fertility of Soils* 24, 39-44.

Kalembasa, S.J., and Jenkinson, D.S. (1973). A comparative study of titrimetric and gravimetric methods for the determination of organic carbon in soil. *Journal of the Science of Food and Agriculture* 24, 1085-1090.

Ladd, J.N., and Amato, M. (1986). The fate of nitrogen from legume and fertiliser sources in soils successively cropped with wheat under field conditions. *Soil Biology and Biochemistry* 18, 417-425.

Ladd, J.N., Oades, J.M., and Amato, M. (1981). Distribution and recovery of nitrogen from legume residues decomposing in soils sown to wheat in the field. *Soil Biology and Biochemistry* 13, 251-256.

Ladd, J.N., Amato, M., and Oades, J.M. (1985). Decomposition of plant material in Australian soils: III. Residual organic and microbial biomass C and N from isotope-labelled legume material and soil organic matter decomposing under field conditions. *Australian Journal of Soil Research* 23, 603-611.

Ladd, J.N., Amato, M., Grace, P.R., and Van Veen, J.A. (1995). Simulation of ^{14}C turnover through the microbial biomass in soils incubated with ^{14}C-labelled plant residues. *Soil Biology and Biochemistry* 27, 777-783.

Ladd, J.N., Van Gestel, M., Jocteur Monrozier, L., and Amato, M. (1996). Distribution of organic ^{14}C and ^{15}N in particle-size fractions of soils incubated with ^{14}C,[15]N-labelled glucose/NH4, and legume and wheat straw residues. *Soil Biology and Biochemistry* 28, 893-905.

McNeill, A.M., Hood, R.C., and Wood, M. (1994). Direct measurement of nitrogen fixation by *Trifolium repens* L. and *Alnus glutinosa* L. using [15]N2. *Journal of Experimental Botany* 45, 749-755.

McNeill, A.M., Zhu, C., and Fillery, I.R.P. (1997). Use of *in situ* ^{15}N-labelling to estimate the total below-ground nitrogen of pasture legumes in intact soil-plant systems. *Australian Journal of Agricultural Research* 48, 295-304.

McNeill, A.M., Zhu, C., and Fillery, I.R.P. (1998). A new approach to quantifying the N benefit from pasture legumes to succeeding wheat. *Australian Journal of Agricultural Research* 49, 427-436.

Nicolardot, B., Molina, J.A.E., and Allard, M.R. (1994). C and N fluxes between pools of soil organic matter: model calibration with long-term incubation data. *Soil Biology and Biochemistry* 26, 235-243.

Oghoghorie, C.G.O., and Pate, J.S. (1972). Exploration of the nitrogen transport system of a nodulated legume using [15]N. *Planta* 104, 35-49.

Palm, C.A., and Rowland, A.P. (1997). A minimum dataset for characterisation of plant quality for decomposition. In 'Driven by Nature: Plant Litter Quality and Decomposition'. (Eds G. Cadisch and K.E. Giller) pp. 379-392. (CAB International, Wallingford UK.)

Palta, J.A., Fillery, I.R.P., Mathews, E.L., and Turner, N.C. (1991). Leaf feeding of (^{15}N) urea for labelling wheat with nitrogen. *Australian Journal of Plant Physiology* 18, 627-636.

Parton, W.J., and Rasmussen, P.E. (1994). Long-term effects of crop management in wheat-fallow: II CENTURY model simulations. *Soil Science Society of America Journal* 58, 530-556.

Pate, J.S. (1973). Uptake, assimilation and transport of nitrogen compounds by plants. *Soil Biology and Biochemistry* 5, 109-119.

Pate, J.S., Peoples, M.B., and Atkins, (1984). Spontaneous phloem bleeding from cryopunctured fruits of a ureide-producing legume. *Plant Physiology* 74, 499-505.

Paustian, K., Parton, W.J., and Persson, J. (1992). Modelling soil organic matter in organic-amended and nitrogen-fertilised long-term plots. *Soil Science Society of America Journal* 56, 476-488.

Peoples, M.B., Faizah, A.W., Rerkasem, B., and Herridge, D.F. (1989). Methods for evaluating nitrogen fixation by nodulated legumes in the field. Monograph No.11 ACIAR, Canberra Australia

Rochester, I..J., Peoples, M.B., Gault, R.R., and Constable, G.A. (1998). Implications of accounting for below-ground N on the calculations of residual returns of fixed N for commercial faba bean crops. In 'Proceedings of the Ninth Australian Agronomy Conference' pp. 493-496. (Wagga Wagga: NSW.)

Ross, D.J. (1992). Influence of sieve mesh size on estimates of microbial carbon and nitrogen by fumigation-extraction procedures in soils under pasture. *Soil Biology and Biochemistry* 24, 343-350.

Russell, C. A., and Fillery, I.R.P. (1996a). *In situ* ^{15}N labelling of lupin below-ground biomass. *Australian Journal of Agricultural Research* 47, 1035-46.

Russell, C. A., and Fillery, I.R.P. (1996b). Estimates of lupin below-ground biomass nitrogen, dry matter, and nitrogen turnover to wheat. *Australian Journal of Agricultural Research* 47, 1047-1059.

Sparling, G. P., and Zhu, C. (1993). Evaluation and calibration of methods to measure microbial biomass C and N in soils from Western Australia. *Soil Biology and Biochemistry* 25, 1793-1801.

Sparling, G. P., Zhu, C., and Fillery, I.R.P. (1996). Microbial immobilisation of ^{15}N from legume residues in soils of differing textures: measurement by persulphate oxidation and ammonia diffusion methods. *Soil Biology and Biochemistry* 28, 1707-1715.

Sparling, G. P., Murphy, D.V., Thomson, R.B., and Fillery, I.R.P. (1995). Short-term net mineralization from plant residues and gross and net N mineralization from soil organic matter after rewetting of a seasonally dry soil. *Australian Journal of Soil Research* 33, 961-973.

Tate, K.R., Ross, D.J. and Feltham, C.W. (1988). A direct extraction method to estimate soil microbial C: effects of experimental variables and some different calibration procedures. *Soil Biology and Biochemistry* 20, 329-335.

Vance, E.D., Brookes, P.C., and Jenkinson, D.S. (1987). An extraction method for measuring soil microbial biomass C. *Soil Biology and Biochemistry* 19, 703-707.

Vanlauwe, B., Nwoke, O.C., Sangina, N., and Merckx, R. (1996). Impact of residue quality on the C and N mineralisation of leaf and root residues of three agroforestry species. *Plant and Soil* 183, 221-231.

Voroney, R.P., Paul, E.A., and Anderson, D.W. (1989). Decomposition of straw and stabilization of microbial products. *Canadian Journal of Soil Science* 69, 63-77.

Warembourg, F.R., Montange, D., and Bardin, R. (1982). The simultaneous use of ^{14}CO$_2$ and ^{15}N$_2$ labelling techniques to study carbon and nitrogen economy of legumes grown under natural conditions. *Physiologia Plantarum* 56, 46-55.

Whitmore, A.P., and Groot, J.J.R. (1994). The mineralisation of N from finely or coarsely chopped residues: measurement and modelling. *European Journal of Agronomy* 3, 103-109.

Wittwer, S.H., Bukovac,M.J., and Tukey, H.B. (1963). Advances in foliar feeding of plant nutrients. In 'Fertiliser technology and usage' (Eds M. H. McVickar, G.L. Bridger and L.B. Nelson.) pp. 429-455. (Soil Science Society of America: Madison, WI.)

Wu, J., Joergensen, R.G., Pommeraning, B., Chaussod, R., and Brookes, P.C. (1990). Measurements of soil microbial biomass C by fumigation-extraction - an automated procedure. *Soil Biology and Biochemistry* 22, 1167-1169.

Chapter 11

Source Identification in Marine Ecosystems
Food Web Studies using $\delta^{13}C$ and $\delta^{15}N$

Albertus J. Smit
Department of Botany, The University of Western Australia, Crawley, WA 6009 Australia.
Email: asmit@cyllene.uwa.edu.au

Key words: isotopic variation, primary source, $\delta^{15}N$, $\delta^{13}C$, food web, trophic structure

1.   INTRODUCTION

The measurement of natural abundance stable isotope signals in various ecosystem components provides a powerfully incisive means for identifying sources of natural or anthropogenic carbon (C) or nitrogen (N) entering an ecosystem, for studying its assimilation and transformation by photoautotrophs at the base of a trophic structure, and for following subsequent incorporation and transport through the food web. Indeed, in some situations, it offers the only feasible tool for studying subtle changes in marine communities resulting from human impacts. Nevertheless, wherever possible it should be used in conjunction with other complementary forms of investigation. Unfortunately, the technique appears limited by many sources of potential isotopic variation likely to occur within the system being investigated, often hampering interpretation of results. This probably led to the coining of *Fretwell's law*:

> "Warning! Stable isotope data may cause severe and contagious stomach upset if taken alone. To prevent upsetting reviewers' stomachs and your own, take stable isotope data with a healthy dose of other hydrologic, geologic, and geochemical information. Then, you will find stable isotope data very beneficial" (cited by Kendall and Caldwell 1998).

M. Unkovich et al. (eds.),
Stable Isotope Techniques in the Study of Biological Processes and Functioning of Ecosystems, 219–245.
© 2001 *Kluwer Academic Publishers. Printed in the Netherlands.*

For the purpose of this chapter 'hydrologic, geologic, and geochemical' may be replaced with more appropriate terms such as 'biologic, ecologic, and biogeochemical'.

If all sources of variation are taken into account in the sampling design and planning of a study, and if used alongside appropriate supporting data, some of the limitations alluded to above may be minimised or circumvented. In any event, it is essential that all potential sources of variation are recognised in respect of experimental design, sample processing and subsequent data analysis and interpretation, so that the true potential of stable isotope techniques at the natural abundance level can be realised.

The aim of this chapter is to discuss the application of $\delta^{13}C$ and $\delta^{15}N$ measurements to the study of trophic relations and source identification in marine systems. To accomplish this, the range and nature of variability of $^{15}N/^{14}N$ and $^{13}C/^{12}C$ in marine systems, will be discussed, together with the factors that result in changes in isotope ratios between and within trophic levels while also considering variability at the level of individual organisms. Elements that have to be taken into account when designing an experiment using natural abundance stable isotopes will then be addressed. The chapter avoids discussion of isotopic fractionation and effects, which can be found in other chapters in this volume, or in reviews such as Fry and Sherr (1984), Owens (1987), Peterson and Fry (1987), Handley and Raven (1992), Preston (1992) and Raven (1992).

2. VARIATIONS RELATING TO PRIMARY SOURCES OF N AND C

For our purpose, the term 'primary sources' will refer to those forms of inorganic N or C that are directly utilised by primary producers from the surrounding environment and are thereafter assimilated into biomass. A primary source can be of either natural (e.g. upwelled 'new' nitrate (NO_3^-) and ammonium (NH_4^+), bicarbonate (HCO_3^-) or dissolved carbon dioxide ($CO_{2(aq)}$), 'recycled' inorganic N or C (from animal activity, ammonification, etc.) or potentially of anthropogenic origin (e.g. NH_4^+ or NO_3^- pollutants from river runoff or sewage disposal). Any such sources can be incorporated into autotrophic biomass which in turn will be transferred to subsequent levels of the trophic structure.

2.1   Nitrogen

As in terrestrial ecosystems, N in the marine environment occurs in both organic and inorganic forms. Dissolved inorganic forms of N (DIN), principally NO_3^- and NH_4^+, typically represent the most important form of N sustaining planktonic and benthic primary production (Dugdale 1967). Nitrate is generally considerably more abundant than NH_4^+, varying in concentration between ~10 nM and 35 µM. NH_4^+ seldom exceeds 2 µM, and is more frequently found at much lower concentrations. Dissolved gaseous nitrogen (N_2) is the most abundant form of DIN in the oceans (~800 – 1000 µM), but of course can be utilised only by N_2 fixing organisms such as blue-green algae (Minagawa and Wada 1986). One important feature of marine ecosystems is that DIN concentrations are extremely variable, both spatially and temporally (Velinsky *et al.* 1989). The complex of physical and biological oceanographic principles that govern the amounts of DIN at global, regional and local scales will not be discussed here.

In addition to variations in concentration, pools of NH_4^+ and NO_3^- may also vary spatially and temporally in ^{15}N abundance (e.g. Goering *et al.* 1990) due to physical and chemical isotope effects (see chapter in this volume by Dawson and Brooks). N_2 fixation and assimilation have been shown to modify the isotope ratio of primary N sources (Hoering and Ford 1960, Macko *et al.* 1982) and biological oxidation-reduction processes such as nitrification and denitrification also introduce shifts in the $\delta^{15}N$ of DIN (Miyake and Wada 1971, Cline and Kaplan 1975). Catabolic biochemical processes such as excretion of organic or inorganic N sources, decomposition and remineralisation within marine ecosystems can also result in production of regenerated N pools with $\delta^{15}N$ values distinctly different from those of the primary N sources (Miyake and Wada 1971, Tieszen *et al.* 1983, Checkley and Miller 1989).

Processes that do not rely on the *in situ* modification or transformation of N sources include those that are brought about through external agencies. For example the DIN of a system can be of natural or anthropogenic origin, and either source can be assimilated by primary producers and incorporated into the trophic structure. Spies *et al.* (1989), Peterson *et al.* (1993) and Rau *et al.* (1981) have demonstrated that, where high levels of anthropogenically-derived N and C are introduced into a system, shifts in $\delta^{15}N$ and $\delta^{13}C$ values can progressively propagate through an entire system. In similar fashion, N entering a marine system as DIN in rainwater (Paerl and Fogel 1994) or strong winds and storms may mix the water column and advect nutrients into the photic zone (Malone *et al.* 1993, Fogel *et al.* 1999), thereby generating changes in $\delta^{15}N$ values of particulate organic N (PON) in the system. The spatial and temporal scales in isotopic variation resulting during execution of these processes may also vary dramatically.

For a discussion on isotope effects and fractionation the reader is referred to the detailed reviews by Owens (1987) and Owens and Watts (1998). For present proposes it suffice to say that there are two fundamental reasons why

^{15}N/^{14}N ratios might vary appreciably among autotrophs: (a) variations are likely to exist in the natural abundance of ^{15}N of DIN and (b) organic N pools in phytoplankton and macrophytes are generally depleted to varying degrees in ^{15}N relative to the DIN in source seawater. As an illustration, Table 1 contains a short list of published δ^{15}N values of various forms and sources of DIN utilised by marine autotrophs.

Due to the variety of mechanisms that may affect primary N sources, it is critical that any source identification study follows the incorporation of these sources into the trophic structure by measuring their initial δ^{15}N values before assimilation into biomass. Two basic methodologies are available for the extraction of DIN for δ^{15}N characterisation of NH_4^+ and/or NO_3^-: (a) distillation and (b) diffusion methods. Steam distillation methods (e.g. Cifuentes *et al.* 1989, Velinsky *et al.* 1989) stem from the technique developed by Bremner and Edwards (1965) for the isolation of NH_4^+ from soil extracts. This technique can be used for meso- or eutrophic waters and sewage effluents (more than ~ 50 μg DIN L^{-1}, depending on the sensitivity of the mass spectrometer). However, it would be extremely difficult, if not impossible, to apply to oligotrophic waters (less than ~ 50 μg DIN L^{-1}) due to constraints imposed by sample size. For seawater with a low DIN concentration one of the currently available diffusion methods would be more appropriate (e.g. Sigman *et al.* 1997, Holmes *et al.* 1998).

2.2   Carbon

The largest active pool in the global C cycle is dissolved inorganic C (DIC) present in the oceans. This results from atmospheric CO_2 participating in equilibrium exchange reactions with the ocean carbonate system. HCO_3^- and CO_2 are the most abundant pools of DIC in the oceans; of these two, HCO_3^- comprises virtually all (more than 99 %) of the total DIC pool (refer to Skirrow 1975 and Benson and Krause 1984 for the stoichiometry of the ocean carbonate system and its effect of isotopic fractionation). Variation in the chemical equilibrium of DIC in marine waters can significantly alter the δ^{13}C signature of source DIC utilised by microalgae and macrophytes.

Superimposed on the major δ^{13}C variations determined by the stoichiometry of carbonate-CO_2, localised patterns of photosynthesis, respiration and decomposition can also modify the ^{13}C natural abundance of seawater DIC. Thus, as photosynthesis discriminates against ^{13}C, the residual DIC tends to be enriched in ^{13}C to values ranging from about +1 to +3 ‰ (Anderson and Arthur 1983). The upwelling of deep-ocean water can also affect the DIC δ^{13}C value to some extent since ocean basins are sites of decomposition of isotopically light organic material, and therefore contain

[13]C-depleted inorganic C carrying values of around 0 ‰ (Anderson and Arthur 1983).

Table 1. δ^{15}N values of a range of primary nitrogen forms obtained from a variety of sources.

N species	Source	Mean[a] ‰	Range[a] ‰	Reference
ammonium	unknown	7	6.5 – 7.5	Miyake and Wada (1967)
ammonium	atmospheric deposition	-5.5	-13 – 2	Paerl *et al.* (1993)
ammonium	atmospheric deposition	-3.1	-12.5 – 3.6	Paerl and Fogel (1994)
ammonium	Chesapeake Bay, varies seasonally	15.9	11.5 – 20.2	Horrigan *et al.* (1990)
ammonium	porewater, Santa Barbara Basin, California	9.7	8.2 – 10.5	Sweeney and Kaplan (1980)
ammonium	porewater, Santa Barbara Basin, California	10.0	8.3 – 12.5	Sweeney and Kaplan (1980)
ammonium	recycled N, oligotrophic system	-3.5		Miyake and Wada (1967)
ammonium	remineralised (estuarine)	13.0	10 – 16	Paerl *et al.* (1993)
ammonium	sewage effluent	8.0	5 – 11	Paerl *et al.* (1993)
DIN			-0.2 – 0.5	Benson and Parker (1961)
DIN	agricultural land runoff	9.0	8 – 10	Paerl *et al.* (1993)
DIN	ETNP	0.7	0.5 – 0.7	Cline and Kaplan (1975)
DIN	fertiliser	0.0	-2 – 2	Paerl *et al.* (1993)
N$_2$, air	atmosphere	0.0		Paerl *et al.* (1993)
nitrate	atmospheric deposition	-1.5	-5 – 2	Paerl *et al.* (1993)
nitrate	central North Pacific	5.5		Michener and Schell (1994)
nitrate	Eastern Tropical North Pacific (ETNP)	5.5	5 – 6	Cline and Kaplan (1975)
nitrate	ETNP	5.8	4.8 – 6.8	Cline and Kaplan (1975)
nitrate	ETNP, >500m	6.5		From Michener and Schell (1994)
nitrate	ETNP, denitrifying zone	19.0		Cline and Kaplan (1975)
nitrate	ETNP, denitrifying zone, 200 – 500m	15.0		from Michener and Schell (1994)
nitrate	ETSP, denitrifying zone	13.0		Liu *et al.* (1987)
nitrate	Northeastern North Pacific	6.1		Miyake and Wada (1967)
nitrate	Northern Atlantic, >200m	5.8		Liu and Kaplan (1989)
nitrate	unknown	5.8		Wada *et al.* (1975)
nitrate	sewage effluent	7.5	5 – 10	Paerl *et al.* (1993)
nitrate + DOM	atmospheric deposition	1.0	-2.0 – 4.7	Paerl and Fogel (1994)
nitrate + nitrite	Chesapeake Bay, varies seasonally	9.2	6 – 12.3	Horrigan *et al.* (1990)

[a] δ vs. atm. N$_2$. δ^{15}N of atmospheric N$_2$ rated at 0 ‰

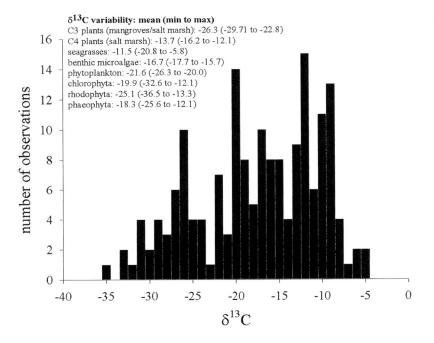

Figure 1. Frequency distributions of $\delta^{13}C$ measured for a variety of marine primary producers sampled from a wide range of coastal and oceanic habitats. Data compiled from a variety of sources.

Although photosynthesis in the marine environment occurs mainly via the C_3 pathway, the resulting $\delta^{13}C$ values of primary producers do not normally resemble those typical of terrestrial C_3 plants. As seen from Figure 1, $\delta^{13}C$ values across a variety of marine macrophytes are considerably lower than those of the primary source of C utilised during photosynthesis. Following from this, marine algae with very negative $\delta^{13}C$ values have been shown to be unable to utilise HCO_3^- as primary inorganic C source, but rather to rely on the diffusion of $CO_{2(aq)}$ into the thallus (Raven *et al.* 1995). Algae of this type often have $\delta^{13}C$ values as low as -33 ‰, partly because $CO_{2(aq)}$ has a more negative $\delta^{13}C$ signature than HCO_3^-, but also because slower diffusion of $CO_{2(aq)}$ in water reduces the extent of fractionation by RuBP carboxylase (O'Leary 1988, Raven and Farquhar 1990). Raven *et al.* (1995) provide an excellent review on the effect of $CO_{2(aq)}$ and HCO_3^- utilisation on $\delta^{13}C$ variability in macrophytes. The situation is further complicated in emergent macrophytes capable of also utilising atmospheric CO_2 (-7.8 ‰). Inputs of this type should lower the $\delta^{13}C$ of the organism by 1 to 4 ‰ (Surif and Raven 1990). Other factors suggested to induce significant effects on plant natural abundance $^{13}C/^{12}C$ ratios include latitude (Rau *et al.* 1982), species composition of phytoplankton communities

(Gearing *et al.* 1984), photon flux density (Thompson and Calvert 1994), pH (Hinga *et al.* 1994), temperature (Johnson 1996) and nutrient availability (Laws *et al.* 1995).

In summary then, it is important to realise that signals for $^{13}C/^{12}C$ and $^{15}N/^{14}N$ may differ appreciably among primary producers for a variety of reasons. These differences in isotope signals may then be transferred with further modification through a particular food chain as the relevant pools of C and N are ingested and assimilated by first order consumers and then carried over indirectly to carnivores. DIC can be extracted for isotopic characterisation using the method of Atekwana and Krishnamurthy (1998). Basically, the procedure involves the conversion of DIC to gaseous CO_2 after treating the water sample with phosphoric acid. As with DIN, it is recommended that studies looking at the incorporation of primary DIC sources into plant and animal biomass should first determine the isotope signal of the unmodified source.

3. VARIATIONS WITHIN INDIVIDUAL ORGANISMS

Most successful isotope studies are underpinned by there being sufficiently large natural isotopic variations between the interrelating components of the ecosystem under study. However, it must be noted that there are also significant variations in $^{15}N/^{14}N$ and $^{13}C/^{12}C$ ratios within each individual plant or animal and that these may be as great or even greater than those exhibited between different organisms. The significance of these intra-organismal variations was first realised by workers such as DeNiro and Epstein (1978), Lyon and Baxter (1978), McConnaughey and McRoy (1979), Tieszen (1978), Tieszen *et al.* (1983) and Stephenson *et al.* (1984). From animal studies, for example, it became evident that certain tissues such as the liver or pancreas consequently showed more negative $\delta^{13}C$ values when compared to other tissues. Tieszen *et al.* (1983) attributed these variations to differences in the biochemical composition of the tissues in question, e.g., tissues containing a high proportion of lipid would be expected to show more negative $\delta^{13}C$ values than those of lower lipid content (Smith and Epstein 1970). When analysing tissues of high lipid content, some authors accordingly prefer to perform a lipid removal treatment prior to analysis (e.g. Bligh and Dyer 1959).

Tieszen *et al.* (1983) also showed that the rate at which C derived from one diet is being replaced with a second diet of different isotope signal depends on the turnover rate, and hence on the metabolic activity of the tissue. More recently, Schmidt *et al.* (1999) found that when earthworms switch their diet to include a food source with a different isotope signal, the corresponding change in the invertebrate manifested itself more rapidly in the excreted mucus than in corresponding bulk body tissue. However, Hesslein *et al.* (1993) found that elemental turnover rates did not differ appreciably between tissue types of certain fish species, leading them to speculate that within-organism variations are not always evident in cold-blooded animals. Nevertheless, studies that aim to gain information on the diet or trophic relations of animals may well be confounded by the tissue type being analysed. Thus, collagen, which has a much longer turnover time than other tissues, would equilibrate with the isotopic composition of the dietary C over a much longer time period than, say, liver or muscle tissue. Time courses of changes in isotopic composition of collagen would therefore provide information on the nature of average diet over a period approaching the lifespan of the organism, whereas that of a tissue that is rapidly turning over such as the liver would largely reflect recent ingestion and metabolism. Taken to the extreme where information is needed on the immediate diet of the organism over the last several hours, it would be highly appropriate to determine stomach contents or recently excreted faeces isotopic composition. Bearing these complications in mind, Tieszen *et al.* (1983) suggest that dietary analyses by means of natural abundance stable isotopes should always include measurements on a range of targeted tissue types. This, of course, is only possible when the organism is large enough to be dissectible. By contrast, smaller animals, such as those frequently encountered in marine environments would have to be analysed whole, and in some cases bulk biomass of a large number of individuals would be needed for an effective and reproducible analysis.

In contrast to the above, another strategy could be to analyse metabolically inactive tissue such as hair, nails, skin, feathers, whiskers, coral skeleton or fish otoliths which would thus reflect the diet of the organism over extensive periods of growth. For example, baleen has been used to reconstruct the seasonal changes of the location of feeding and diet of the bowhead whale (Schell *et al.* 1989). Similarly, stable isotope techniques applied to feathers enabled Hobson and Wassenaar (1997) to establish a link between the breeding and wintering grounds of certain migrating birds. The use of metabolically inactive tissues that can be removed from an animal non-destructively carries the added benefit of allowing sampling of endangered or protected species (Hobson *et al.* 1996).

Differences in growth rates between different parts of the same aquatic plant have also been shown to introduce intra-organism variations. Thus, Smit (1998) found a difference of up to 3.3 ‰ in $\delta^{15}N$ between the young lateral branches and the main axis of the Rhodophycean alga *Gracilaria gracilis*. Although the lateral branches were not necessarily of different tissue type or of biochemical composition than that of the main axis, the differences in $\delta^{15}N$ values appeared to stem from differences in growth rates (Smit and Bolton 1999) and N uptake rates (A. J. Smit, unpubl.). By analogy, lateral branches (or young tissue) in certain seaweeds may be considered as behaving much like metabolically active tissue in animals, and in this connection it is well known that growth and N uptake rates in finely branched seaweeds are higher than those of coarser thallus construction (Littler 1980, Littler and Arnold 1982). The same is then likely to apply to fine lateral branches and the thick main axis on one thallus. The long-term stability of the $\delta^{15}N$ signal of fast growing tissue should of course be of lesser duration than that of the older, slower growing tissue since N pools of the former are being continually diluted by N taken from the environment. Of course, differences in $\delta^{15}N$ between tissues of different growth rates would not be evident were the macrophyte to be growing in an environment dominated by a single N source of constant isotopic composition. Contrasting with this, an environment that receives transient pulses of isotopically different N sources, such as would occur in some polluted coastal environments, would generate substantial differences in $\delta^{15}N$ signals in recipient seaweeds and these should be best displayed in rapidly growing algae (Gartner, unpubl.). In other words, rapidly growing tissue should respond almost immediately to changes in $\delta^{15}N$ of the fluctuating DIN environment, whereas $\delta^{15}N$ in the older, slower growing tissue should provide an index of the average DIN taken up since the development of that tissue in the past. In rapidly growing organisms, elemental turnover (and hence changes in isotope signals) is a function of mass gain, whereas in adult or slow growing individuals, the determinant is time or maintenance metabolism (Hesslein *et al.* 1993).

It is important to realise that growth rates of an organism will also determine how rapidly isotopic change occurs in their bodies as they change diet. For example, post-larval marine invertebrates tend to have a very fast growth rates allowing their isotope ratio to rapidly respond to a shift in the isotope composition of their diet, given that the second diet is significantly different in its isotope signal from the first (Fry and Arnold 1982). In larger animals, the growth rate slows down, so isotope turnover will be more related to the maintenance of metabolism rather than to the rate of growth. This is exemplified in animals deprived of food (e.g. Hobson *et al.* 1993, Scrimgeour *et al.* 1995, Gannes *et al.* 1997). Such animals become enriched in ^{15}N and ^{13}C following starvation as isotopically light C and N are lost through catabolic processes - essentially this is a form of autolysis as they feed off their own tissue. This phenomenon has been observed in animals such as birds, fish and invertebrates, but is apparently not universally applicable since it does not occur in larval krill or earthworms (Frazer *et al.*

1997, Schmidt *et al.* 1999). It has also been suggested that among adult organisms, large animals generally showed slower C or N turnover rates than smaller animals (Tieszen *et al.* 1983). Turnover rate is directly related to metabolic rate, as it has been shown that small animals have a faster metabolic rate than large animals (Schmidt-Nielsen 1983).

Most trophic relations studies assume a constant fraction between food source and consumer (see Section 5, below). However, a study on captive seals held on a constant diet for two years (Hobson *et al.* 1996) showed that isotopic fractionation depends on the tissue type sampled. Most tissue types exhibited a consistent fractionation in the commonly accepted $2 - 3$ ‰ range, but blood samples always showed a significantly lower ^{15}N enrichment of 1.7 ‰. Moreover, fractionation values for a specific tissue type may vary between species (e.g. Hobson and Clark 1992).

4. VARIATIONS BETWEEN INDIVIDUALS OF THE SAME SPECIES

Apart from the utilisation of isotopically distinct primary C and N sources (Section 1), several factors are known to influence the isotopic variability between individuals of the same species. Grice *et al.* (1996) have demonstrated that light intensity greatly affects $\delta^{13}C$ of several seagrass species, with exposure to full sunlight resulting in a 4 ‰ enrichment in ^{13}C. They attributed this response to the increased uptake of ^{13}C from the DIC pool or increased internal recycling of CO_2 in the lacunae due to increased lacunal size. Similarly, Wefer and Killingley (1986) demonstrated that the Chlorophycean macroalga, *Halimeda incrassata*, growing in shallow water was more enriched in ^{13}C compared to those deeper down. The effect of light intensity on the fractionation of $^{13}C/^{12}C$ is discussed in detail by Hemminga and Mateo (1996).

Temperature and salinity may also affect the isotopic composition of macrophytes and microalgae, but only indirectly in such instances by altering the isotope ratios of the source C by affecting the solubility constant of CO_2, and hence its concentration in seawater (Mook *et al.* 1974). The change in $\delta^{13}C$ of source C with temperature has also been used to explain the apparent correlation of $\delta^{13}C$ of primary producers with latitude (Rau *et al.* 1989, Hemminga and Mateo 1996).

C and N isotope compositions of photoautotrophs and consumers have also been found to vary with season (Goering *et al.* 1990). In most cases seasonal variations in temperature, light intensity and/or salinity collectively affect the δ-values of primary sources. However, seasonal changes in the abundances of different types of organisms are known to exercise similar effects. As an example of the latter, Gearing *et al.* (1984) have shown that the δ^{13}C value of estuarine phytoplankton populations varies seasonally and attribute this change to the variation in types of phytoplankton present on a seasonal basis. Other examples of seasonal fluctuations in isotope ratios are detailed and evaluated by Parker (1964), Sackett *et al.* (1965), Deuser (1970), Fontugne and Duplessy (1978) and Smith and Kroopnick (1981).

Isotope compositions are found to vary appreciably between organisms of different age classes, and also seasonally within organisms of the same age. For example, Fry and Arnold (1982), Hesslein *et al.* (1993) and Sierzen *et al.* (1996) have demonstrated that fish and shrimp have isotope ratios that are determined by the dominant food source utilised, and that, as diets vary with age, isotope composition consistently varies accordingly. As a result, where isotope ratios vary between localities that an organism exploits at different times, data can be used to study the migratory behaviour or interpopulation variation in diet of animals (France 1995, Chamberlain *et al.* 1997, Hobson and Wassenaar 1997, Smit *et al.* 1998, Smit, unpubl.).

5.   VARIATIONS DUE TO ASSIMILATION AND TROPHIC TRANSFER

The premise underlying food web studies utilising ^{13}C/^{12}C and ^{15}N/^{14}N ratios is that isotopic composition of different consumer organisms should reflect those of their respective diets to within a few ‰, with the general proviso that a consumer is appreciably enriched in ^{13}C or ^{15}N relative to its current food source (Peterson and Fry 1987). Specifically, C isotope signals measured in animals vary to within +0.5 to +1 ‰ of their food source (Haines 1976, DeNiro and Epstein 1978, Haines and Montague 1979, McConnaughey and McRoy 1979, Fry *et al.* 1978, Tieszen *et al.* 1983, Checkley and Entzeroth 1985), whereas those of N vary to within -0.3 to +3.4 ‰ relative to that of the food source (DeNiro and Epstein 1981, Macko *et al.* 1982, Minagawa and Wada 1984, Schoeninger and Wada 1984). These fractionation values are used in both terrestrial and marine studies.

The general conservatism in propagation of ^{13}C/^{12}C signals from primary producer to consumer comprises a most useful element when tracing C flow through food webs, especially if there are large differences in δ^{13}C of the sources of C being consumed, e.g. the differences between seagrass *vs.* seaweed sources. Figure 2a, c and e show frequency histograms for a variety of primary producer and consumer organisms collected from a seagrass meadow of coastal Western Australia and illustrate the enrichment in ^{13}C that can occur with increasing trophic level. Although the data incorporate a diverse range of organisms belonging to several feeding groups, the overall conclusion is that invertebrate and fish groups belong to separate trophic levels as suggested in terms of the mean $\delta^{13}C$ values. The magnitude of the enrichment in ^{13}C with increasing trophic level is of the same magnitude as other studies (see above), although the measured difference is not statistically significant. This result readily demonstrates the need to isolate the individual sources of variation from the analysis, and in this instance to remove data for some of the carnivorous invertebrates (e.g. anemones and some polychaetes). The latter appear to account for the major element of the large variation encountered among the otherwise predominantly herbivorous invertebrates.

To estimate the percentage contribution of each food source consumed by the consumer organism, a multiple-source mixing model is typically applied (e.g. as developed by Ben-David *et al.* 1997, the model is shown in Section 6c, below). Such models assumes that all sources of food can potentially be preyed upon to at least some extent and subsequently assimilated into animal biomass. It then evaluates the extents of dietary mixing of food sources of different isotopic signal, assuming that the isotope ratio measured in the consumer represents a compromise value for the average food source consumed. Unfortunately, there is often no way of knowing whether all available food items are really being consumed, but insight can be gained through gut content analysis (e.g. Smit, unpubl.). However, such gut analyses still fail to provide unequivocal information on whether the food source is actually being assimilated or not. An essential prerequisite, of course, is that significant differences must exist in the δ-values of the ingested food sources. For example, applying this mixing model to the data in Figure 2, it becomes apparent that members of Rhodophyta (group 1) and Phaeophyta are the dominant (but not exclusive) source of dietary C and N to the invertebrate consumers in the food web (Smit, unpubl.). Used in this way, a mixing model approach combined with stomach content analyses removes a great deal of the subjectivity normally encountered when evaluating the role of potential food sources.

In contrast to the wide variation typically exhibited in $\delta^{13}C$ across primary producers (Figure 2 b, d and f), corresponding $\delta^{15}N$ signals tend to be much less variable. This possibly reflects that in most situations a common source of DIN is available to all primary producers. In this case, small variability in $\delta^{15}N$ between primary producers limits the use of stable isotope techniques in tracing N sources, but the stepwise enrichment in ^{15}N from producer to grazer/predator levels still makes N isotopes useful for determining the total number of trophic levels in the food web. It is therefore useful to analyse specimens for both $^{15}N/^{14}N$ and $^{13}C/^{12}C$ ratios (e.g. Kwak and Zedler 1997). Current model isotope ratio mass spectrometers (IRMS) allow one to achieve this using only one sample (dual analysis), and in the process adding very little to the overall preparation time while minimising costs of analyses.

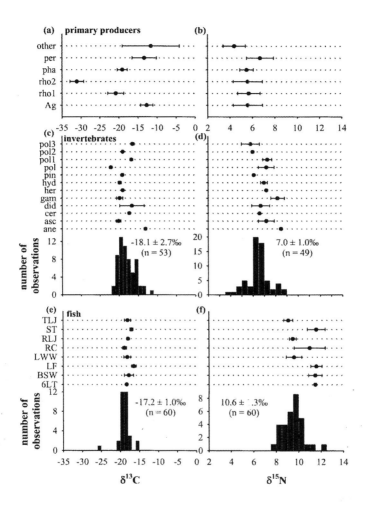

Figure 2. Trophic relations in a Western Australian seagrass habitat. $\delta^{15}N$ and $\delta^{13}C$ isotope ratio whisker plots (mean ± SD) for all groups of primary producers (a-b), invertebrates (c-d) and fish (e-f), together with frequency distribution histograms for pooled invertebrate and fish data are presented. Primary producer codes: sed = sediment; other = detrital material of unknown origin; per = periphyton; pha = Phaeophyta; rho2 = Rhodophyta group 2; rho1 = Rhodophyta group 1; Ag = *Amphibolis griffithii*. Invertebrate codes: ane = anemone; asc = ascidian; cer = *Cerapus* sp.; did = *Didimyus* sp.; hyd = hydroid; pin = *Pinna bicolor*; pol = *Polycarpa clavata*; pol1 = polychaete1; pol2 = polychaete2; pol3 = polychate3. Fish codes: TLJ = *Acanthaluteres vittiger*; ST = *Pelsartia humeralis*; RLF = *Scobonichthys granulatus*; RC = *Odax acroptilus*; LWW = *Siphonognathus radiatus*; LF = *Leviprora inops*; BSW = Notolabrus parilus; 6LT = *Pelates sexlineatus*. (Data from Smit, unpubl.)

The mechanism for C isotopic enrichment along the trophic structure is thought to be dependent on the differential assimilation of major biochemical components of different diets and catabolic activities resulting in selective excretion or catabolism of light isotopes (Malej *et al.* 1993). Recent laboratory studies have indicated that C isotopic fractionation is not constant (see references above), but depends on the nutritional quality of the diet (Fantle *et al.* 1999). For example, based on compound specific isotope analysis (CSIA) and controlled laboratory experiments involving blue crabs, Fantle *et al.* (1999) developed a conceptual model explaining possible relationships between isotopic fractionation and food quality. According to the model, fast growing crabs feed on a food source that is nutritious enough to provide both the energy for metabolic activities, and amino acids for the biosynthesis of new tissue. As the largest amount of C from the diet is incorporated into tissue, the $\delta^{13}C$ of the newly synthesised tissue could be expected to be similar to that of the food. In contrast, a slowly growing individual sustained by a protein-poor diet would be likely to fractionate C to a greater extent and accordingly carry a heavier $\delta^{13}C$ value than that of its diet. This is because a low quality food source would not provide enough energy to sustain basic metabolic needs, and the crab's own lipid and protein stores would therefore have to be catabolised to provide the required energy.

6.   INTERPRETATION AND ANALYSIS OF DATA

6.1   Trophic position histograms

Many stable isotope studies that have attempted to analyse trophic structure or trace the flow of a particular food source through different consumer levels have made use of critically small numbers (typically <4 – 5) of replicate samples. Lack of sufficient replication would thus fail to cover the full range of variability within a species, and therefore possibly over-simplify more complex patterns in the variation of δ-values. Monteiro *et al.* (1991) aimed to overcome this limitation by using a methodology which he termed the Trophic Position Isotope Spectrum (TPIS). This method utilises a frequency histogram constructed from δ-values determined from a large number of individuals per species (representing a certain size class, season, habitat or any other variable that might result in isotopic variation). The isotope signals form modes in the frequency distribution near the δ-values representative of the dominant source of food utilised by the organism. It can be employed to gain greater understanding into the variance components in systems that are temporally and spatially heterogeneous (Owens 1987,

Goering *et al.* 1990). TPIS representations may be useful in assessing the importance of multiple food sources available to a range of consumers. This procedure is underpinned by the basic principle of isotope ecology, *i.e.* that each δ-value provides information of the dietary history of an individual organism integrated over the turnover time of the tissue sampled. Differences in δ-values between organisms are therefore the result of ontogenetic dietary differences over the lifetime of the organism (but weighted to the recent past). For example Figure 3 presents a frequency histogram of 14 individual gammarid amphipods with a modal $\delta^{13}C$ value of -19.7 ‰; this value is enriched by ~ 2.8 ‰ with respect to the mixed algal epiphyte signal. In this case, one would therefore conclude that, of all the food sources sampled, epiphytes comprise the most likely source of dietary C.

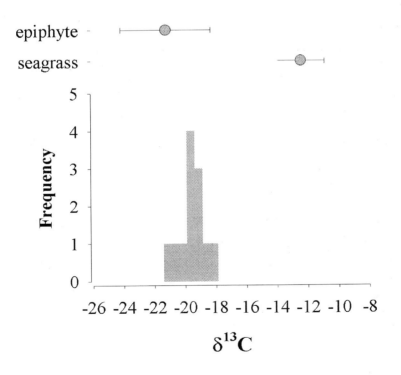

Figure 3. Frequency distribution of 14 gammarid amphipods collected from a seagrass meadow; $\delta^{13}C$ signals (± SD) for seagrass (*A. griffithii*) and mixed seagrass epiphytes are also indicated. Here, the amphipods are enriched by ~ 2 ‰ with respect to epiphytes.

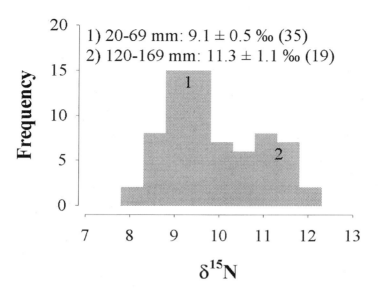

1) 20-69 mm: 9.1 ± 0.5 ‰ (35)
2) 120-169 mm: 11.3 ± 1.1 ‰ (19)

Figure 4. Frequency distribution of 71 *Siphonognathus radiatus* individuals. The modes of the frequency distributions correspond to the two size classes indicated; the number in partenthesis indicates *n* (note: statistically, a third size class of 70 – 119 mm can be distinguished (10.3 ± 0.7 ‰, *n* = 47), but it is not apparent on the frequency histogram) (Smit, unpubl.).

A second example of this approach is provided in Figure 4, which represents a frequency histogram summarising the δ^{15}N distribution of 71 individuals of long-rayed weedwhiting, *Siphonognathus radiatus* (Smit *et al.* 1998). For the sake of this example, we will assume there is no *a priori* knowledge on the origin of the individuals, their size distribution, or on any other factor that might be responsible for isotopic variation. The histogram is clearly bi-modal, and we therefore conclude that values for individuals segregate based on differences in diet. However, in this particular study, sizes of individual fish were measured, whereupon it was found that the first mode of the histogram corresponded to animals ranging from 20 to 69 mm in length, the second mode to animals of 120 to 169 mm in length. It is also known from gut content analysis that larger *S. radiatus* individuals consume a larger proportion of molluscs (mainly mytillids) compared to the smaller size class, with the balance of the diet in both cases being comprised of crustaceans (gammarid amphipods and harpacticoid copepods). Unfortunately, isotope signals for these food sources were not measured, but other results indicated that gammarid amphipods and certain bivalves carried δ^{15}N values of approximately 5.5 and 6.8 ‰, respectively. This suggests that ontogenetic shifts in dietary preference may well have been responsible for the observed drift in $\delta^{15}N$ with fish length. Note that in the above example the argument for a change in diet with fish size class is somewhat weakened by the fact that food and consumer organisms were not sampled in the same study. Interannual, seasonal and spatial variation of the isotope signals may have acted as confounding factors, so the conclusion given above is based only on indirect evidence.

Conducted fully to include seasonal samples, the use of frequency histograms as alluded to above, may be viewed as an elegant visually based method when applied to systems with only a few trophic levels and a small number of species. However, the high level of replication makes the technique extremely costly if extended to complex ecosystems with multiple interactions between many possible food sources and diverse consumer organisms distributed across several trophic levels.

6.2 $\delta^{15}N$ *vs.* $\delta^{13}C$ plots

An alternative method for presenting isotope data to aid in the interpretation of trophic flows of C and N is one involving comparative plots of values for $\delta^{15}N$ and $\delta^{13}C$. For example, Figure 5 presents data for several food and consumer organisms collected from a *Posidonia coriacea* seagrass meadow. The two groups of Rhodophyta differ greatly in terms of $\delta^{13}C$, apparently due to differences in their DIC acquisition mechanisms (refer to Section 2b), but there is no corresponding variation in terms of $\delta^{15}N$, presumably due to the utilisation of a common pool of DIN. The Phaeophyta have similar isotope signals to the Rhodophyta Group 1, but seagrass emerges as significantly depleted in ^{15}N and enriched in ^{13}C relative to the other primary producers. Statistically, the primary producers can thus be grouped into three food sources: Rhodophyta Group 2, Rhodophyta Group1 plus Phaeophyta, and the seagrass. Note the main requirement for the application of natural abundance stable isotopes as tracer of C and N flow pathways is that significant differences exist in isotopic signatures between food sources.

The caprellid and gammarid amphipods are known to be important grazers in many marine ecosystems (Barnes, 1987). As shown in Figure 5, the caprellid amphipods are enriched in both ^{13}C and ^{15}N with respect to the Rhodophyta Group 1 (approximately +1.7 and +2.3 ‰, respectively). Similarly, the gammarid amphipods exhibit a similar relationship. Due to the standard deviations around the mean Phaeophyta signal, and the fact that they are not significantly different from the closely associated rhodophycean group, these algae cannot be excluded as possible food source. Neither does the Rhodophyta Group 2 nor the seagrass seem to feature as a food source.

In the same system, the mollusc *Oliva lignaria* is carnivorous, as reflected in its heavy $\delta^{15}N$ value and to a lesser extent by its $\delta^{13}C$ signal, thus placing it at about one trophic level above the amphipods.

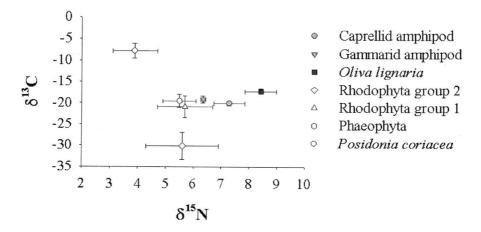

Figure 5. $\delta^{15}N$ vs. $\delta^{13}C$ plot of food sources and consumers from a Western Australian seagrass meadow (error bars represent 1 SD).

6.3   Mixing models

The previous example can be extended in its analysis by employing the mixing model approach (e.g. Ben-David *et al.* 1997). The analysis requires mean $\delta^{13}C$ and $\delta^{15}N$ values of each food source when estimating the contribution of the sources to the invertebrate feeder. The mean food δ-values (*A, B* and *C*) are corrected for the fractionation between the food and the invertebrate, yielding *A', B'* and *C'*. Although the model can incorporate the effect of variable fractionation between producers and consumers (e.g. Fantle *et al.* 1999), a constant 2.1 and 1.0 ‰ for $^{15}N/^{14}N$ and $^{13}C/^{12}C$ were used, respectively. The Euclidean distance between the invertebrate δ-values and the corrected food δ-values are then computed according to:

$$Z_A = \sqrt{(\delta^{13}C_{A'} - \delta^{13}C_I)^2 + (\delta^{15}N_{A'} - \delta^{15}N_I)^2} \qquad (1)$$

where Z_A is the Euclidean distance between the invertebrate and food *A'*, $\delta^{13}C_{A'}$ and $\delta^{15}N_{A'}$ the $\delta^{13}C$ and $\delta^{15}N$ signatures for food *A'*, and $\delta^{13}C_I$ and $\delta^{15}N_I$ the $\delta^{13}C$ and $\delta^{15}N$ signatures for the invertebrate, *I*. The same equation is then used to calculate the Euclidean distance between the remaining food sources, *B* and *C*. The contribution of each food source to the diet of the consumer is assumed to be inversely related to the Euclidean distance between the corrected signature of the source and the consumer, as represented below:

$$\%X = \frac{Z_X}{Z_A^{-1} + Z_B^{-1} + Z_C^{-1}} \times 100 \tag{2}$$

where Z_X is the Euclidean between any of the corrected food sources (*A'*, *B'* or *C'*) and the consumer.

The same general conclusions can be drawn from applying this model to the gammarid amphipod data in Figure 5. However, because this particular model incorporates all potential food sources, dietary mixing is now also taken into account. This becomes apparent in the increased importance of the other available food sources such as seagrass and the very ^{13}C-depleted rhodophycean algae (Table 2). Nevertheless, the phaeophycean and rhodophycean (Group 1) algae remain the main component of their diet. The principal value of this method is that it removes much of the subjectivity inherent in the previous approach, but, despite this advantage, compliance to important assumptions or prerequisites must still apply (Ben-David *et al.* 1997). Firstly, the model requires that there are significant differences among the food sources in terms of the δ-values entered into the respective equations. This is true for the $δ^{13}$C data, but differences are small in terms of $δ^{15}$N. Secondly, it assumes that all food sources consumed by invertebrate consumers are actually being assimilated into biomass. There is no way of validating this assumption without stomach content analysis. Even with stomach content data, digestion and assimilation cannot truly be verified. In fact, Ben-David *et al.* (1997) suggest that results obtained from mixing models are used as an index of potential food assimilation, rather than as an indication of proportions of the sources actually being eaten.

Table 2. Contribution (%) from mixing model calculations applied to the mean gammarid amphipod δ-values in Figure 5.

Food source	% contribution
Rhodophyta Group 1	46.7
Rhodophyta Group 2	9.3
Phaeophyta	37.7
Posidonia coriacea	6.4

6.4   Independence of replicates

The problem of lack of independence of data (Underwood, 1997) is great in stable isotope ecology studies. The reason for non-independence in such studies is probably based on the nature of many of the samples analysed, *i.e.*, their size. For example, it is often not possible to analyse individual phyto- or zooplankters. As mentioned before, many organisms require the use of a composite sample, often with as many as millions of individuals in the case of plankton, to provide sufficient material for analysis. In other cases, the number of individuals per sample is less severe (say five isopods to provide one δ-value), but replicate samples often remain non-independent. It therefore falls on the scientist to create independence through careful experimental design.

The first and probably most obvious problem concerns pooling individuals to provide one replicate sample since this substantially decreases the variance around the mean due to a positive correlation between replicates. Secondly, it increases the chance of making a Type I error (rejection of the null hypothesis when it is true) when testing for statistical differences among samples. Further discussion of the problem of non-independence among replicates and samples is provided by Underwood (1997).

When it is not possible to obtain independent replicates from one population, the only alternative may be to obtain a single sample from a population, but to sample many (replicate) independent populations. For example, suppose one wants to compare or contrast zooplankton isotope signals from two habitats: sheltered sandy embayments and exposed rocky capes. One might start by defining a stretch of coastline that contains three sandy bays isolated from one another by rocky headlands. For each of the two habitats three independent replicates samples can then be obtained, thereby making the statistical comparison of isotope signatures between the two habitats possible. In most instances, the solution to non-independence involves nothing more than careful spatial replication of the sampling design.

6.5   Definition of a population

When conducting any study it is important that an explicit definition is provided of the populations involved. Thus, a general phytoplankton sample containing cells of less than 65 μm in size would probably include the following groups: netplankton (<65 μm), nanoplankton (2 – 20 μm), picoplankton (<2 μm) and the smaller bacterioplankton. Since these groups play diverse ecological roles, and may include photoautotrophs, N_2-fixers and heterotrophs, isotopic signals of each group should vary appreciably. Highly detailed results can be obtained through further size-fractionation of the total phytoplankton community. In principle, the types of questions that can be answered by stable isotope techniques are extremely broad, but in practice, resolution is limited only by the amount of sampling detail possible using current techniques.

7.   CONCLUSION

This review has shown that any of a wide variety of factors may be held responsible for generating considerable isotopic variability in $^{15}N/^{14}N$ and $^{13}C/^{12}C$ signals of the primary C and N sources in tissue components of individual organisms and different individuals within and between populations and communities. Variability of this nature may be viewed as either a limitation or an asset when attempting to understand the functioning of marine ecosystems. In any study, careful high-resolution sampling of potential or actual sources of variation should be defined and accounted for at each scale at which the system under study is to be considered. These scales might range from the individual organism (e.g. between tissue variability, biochemical makeup, etc.) up to the population itself (e.g. intra-species variability, age, growth rate, nutritional history, etc.). Such stratification may then continue up to the scale of the entire ecosystem (e.g. inter-population, intra- and inter-community variability), and in the process encompass effects of seasonality and location as additional factors to be considered in a particular analysis. If desired, variability between entire ecosystems can even be assessed.

REFERENCES

Anderson T. F. and Arthur M. A. (1983) Stable isotopes of oxygen and carbon and their application to sedimentologic and paloenvironmental problems. In 'Stable Isotopes in Sedimentary Geology.' (Eds M.A. Arthur, T.F. Anderson, J. Veizer L.S. Land.) pp. 1.1-1.151. (Society of Economic Palaeontologists and Mineralogists: Tulsa, Oklahoma.)

Atekwana, E. A. and Krishnamurthy, R. V. (1998). Seasonal variations of dissolved inorganic carbon and $\delta^{13}C$ of surface waters: applications of a modified gas evolution technique. *Journal of Hydrology* 205, 265-278.

Barnes R. D. (1987) Invertebrate zoology. Saunders College Publishing, 891 pp.

Ben-David M. R., Flynn R. W. and Schell D. M. (1997) Annual and seasonal changes in diets of martens: evidence from stable isotope analysis. *Oecologia* 111, 280-291.

Benson B. B. and Krause D. Jr. (1984) The concentration and isotopic fractionation of oxygen dissolved in freshwater and seawater in equilibrium with the atmosphere. *Limnology and Oceanography* 29, 620-632.

Benson B. B. and Parker P. D. M. (1961) Nitrogen/argon and nitrogen isotope ratios in aerobic seawater. *Deep-Sea Research* 7, 165-179.

Bligh E. G. and Dyer W. J. (1959) A rapid method of total lipid extraction and purification. *Canadian Journal of Biochemistry and Physiology* 37, 911-917.

Bremner J. M and Edwards A. P. (1965) Determination and isotope-ratio analysis of different forms of nitrogen in soils: I. Apparatus and procedure for distillation and determination of ammonium. *Proceedings of the Soils Science Society* 29, 504-507.

Chamberlain C. P., Blum J. D., Holmes R. T., Feng X., Sherry T. W. and Graves G. R. (1997) The use of isotope tracers for identifying populations of migratory birds. *Oecologia* 109, 132-141.

Checkley D. M. and Entzeroth L. C. (1985) Elemental and isotopic fractionation of carbon and nitrogen by marine, planktonic copepods and implications to the marine nitrogen cycle. *Journal of Plankton Research* 7, 553-568.

Checkley Jr. D. M. and Miller C. A. (1989) Nitrogen isotope fractionation by oceanic zooplankton. *Deep-Sea Research* 36, 1449-1456.

Cifuentes L. A., Fogel M. L., Pennock J. R. and Sharp J. H. (1989) Biochemical factors that influence the stable nitrogen isotope ratio of dissolved ammonium in the Delaware Estuary. *Geochimica et Cosmochimica Acta* 53, 2713-2421.

Cline J. D. and Kaplan I. R. (1975) Isotopic fractionation of dissolved nitrate during denitrification in the eastern tropical North Pacific Ocean. *Marine Chemistry* 3, 271-299.

DeNiro M. and Epstein S. (1978) Influence of diet on the distribution of carbon isotopes in animals. *Geochimica et Cosmochimica Acta* 42, 495-506.

DeNiro M. J. and Epstein S. (1981) Influence of diet on the distribution of nitrogen isotopes in animals. *Geochimica et Cosmochimica Acta* 45, 341-353.

Deuser W. G. (1970) Isotopic evidence for diminishing supply of available carbon during diatom blooms in the Black Sea. *Nature* 225, 1069-1071.

Dugdale R. C. (1967) nutrient limitation in the sea: dynamics, identification, and significance. *Limnology and Oceanography* 12, 685-695.

Fantle M. S., Dittel A. I., Schwalm S. M., Epifanio C. E. and Fogel M. L. (1999) A food web analysis of the juvenile blue crab, *Callinectes sapidus*, using stable isotopes in whole animals and individual amino acids. *Oecologia* 120, 416-426.

Fogel M. L., Aquilar C., Cuhel R., Hollander D. J. Willey J. D. and Paerl H. W. (1999) Biological and isotopic changes in the coastal waters induced by Hurricane Gordon. *Limnology and Oceanography* 44, 1359-1369.

Fontugne M. and Duplessy J. C. (1978) Carbon isotope ratio of marine plankton related to surface water masses. *Earth Planetary Science Letters* 41, 365-371.

France R. (1995) Stable isotope survey of the role of macrophytes in the carbon flow of aquatic food webs. *Vegetatio* 124, 67-72.

Frazer T. K., Ross R. M., Quetin L. B. and Montoya J. P. (1997) Turnover of carbon and nitrogen during growth of larval krill, *Euphausia suberba* Dana: a stable isotope approach. *Journal of Experimental Marine Biology and Ecology* 212, 259-275.

Fry B. and Arnold C. (1982) Rapid $^{13}C/^{12}C$ turnover during growth of brown shrimp (*Penaeus aztecus*). *Oecologia* 54, 200-204.

Fry B., Jeng W., Scalan R. S., Parker P. L. and Baccus J. (1978) $\delta^{13}C$ food web analysis of a Texas sand dune community. *Geochimica et Cosmochimica Acta* 42, 1299-1302.

Fry B. and Sherr E. B. (1984) $\delta^{13}C$ measurements as indicators of carbon flow in marine and freshwater ecosystems. *Contributions in Marine Science* 27, 13-47.

Gannes L. Z., O'Brien D. M. and Del Rio C. M. (1997) Stable isotopes in animal ecology: assumptions, caveats, and a call for more laboratory experiments. *Ecology* 78, 1271-1276.

Gearing J. N., Gearing P. J., Rudnick D. T., Requejo A. G. and Hutchins M. J. (1984) Isotopic variability of organic carbon in a phytoplankton-based temperate estuary. *Geochimica et Cosmochimica Acta* 48, 1089-1098.

Goering J., Alexander V. and Haubenstock N. (1990) Seasonal variability of stable carbon and nitrogen isotope ratios of organisms in a north Pacific bay. *Estuarine, Coastal and Shelf Science* 30, 239-260.

Grice A. M., Loneragan N. R. and Dennison W. C. (1996) Light intensity and the interactions between physiology, morphology and stable isotope ratios in five species of seagrass. *Journal of Experimental Marine Biology and Ecology* 195, 91-110.

Handley L. L. and Raven J. A. (1992) The use of natural abundance isotopes in plant physiology and ecology. *Plant, Cell and Environment* 15, 965-985.

Haines E. B. (1976) Stable carbon isotope ratios in the biota, soils and tidal water of a Georgia salt marsh. *Estuarine, Coastal and Shelf Science* 4, 609-616.

Haines E. B. and Montague C.L. (1979) Food sources of estuarine invertebrates analyzed using $^{13}C/^{12}C$ ratios. *Ecology* 60, 48-56.

Hemminga M. A. and Mateo M. A. (1996) Stable carbon isotopes in seagrasses: variability in ratios and use in ecological studies. *Marine Ecology Progress Series* 140, 285-298.

Hesslein R. H., Hallard K. A., Ramlal P. (1993) Replacement of sulfur, carbon, and nitrogen in tissue of growing broadfish (*Coregomus nasus*) in response to a change in diet traced by $\delta^{34}S$, $\delta^{13}C$ and $\delta^{15}N$. *Canadian Journal of Fisheries and Aquatic Science* 50, 2071-2076.

Hinga K. R., Arthur M. A., Pilson M. E. Q. and Whitaker D. (1994) Carbon-isotope fractionation by marine-phytoplankton in culture - the effects of CO_2 concentration, pH, temperature, and species. *Global Biogeochemical Cycles* 8, 91-102.

Hobson K. A. and Clark R. G. (1992) Assessing avian diets using stable isotope analysis. II. Factors influencing diet-tissue fractionation. *Condor* 94, 189-197.

Hobson K. A. and Wassenaar L. I. (1997) Linking breeding and wintering grounds of neotropical migrant songbirds using stable hydrogen isotopic analysis of feathers. *Oecologia* 109: 142-148.

Hobson K. A., Alisauskas R. T. and Clark R. G. (1993) Stable-nitrogen isotope enrichment in avian tissues due to fasting and nutritional stress: implications for isotopic analysis of diet. *Condor* 95, 388-394.

Hobson K. A., Schell D. M., Renouf D. and Noseworthy E. (1996) Stable carbon and nitrogen isotopic fractionation between diet and tissues of captive seals: implications for dietary reconstruction involving marine mammals. *Canadian Journal of Fisheries and Aquatic Science* 53, 528-533.

Hoering T. C. and Ford H. T. (1960) Isotopic effect in the fixation of nitrogen by *Azotobacter*. *Journal of the American Chemical Society* 82, 376-378.

Holmes R. M., McClelland J. W., Sigman D. M., Fry B. and Peterson B. J. (1998) Measuring $^{15}N-NH_4^+$ in marine, estuarine and fresh waters: An adaptation of the ammonia diffusion method for samples with low ammonium concentrations. *Marine Chemistry* 60, 235-243.

Horrigan S. G., Montoya J. P., Nevins J. L. and McCarthy J. J. (1990). Natural isotopic composition of dissolved inorganic nitrogen in the Chesapeake Bay. *Estuarine, Coastal and Shelf Science* 30, 393-410.

Johnson A. M. (1996) The effect of environmental variables on 13C discrimination on two marine phytoplankton. *Marine Ecology Progress Series* 132, 257-263.

Kendall C. and Caldwell E. A. (1998) Fundamentals of isotope geochemistry. In 'Isotope Tracers in Catchment Hydrology.' (Eds C. Kendall and J. J. McDonnell) 839 pp. (Elsevier: Amsterdam.)

Kwak T. J. and Zedler J. B. (1997) Food web analysis of southern California coastal wetlands using multiple stable isotopes. *Oecologia* 110, 262-277.

Laws E. A., Popp B. N., Bidigare R. R., Kennicutt M. C. and Macko S. A. (1995) Dependance of phytoplankton carbon isotope composition on growth rate and [CO_2]aq: theoretical considerations and experimental results. *Geochimica et Cosmochimica Acta* 59, 1131-1138.

Littler M. M. (1980) Morphological form and photosynthetic performance of marine macroalgae: tests for a functional/form hypothesis. *Botanica Marina* 22, 161-165.

Littler M. M. and Arnold K. E. (1982) Primary productivity of macroalgal functional form groups from southwestern North America. *Journal of Phycology* 18, 307-311.

Liu K. –K. and Kaplan I. R. (1989) The eastern tropical Pacific as a source of ^{15}N-enriched nitrate in seawater off southern California. *Limnology and Oceanography* 34, 820-830.

Liu K. –K., Shaw P. –T and Kaplan I. R. (1987) Modeling of nitrogen isotopic variation of nitrate within the denitrification zone in the eastern tropical South Pacific. *EOS* 68, 1714.

Lyon T. and Baxter M. (1978) Stable carbon isotopes in human tissues. *Nature* 273, 750-751.

Macko S. A., Lee W. Y. and Parker P. L. (1982) Nitrogen and carbon isotope fractionation by two species of marine amphipods: laboratory and field studies. *Journal of Experimental Marine Biology and Ecology* 63, 145-149.

Malej A., Faganeli J. and Pezdi J. (1993) Stable isotope and biochemical fractionation in the marine pelagic food chain: the jellyfish *Pelagia noctiluca* and net zooplankton. *Marine Biology* 116, 65-570.

Malone T. C., Pike S. E. and Conley D. J. (1993) Transient variations in phytoplankton productivity at the JGOFS Bermuda time series station. *Deep-Sea Research* 40, 903-924.

McConnaughey T. and McRoy C. P. (1979) ^{13}C label identifies eelgrass (*Zostera marina*) carbon in an Alaskan estuarine food web. *Marine Biology* 53, 263-269.

Minagawa M. and Wada E. (1984) Stepwise enrichment of ^{15}N along food chains: further evidence and the relation between $\delta^{15}N$ and animal age. *Geochimica et Cosmochimica Acta* 48, 1135-1140.

Minagawa M. and Wada E. (1986) Nitrogen isotope ratios of red tide organisms in the east China Sea: a characterization of biological nitrogen fixation. *Marine Chemistry* 19, 245-259.

Michener R. H. and Schell D. M. (1994) Stable isotope ratios as tracers in marine aquatic food webs. In 'Stable Isotopes in Ecology and Environmental Science.' (Eds K. Lajtha and R. H. Michener.) (Blackwell Scientific Publications: Oxford.)

Miyake Y. and Wada E. (1967) The abundance ratio of ^{15}N/^{14}N in marine environments. *Records of Oceanographic Works, Japan* 7, 37-53.

Miyake Y. and Wada E. (1971) The isotope effect of nitrogen in biochemical oxidation-reduction reactions. *Records of Oceanographic Works Japan* 9, 37-57.

Monteiro P. M. S., James A. G., Sholto-Douglas A. D. and Field J. G. (1991) The $\delta^{13}C$ trophic position isotope spectrum as a tool to define and quantify carbon pathways in marine food webs. *Marine Ecology Progress Series* 78, 33-40

Mook W. G., Bommerson J. C. and Staverman W. H. (1974) Carbon isotope fractionation with diffusion of carbon dioxide and gaseous carbon dioxide. *Earth Planet Science Letters* 22, 169-176.

O,Leary M. H. (1988) Carbon isotopes in photosynthesis. BiopScience 38, 328-336.

Owens, N. J. P. (1987) Natural variations in ^{15}N in the marine environment. *Advances in Marine Biology* 24, 389-451.

Owens N. J. P. and Watts L. J. (1998) ^{15}N and the assimilation of nitrogen by marine phytoplankton: the past, present and future? In 'Stable Isotopes: Integration of Biological,

Ecological and Geochemical processes.' (Ed H Griffiths.) pp. 257-284. (Bios Scientific Publishers, Ltd.: Oxford.)

Paerl H. W. and Fogel M. L. (1994) Isotopic characterization of atmospheric nitrogen inputs as sources of enhanced primary production is coastal Atlantic Ocean waters. *Marine Biology* 119, 635-645.

Paerl H. W., Fogel M. L. and Bates P. W. (1993) Atmospheric nitrogen deposition in coastal waters: implications for marine primary production and flux C flux. In 'Trends in Microbial Ecology.' (Eds R. Guerrero and C. Pedrós-Alió.) pp. 459-464. (Spanish Society of Microbiology: Barcelona.)

Parker P. L. (1964) The biogeochemistry of the stable isotopes of carbon in a marine bay. *Geochimica et Cosmochimica Acta 28:* 1155-1164.

Peterson B. J. And Fry B. (1987) Stable isotopes in ecosystems studies. *Annual Review of Ecology and Systematics* 18, 293-320.

Peterson B., Fry B. and Deegan L. (1993) The trophic significance of epilithic algal production in a fertilized tundra river ecosystem. *Limnology and Oceanography* 38, 872-878.

Preston T. (1992) The measurement of stable isotope natural abundance variations. *Plant, Cell and Environment* 15, 1091-1097.

Rau G. H., Sweeney R. E. and Kaplan I. R. (1982) Plankton 13C:12C ratio changes with latitude: differences between northern and southern oceans. *Deep-Sea Research* 29, 1035-1039.

Rau G. H., Sweeney R. E., Kaplan I. R., Mearns A. J. and Young D. R. (1981) Differences in animal ^{13}C, ^{15}N and D abundance between a polluted and and unpolluted coastal site: likely indicators of sewage uptake by a marine food web. *Estuarine, Coastal and Shelf Science.* 13, 701-707.

Rau G. H., Takahashi T. and Des Marais D. J. (1989) Latitudinal variations in plankton $\delta^{13}C$: implications for CO_2 and productivity in past oceans. Nature 341, 516-518.

Raven J. A. (1992) Present and potential uses of the natural abundance of stable isotopes in plant science, with illustrations from the marine environment. *Plant, Cell and Environment* 15, 1083-1091.

Raven J. A. and Farquhar G. D. (1990) The influence of N metabolism and organic acid synthesis on the natural abundance of C isotopes in plants. *New Phytologist* 116, 505-529.

Raven J. A., Walker D. I., Johnson A. M., Handley L. L. and Kübler J. E. (1995) Implications of ^{13}C natural abundance measurements for photosynthetic performance by marine macrophytes in their natural environment. *Marine Ecology Progress Series* 123, 193-205.

Sackett W. M., Eckelmann W. R. and Bender M. L. (1965) Temperature dependence of carbon isotope composition in marine plankton and sediments. *Science* 148, 235-237.

Schell D. M., Saupe S. M. and Haubenstock N. (1989) Bowhead whale (*Balaena mysticetus*) growth and feeding as estimated by $\delta^{13}C$ techniques. *Marine Biology* 103, 433-443.

Schoeninger M. J. and Wada M. J. (1984) Nitrogen and carbon isotope composition of bone collagen from marine and terrestrial animals. *Geochimica et Cosmochimica Acta* 48, 625-639.

Schmidt O., Scrimgeour C. M. and Curry J. P. (1999) Carbon and nitrogen stable isotope ratios in body tissue and mucus of feeding earthworms (*Lumbricus festivus*). *Oecologia* 118, 9-15.

Schmidt-Nielsen, K. (1983) Animal Physiology: Adaptation and Environment. 3rd Ed. 619 pp. (Cambridge University Press: Cambridge.)

Scrimgeour C. M., Gordon S. C., Handley L. L. and Woodford J. A. T. (1995) Trophic levels and anomalous $\delta^{15}N$ of insects on raspberry (*Rubus idaeus* L.). *Isotopes in Environmental Health Studies* 31, 107-115.

Sierzen M. E., Keough J. R. and Hagley C. A. (1996) Trophic analysis of ruffe (*Gymnocephalus cernuus*) and white perch (*Morone americana*) in a Lake Superior coastal food web, using stable isotope techniques. *Journal of Great Lakes Research* 22, 436-443.

Sigman D. M., Altabet M. A., Michener R., McCorkle D. C., Fry B. and Holmes R. M. (1997) Natural-abundance-level measurement of nitrogen isotopic composition of oceanic nitrate: an adaptation of the ammonia diffusion method. *Marine Chemistry* 57, 227-242.

Skirrow G. (1975) The dissolved gases - carbon dioxide. In 'Chemical Oceanography.' Vol. 2, 2nd edn. (Eds J.P. Wiley and G. Skirrow.) pp. 1-192. (Academic Press: London.)

Smit A. J. (1998) Nitrogen environment, ecophysiology and growth of *Gracilaria gracilis* from Saldanha Bay, South Africa. Ph.D. thesis, The University of Cape Town, South Africa, 158 pp.

Smit A. J. and Bolton J. J. (1999) Organismic determinants and their effect on growth and regeneration in *Gracilaria gracilis*. *Journal of Applied Phycology* 11, 293-299.

Smit A. J., Brearley A., Hyndes G. and Lavery P. (1998). Shells and Dredging Environmental Management Programme. *Project S1: Ecological Significance of Seagrasses. Task 11: Trophic structure and linkages.* (Cockburn Cement Limited: Perth Western Australia)

Smith B. N. and Epstein S. (1970) Biogeochemistry of stable isotopes of hydrogen and carbon in salt marsh biota. *Plant Physiology* 46, 738-742.

Smith S. V. and Kroopnick P. (1981) Carbon-13 isotopic fractionation as a measure of aquatic metabolism. *Nature* 294, 252-253.

Spies R. B., Kruger H., Ireland R. and Rice Jr., D. W. (1989) Stable isotope ratios and contaminant concentrations in a sewage-distorted food web. *Marine Ecology Progress Series* 54, 157-170.

Stephenson R. L., Tan F. C. and Mann K. H. (1984) Stable carbon isotope variability in marine macrophytes and its implications for food web studies. *Marine Biology* 81, 223-230.

Surif M. B. and Raven J. A. (1990) Photosynthetic gas exchange under emersed conditions in eulittoral and normally submerged members of the Fucales and Laminariales: interpretation in relation to C isotope and N and water use efficiency. *Oecologia* 82, 68-80.

Sweeney R. E. and Kaplan I. R. (1980) Natural abundance of ^{15}N as a source indicator for near-shore marine sedimentary and dissolved nitrogen. *Marine Chemistry* 9, 81-94.

Tieszen K. R. (1978) Carbon isotope fractionation in biological material. *Nature* 276, 97-98.

Tieszen L. L., Boutton T. W., Tesdahl K. G. and Slade N. A. (1983) Fractionation and turnover of stable carbon isotopes in animal tissues: implications for $\delta^{13}C$ analysis of diet. *Oecologia* 57, 32-37.

Thompson P. and Calvert S. (1994) Carbon isotope fractionation by a marine diatom: the influence of irradiance, daylength, pH and nitrogen source. *Limnology and Oceanography* 40, 673-679.

Underwood A. J. (1997) Experiments in ecology: their logical design and interpretation using analysis of variance. 504 pp. (Cambridge University Press: United Kingdom.)

Velinsky D. J., Pennock J. R., Sharp J. H., Cifuentes L. A. and Fogel M. L. (1989) Determination of the isotopic composition of ammonium-nitrogen at the natural abundance level from estuarine waters. *Marine Chemistry* 26, 351-361.

Wada E., Kadonaga T. and Matsuo S. (1975) ^{15}N abundance in nitrogen of naturally occurring substances and global assessment of denitrification from isotopic viewpoint. *Geochemical Journal* 9, 139-148.

Wefer G. and Killingley J. S. (1986) Carbon isotopes in organic matter from a benthic alga *Halimeda incrassata* (Bermuda): effects of light intensity. *Chemical Geology* 59, 321-130.

Chapter 12

$\delta^{13}C$ as an Indicator of Palaeoenvironments
A Molecular Approach

Kliti Grice
Centre for Petroleum and Environmental Organic Geochemistry, APCRC/CEMS/ School of Applied Chemistry, Curtin University of Technology, GPO Box U1987, Perth, WA 6001 Australia. Email: K.Grice@alpha1.curtin.edu.au

Key words: biomarkers, compound specific isotopes, palaeoenvironments, palaeoclimates, petroleum, organic matter, carbon isotopes, molecules

1.   INTRODUCTION

Reconstruction of ancient climates (palaeoclimates) is a major goal in areas of the Earth sciences seeking insight into the history of the Earth. On the one hand a petroleum organic geochemist is able to use such information to provide valuable insight on palaeoclimatic and palaeoenvironmental conditions likely to generate rocks offering good sources of petroleum, thereby leading to more effective exploration strategies. On the other hand, an environmental geoscientist might apply palaeoclimate information to predict the likelihood of possible future ice-house and greenhouse conditions.

Isotope ratio monitoring gas chromatography mass spectrometers (irm-GCMS), whereby a GC is linked to an isotope ratio mass spectrometer *via* a combustion interface (Matthews and Hayes 1978, Hayes 1983) comprise an important tool for determining $^{13}C/^{12}C$ of individual organic components in complex mixtures (e.g. Hayes *et al.* 1990). This technique is commonly referred to as compound specific isotope analysis (CSIA).

As organic matter (OM) depends upon the relative proportions and isotopic compositions of marine and terrigenous OM, isotopic effects associated with organisms producing the OM, associated palaeo-

247

M. Unkovich et al. (eds.),
Stable Isotope Techniques in the Study of Biological Processes and Functioning of Ecosystems, 247–279.
© 2001 *Kluwer Academic Publishers. Printed in the Netherlands.*

geographical, palaeoenvironmental and palaeoclimatic changes, and the nature and degree of genetic alteration and reworking of OM can be predicted with some confidence from $^{13}C/^{12}C$ analysis. The basis for this approach is that each individual organic compound in OM inherit a specific $^{13}C/^{12}C$ signal resulting from its natural origins and its post-depositional transformation pathways. In organic geochemistry, CSIA has been applied to the analysis of components called biomarkers (molecular fossils) found in complex mixtures of petroleum, sedimentary OM, and related samples. These components are related to biochemicals in algae, bacteria and higher plants (e.g. Brassell and Eglinton 1986) and organisms processing already fixed carbon, methanotrophs (Summons *et al.* 1994), heterotrophs (Grice *et al.* 1998a and references therein) *etc.* (Table 1). As a result their characteristic isotopic compositions ($^{13}C/^{12}C$) have proven useful for establishing pathways of carbon flow in ancient environments (Hayes *et al.* 1990), in interpreting depositional conditions in palaeowaters (e.g. Collister *et al.* 1992, 1994a, 1994b, Hollander *et al.* 1993, Schoell *et al.* 1994a, b, Grice *et al.* 1996a, b, 1997, Schouten *et al.* 1997a, Grice *et al.* 1998b, c, d) and for assessing ambient CO_2 concentrations in palaeoenvironments (e.g. Jasper and Hayes 1990, Jasper *et al.* 1994, Bidigare *et al.* 1997, Huang *et al.* 1999, Kuypers *et al.* 1999).

The bold numbers in brackets (below) refer to chemical structures, for example, botryococcane (**1**) isorenieratane (**2**) and methyl *iso*-butyl maleimide (**3**), shown in the appendix. Based on structural considerations, biomarkers in complex mixtures of petroleum or sedimentary OM, prove to be specific in terms of source organisms. Thus, botyrococcane derives from botryococcene (**4**) biosynthesised by the fresh/brackish water alga *Botryococcus braunii*. Isorenieratane is derived from the carotenoid isorenieratene (**5**) and methyl *iso*-butyl maleimide is a degradation product of bacteriochlorophylls *c, d* or *e* (e.g. **6**) uniquely biosynthesised by the anoxygenic photosynthetic Chlorobiaceae (e.g. Fig. 1).

Unfortunately, many biomarkers may be of multiple origin; for example, the C_{20} isoprenoid phytane (**7**), might originate from ether-linked membrane lipids of methanogenic/halophilic archaea (e.g. **8**, Fig. 3) or might derive from the phytyl side chain of chlorophyll *a* (**9**) in phytoplankton and higher plants (Fig. 3). Also methyl ethyl maleimide (**10**) is a non-specific biomarker in that it is a general degradation product from chlorophyll *a* (**9**), the major chlorophyll of many photosynthetic organisms (Fig. 2).

In certain instances, CSIA, when studied alongside the incorporation of reduced sulfur specific organosulfur compounds, has aided geochemists in evaluating the sources of unambiguous biomarkers (e.g. Grice *et al.* 1998b). However, the interpretation of stable carbon isotopic compositions of biomarkers continues to rely on limited CSIA of lipids (e.g. Monson and

Hayes 1982, Summons *et al.* 1994, Collister *et al.* 1994a, Schouten *et al.* 1998, van der Meer *et al.* 1998) whether in extant organisms or in extrapolations to molecular level of conventional studies on bulk carbon isotopic composition of organisms grown under varying conditions.

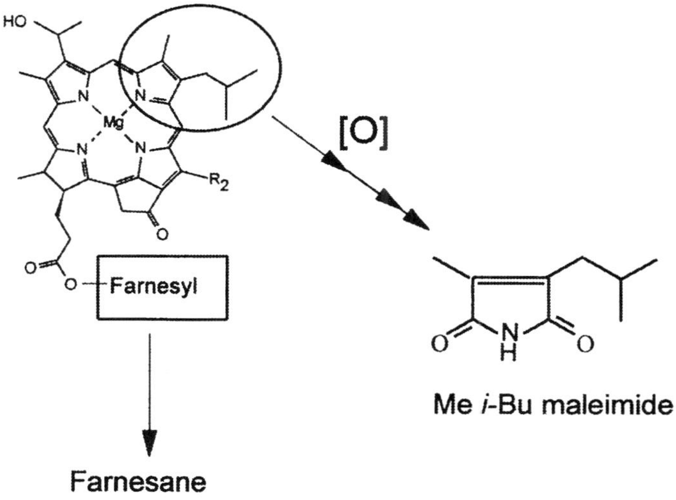

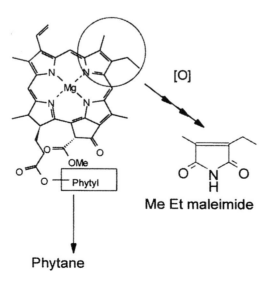

Figure 1. Origin of methyl *iso*-butyl maleimide (*ca.* -15‰) and farnesane (-18‰) from a bacteriochlorophyll *d* in *Chlorobiaceae* (green sulfur bacteria, using reversed TCA cycle).

Figure 2. Origin of methyl ethyl maleimide (*ca.* -27‰) and phytane (*ca.* -31‰) from chlorophyll *a* in for example, phytoplankton (algae/cyanobacteria using a C3 pathway).

**Biochemical
(organism)**

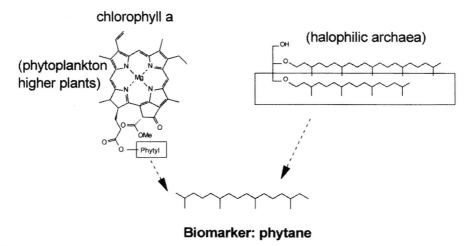

Biomarker: phytane

Figure 3. Origin of phytane from archeae or phytoplankton.

2. EXPERIMENTAL

2.1 Instrumentation

Conventional and more advanced GC-MS (gas chromatography-mass spectrometry) techniques have proven useful in identifying the carbon skeletons of components and the position of their functional groups in natural organic mixtures of soils, sediments and petroleum. However such studies, based on structure alone, are not always definitive in distinguishing their natural product precursors. Nevertheless stable carbon isotope analysis may still allow one to establish the natural sources of compounds. While conventional bulk stable isotope technique permits determination of the $^{13}C/^{12}C$ of a homogenous whole tissue sample, this approach is not ideally suited for individual compounds in a mixture, as lengthy laboratory procedures are required to separate and purify components. Matthews and Hayes (1978) were the first to recognise the value of coupling a combustion furnace to a gas chromatograph (GC) when assessing the isotopic compositions of individual components of complex mixtures. Isotope ratio monitoring gas chromatography mass spectrometers (irm-GCMS) allowing for the effective use of the above approach became commercially available in the 1990s (Matthews and Hayes 1978, Hayes 1983). In essence, as each component elutes from the GC column it is oxidised (CuO, 900°C) to CO_2 and H_2O and NO_X/N_2 (if nitrogen is present). The H_2O is then removed cryogenically (-100°C) or by using a Nafion membrane and the CO_2 then enters the mass spectrometer. Irm-GCMS instruments are now normally fitted with a continuous flow dual inlet consisting of two vitreous capillaries which carry helium gas continuously into the ion source. The precision for $^{13}C/^{12}C$ analysis is 0.2‰ or better, with typically only small amounts of sample (20ng) being required for each component when using an on-column injector. More recent developments include irm-GCMS instruments capable of successfully measuring $^{15}N/^{14}N$, $^{18}O/^{16}O$ (e.g. Merrit 1993, Brand *et al.* 1994) and D/H (Burgoyne and Hayes 1998, Sessions *et al.* 1999). Their use is considered outside the scope of this chapter and book.

2.2 Extraction and separation procedures

Biomarkers in sediments have of course to be extracted and concentrated prior to analysis. A suitable analytical protocol is depicted in Figure 4. Extractable OM is obtained from ground sediments using ultrasonic or Soxhlet extraction with high purity solvents (dichloromethane:methanol, 4:1). Since petroleum and extractable OM are complex mixtures, liquid chromatography techniques are generally used to separate oils and extracts into fractions (saturated, aromatic and polar hydrocarbons). Fractions are separated by virtue of their different polarities, with the polarity of the solvent increased to facilitate progressive separation of different fractions. Prior to CSIA, components in fractions are identified by GC-MS analysis and use of standards. Even after petroleum and extracts are fractionated, some fractions may still be extraordinarily complex and thus present impediments to successful use of CSIA. Before accurate isotope ratios can be obtained, components have to be GC-amenable and GC peaks of interest have to be resolved to the baseline. For less prominent components, highly developed skills are employed in separation techniques such as molecular sieving to isolate fractions based on molecular size (e.g. Armanios *et al.* 1993, Ellis *et al.* 1994, Grice 1995) and/or preparative thin layer chromatography (tlc) techniques.

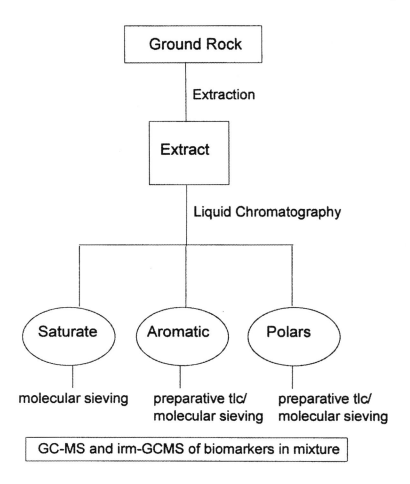

Figure 4. Analytical scheme.

3.   ISOTOPIC MOLECULAR INDICATORS OF PALAEOENVIRONMENTS

3.1   Phytoplankton biomarkers

When reconstructing likely ancient environments of deposited materials based on CSIA of biomarkers it is important to establish an " isotopic reference point" against which values for various separated components can then be compared. Sedimentary steranes (e.g. **11**, and algal phytane, **7**) are particularly useful components in this respect since they occur widely and are derived from algal sterols and chlorophyll *a* (**9**), respectively. Hence, $\delta^{13}C$ values of steranes and phytane can provide a definitive "reference point" for lipid components biosynthesised by the primary producers in the euphotic zone. For example in sediment samples from the Permian Kupferschiefer, the $\delta^{13}C$ values for phytane (**7**) were found to be similar to the algal-derived cholestane (**11**) from the same samples, indicating an algal origin for both these components. However, other contributors to phytane were not excluded, for example, cyanobacteria in which chlorophyll *a* (**9**) is the sole photosynthetic pigment (Grice *et al.* 1996a). In the same samples, methyl ethyl maleimide (**10**), a degradation product predominantly derived from the macrocycle of chlorophyll *a* (**9**) was found to be 3.5 – 4.5‰ more enriched in ^{13}C than phytane from the phytyl side chain, an order of difference to be expected from a common chlorophyll origin (Grice *et al.* 1996b, 1997) (see biosynthesis section below and Fig. 2). A difference of *ca.* 4.5‰ has also been observed between sedimentary porphyrins (derived from photosynthetic pigments) and phytane in sediments (Hayes *et al.* 1990).

When using CSIA of biomarkers in sedimentary OM to reconstruct ancient environments of deposition, it has generally been assumed that stable carbon isotopic fractionation effects associated with the biosynthesis of sterols and isoprenoids (e.g. phytol) as well as cell sizes/cell geometry (Popp *et al.* 1998) of the eukaryotic community, may be safely averaged out for sediment samples covering tens to hundreds of years. Furthermore, it has been established that zooplankton feeding on phytoplankton do not alter the stable carbon isotopic compositions of lipids such as the sterols, which these organisms consume, metabolise to cholesterol, retain, and/or egest (eg. Grice *et al.* 1998a). Nevertheless, zooplankton are selective feeders, so might well select particular classes or sizes of phytoplankton of a distinct $\delta^{13}C$ composition. Thus, $\delta^{13}C$ signals of steroids (sterols and steranes) in sediments would probably be representative of the mixture of the phytoplankton population, even though the individual populations might carry different $\delta^{13}C$ compositions. This restriction should be borne in mind when using $\delta^{13}C$ of sterols and steranes as "isotopic reference points". Furthermore, recent studies of CSIA of lipids in algae (see below) appear to complicate the interpretation of $\delta^{13}C$ of phytoplankton biomarkers, since different algae are known to biosynthesise lipids with quite different carbon isotopic compositions (Schouten *et al.* 1998). Although differences between phytane and cholestane in OM may not be due to different source organisms, Schouten *et al.* (1998) still suggests that cholestane is one of the most suitable "isotopic reference points" of the primary producers living in the euphotic zone of ancient water columns.

In the following sections and Table 1, a number of examples of biochemicals specific for certain organisms, and their biomarkers found in OM will be considered alongside their $\delta^{13}C$ signals to provide a background on the use of this information to reconstruct ancient environments of deposition.

3.1.1   *Botryococcus braunii*

The green unicellular microalga *Botryococcus braunii* (*B. braunii*) is widely distributed in freshwater, brackish lakes and reservoirs at temperate, tropical and arctic latitudes (Tyson 1995 and references therein). Although morphologically similar, three distinct races (A, B and L) of this colonial alga are recognised according to their hydrocarbon composition. The A race exclusively biosynthesises odd carbon numbered *n*-alkadienes and trienes in the C_{25} to C_{31} range (e.g. **12**, Metzger *et al.* 1986, 1991), whereas the B race produces C_{30} to C_{37} branched isoprenoidal hydrocarbons, termed botryococcenes (e.g. **4**, Maxwell *et al.* 1968). Certain strains of the B race also biosynthesise cyclobotryococcenes (e.g. **13**, Metzger *et al.* 1985, David *et al.* 1988 and Huang *et al.* 1988). The L race yields a single C_{40} isoprenoid hydrocarbon, lycopa-14(*E*),18(*E*)-diene (**14**, Metzger and Casadevall 1987, Metzger *et al.* 1991). Botryococcenes exclusively found in the B race can give rise to specific botryococcane biomarkers (e.g. **1**). Unlike botryococcanes, lycopane (**15**) is frequently found in sediments. However, it is not specific to precursor lycopadiene of the L race, as it can also be derived from reduction of the C_{40} carotenoid, lycopene (**16**), a compound also found in a variety of other micro-organisms and higher plants (Schmidt 1978 and references therein) and in certain organisms of marine origin (e.g. Brassell *et al.* 1981, Wakeham *et al.* 1993). Furthermore, long-chain *n*-alkanes (e.g. **17**), with their predominance of odd-over-even carbon numbers, can be derived from reduction of the *n*-alkadienes and trienes of race A (e.g. Metzger *et al.* 1991). This might also indicate a higher plant origin (e.g. Eglinton and Hamilton 1963).

Botryococcanes have been reported to be significantly enriched in ^{13}C compared with phytoplankton biomarkers in both sediments (Boreham *et al.* 1994, Huang *et al.* 1995, Grice *et al.* 1998b, Huang *et al.* 1999) and petroleum (Dowling *et al.* 1995). High $\delta^{13}C$ values have also been noted in culture studies of *Botryococcus braunii* (Summons *et al.* 1996), but the reasons for this remain unclear. Boreham *et al.* (1994) have suggested that the isotopic fractionation accompanying photosynthesis in *B. braunii* may not be fully expressed due to limiting internal pCO_2, perhaps due to the thick outer walls of these microalgae limiting diffusion rates for CO_2 and thereby ^{13}C enriching biomass. However, since *B. braunii* can also utilise bicarbonate (Huang *et al.* 1999 and references therein), heavier $\delta^{13}C$ values might also be expected from this source. A biomarker for *B. braunii* was recently found to be significantly enriched in ^{13}C (having a value of around - 5‰) at the last glacial maximum, contrasting with a value of -30‰ at the beginning of the Holocene (Huang *et al.* 1999). The former isotopically heavy value was attributed to low atmospheric pCO_2 leading to depletion of dissolved CO_2 and therefore forcing *B. braunii* to recourse to bicarbonate in its lake habitat. Another interesting example of unusual *B. braunii* biomarkers has been reported in some hypersaline sediments of Miocene/Pliocene age from the Mount Sdom Formation, Dead Sea, Israel (Grice *et al.* 1998b). Here, a novel series of organosulfur compounds (OSC) derived from functionalised lipids of races B and L of *B. braunii* algae have been identified. One sample was considered to be mainly comprised of lipids derived from race B and possibly also from A and L, whereas the other of lipids came from *B. braunii* races B and L. Overall, B and L race derived components were 13 – 20‰ and 5 – 7‰, respectively more enriched in ^{13}C than comparable phytoplanktonic biomarkers of marine origin. Stable carbon isotopic data of the *B. braunii* components pointed to differing seasonalities of bloom development in each race. In one case, the biomarker isorenieratane (**2**) was present and its carbon isotopic composition therefore consistent with an origin from Chlorobiaceae (see green sulfur bacteria below). In conclusion there would have been periods when the water column was highly stratified, with anoxic waters extending up into the photic zone. The CSIA data thus provided compelling evidence for the existence of freshwater algae in this ancient hypersaline euxinic environment.

3.1.2 Cyanobacteria

Monomethylbranched alkanes, in particular 7-methylheptadecane (e.g. **18**), and 8-methylheptadecane and 7,11-dimethylheptadecane are known to be biosynthesised by cyanobacteria (e.g. Summons *et al.* 1998). For example, it has been reported by Summons *et al.* (1998) that cyanobacteria are able to synthesise normal or branched alkanes of different $\delta^{13}C$ compositions depending upon the environmental conditions. Cyanobacteria also biosynthesise extended bacteriohopanetetrol derivatives (e.g. **19**). $\delta^{13}C$ values of extended hopanoids (e.g. **20**, > C_{30}) determined in sediments (Grice *et al.* 1996b 1997, Schouten *et al.* 1997a) generally have similar $\delta^{13}C$ values to algal-derived cholestane and/or phytane, indicating a possible origin from cyanobacteria inhabiting upper regions of a water column. However, Schoell *et al.* (1994a) found $\delta^{13}C$ differences between the cyanobacterial hopanoids (-29.5 to -31.5‰) and algal-derived steranes (-25‰) in the Middle Miocene which were interpreted as marking the interglacial-glacial transition from a well mixed to a highly thermally stratified ocean. Schoell *et al.* (1994a) attributed these isotopic differences to cyanobacteria living in deeper parts of the water column of a stratified oceanic environment utilising a more depleted inorganic carbon source than the algae inhabiting the upper water column

3.1.3   Archeae

Archeae are a group of primitive organisms that consist of methanogenic bacteria, extreme halophilic bacteria, and certain thermoacidophilic bacteria. OM derives not only from primary producers, but also from secondary producers. It has been suggested that glycerol diether membrane lipids biosynthesised by archeae give rise to regular isoprenoid hydrocarbons upon diagnosis. Biphytane (**21**) and PME (2,6,10,15,19-pentamethyleicosane, **22**) are highly specific to methanogenic bacteria (methane producers) (e.g. Brassell *et al.* 1981, Schouten *et al.* 1997b and references therein). Biphytane (**21**) has been found to be enriched in ^{13}C by about $4 - 5$‰ compared to phytoplankton biomarkers in pelagic sediments and in the water column (Hoefs *et al.* 1997). This difference has been attributed to either, (i) the use of dissolved inorganic carbon as carbon source, but with biosynthetic pathways within methanogens discriminating less against ^{13}C than in algae using the C3 pathway, or (ii) Archeae obtaining their carbon source from low-molecular weight organic substrates (acetate/methylated amines) generated during decomposition of algal-derived particulate or dissolved OM. The methane produced by methanogens is isotopically light compared to their bulk biomass. δ^{13}C of PME (**22**) in some Cretaceous black shales has been found to have a similar δ^{13}C value to phytane (Vinke *et al.* 1998). However, other presumed algal biomarkers turn out to be lighter in ^{13}C by about 10‰, indicating that phytane is probably related to the same source as PME (an archeael origin). The presence of biphytane (**21**) and/or PME (**22**) which are enriched in ^{13}C relative to phytoplankton markers in OM also clearly indicates the presence of methanogens, and probable anoxia in the depositional environment.

The isotopically light methane produced by methanogens can be further fractionated by methanotrophic and methylotrophic bacteria leading to biomass become extremely depleted in ^{13}C (*ca.* -90 ‰). Bacteriohopanetetrol derivatives biosynthesised by these organisms give rise to sedimentary hopanoids which are similarly highly depleted in ^{13}C (Freeman *et al.* 1990). For example, the presence of ^{13}C-depleted hopanoids in non-marine sediments has been related to methane cycling in the Eocene Messel Shale (*ca.* -65‰; Freeman *et al.* 1990), the Green River Formation (*ca.* -80‰; Collister *et al.* 1992) and in a recent lake sediment (*ca.* -50‰;

Spooner *et al.* 1994). Phytane in Messel Shale was also found to be depleted in ^{13}C by *ca.* 5‰ compared against C_{19} regular isoprenoid, pristane (**23**). This was consistent with an origin from ether lipids of methanotrophic bacteria, whereas pristane (*ca.* -25‰) was consistent with derivation from phytol of algal chlorophyll *a* (Freeman *et al.* 1990). The recovery of highly ^{13}C depleted hopanoids and phytane in OM points strongly to the presence of a full methane cycle and reducing conditions in these palaeoenvironments.

Halophilic archeae biosynthesise two main glycerol diether membrane-bound lipids (e.g. De Rosa *et al.* 1982, Teixidor *et al.* 1993), one with two regular isoprenoid chains C_{20}, the other with a C_{20} and C_{25} regular isoprenoid chain (**24**). Both of these are possible sources of the C_{21} to C_{25} regular isoprenoids found in oils (Albaiges 1980) and other sediments deposited in hypersaline lakes (e.g. Jiamo *et al.* 1988, Grice *et al.* 1998c) and hypersaline marine environments (ten Haven *et al.* 1988, Keely *et al.* 1993, Sinninghe Damste *et al.* 1993a). Recently, δ^{13}C analyses of these individual isoprenoids extracted from some Dead Sea halite deposits (Miocene/Pliocene) and other hypersaline deposits (Grice *et al.* 1998c) have shown them to be enriched in ^{13}C by up to 7 ‰ compared to biomarkers of presumed phytoplanktonic origin. This implies a source other than algae and cyanobacteria and tentatively assigns them collectively as halophilic archaea (haloarchaea). Based on biomarker distributions, δ^{13}C contents and mineral compositions, these sediments thus appear to have been deposited in a stratified water body with concentrated brine at depth. Continual evaporation and deposition of salts would have been favourable conditions for the growth of such halophilic archeal communities. The ^{13}C enriched isoprenoids are attributed to haloarchaea (salt tolerant chemoheterotrophs), using simple organic compounds such as amino acids and carbohydrates as their carbon sources (e.g. Tindall 1992 and references therein).

The isotopic composition of lipids biosynthesised by a heterotrophic organism is dependent on the carbon isotopic composition of its food source (DeNiro and Epstein 1977, 1978). Lipids are generally depleted in ^{13}C compared with other products of biosynthesis such as carbohydrates and proteins (DeNiro and Epstein 1977). Thus lipids biosynthesised by a halophile should be slightly depleted in ^{13}C compared with the biomass of the primary producer. In the present day Dead Sea, for example, *Dunaliella* sp. represents the main primary producer and halophiles readily feed off their biosynthetic products, particularly glycerol (Oren 1993). This heterotrophic relationship becomes prominent after the development of *Dunaliella* blooms, when glycerol becomes rapidly taken up by archaeal communities whereupon it is subject to rapid turnover. It is thought that during the Miocene/Pliocene a similar primitive ecosystem existed in the Dead Sea basin. Furthermore, high growth rates during development of

Dunaliella blooms may explain why the lipids which they biosynthesise are considerably enriched in ^{13}C. Indeed a number of authors have demonstrated that high growth rates result in less discrimination against ^{13}C, leading to an enrichment in ^{13}C of biomass (e.g. Laws *et al.* 1995), so it is likely that *Dunaliella,* one of few primary producers capable of growing rapidly under stressed salt conditions, may be responsible for producing isotopically heavy products of biosynthesis. The consumption of this isotopically heavy glycerol by halophiles would in turn lead to a ^{13}C enrichment of their biomass, although the isotopic discrimination effects against ^{13}C by the enzymes associated with the biosynthesis of ether lipids may well perturb this situation. Further research is clearly still required to establish why heavy isotopic compositions of biomarkers occur in hypersaline environments. Nevertheless the presence in OM of isotopically heavy regular isoprenoids $>C_{20}$ would still point to the existence of hypersaline conditions of deposition in many palaeoenvironments.

3.1.4   Photosynthetic sulfur bacteria

Isorenieratane (**2**) and some related aromatic compounds derived from isorenieratene (**5**) in Chlorobiaceae have been widely used as biomarkers for photic zone euxinia (presence of hydrogen sulfide, absence of oxygen) in geological samples. These compounds are typically enriched in ^{13}C by about 15‰ relative to other lipid biomarkers such as obtained, for example, from phytoplankton (e.g. Summons and Powell 1986, 1987b, Sinninghe Damsté *et al.* 1993b, Hartgers *et al.* 1994a, b, Grice *et al.* 1996a 1997). A similar distinct isotopic difference has been observed in compounds such as chlorobactane (**25**) derived from chlorobactene (**26**, Grice *et al.* 1998d) and for the highly specific bacteriochlorophylls of Chlorobiaceae (e.g. **6**, Grice *et al.* 1996a, b). Highly specific bacteriochlorophylls give rise for example to methyl *iso*-butyl maleimide (**3**) and farnesane (**27**) derived from the bacteriochlorophyll macrocycle and lipid farnesyl side chain, respectively (Fig. 1). Farnesane (**27**) has been found to be depleted in ^{13}C by about 3‰ compared to methyl *iso*-butyl maleimide (**3**), a finding consistent with an origin from bacteriochlorophyll (Grice *et al.* 1996a, b). Therefore, the presence of isotopically heavy isorenieratane (**2**) or chlorobactane (**25**) (and aromatic components therefrom), or methyl *iso*-butyl maleimide (**3**) or farnesane (**27**) in OM is evidence for existence of high activity of Chlorobiaceae (and microbial sulfur cycle) which fix CO_2 *via* the reversed tricarboxylic acid cycle (Evans *et al.* 1966). Their presence would be expected to lead to an enrichment in ^{13}C of biomass (Quandt *et al.* 1977, Sirevag *et al.* 1977) of ancient water columns, while at the same time indicate the presence of stratified water bodies with euxinic zones extending into the photic zone.

$\delta^{13}C$ values of purple sulfur bacterial (Chromatiaceae)-derived compounds, in particular ones derived from okenone (**28**) (Schaeffer *et al.* 1997), are significantly depleted in ^{13}C namely of $\delta^{13}C$ of *ca.* -45‰ compared to phytoplankton markers at *ca.* -35‰. These data are consistent with Chromatiaceae living in a deeper part of the water column where they would probably utilise more depleted inorganic carbon sources than would phytoplankton inhabiting an upper water column.

3.1.5   Dinoflagellates

Biomarkers, 4-methyl 24-ethyl cholestane and dinosteranes (e.g. **29**) are generally considered to originate from the 4-methyl 24-ethyl sterol and dinosterol (e.g. **30**) precursors in dinoflagellates (e.g. Summons *et al.* 1987a, 1992). However, the possibility of methanotrophic bacterial sources applying to some of these compounds has also been suggested (Bird *et al.* 1971). The $\delta^{13}C$ values of the C_{30} 4-methylsteroids of some Eocene-Paleocene Chinese hypsersaline lacustrine sediments (Grice *et al.* 1998d) have been found to be enriched in ^{13}C by *ca.* 8‰ compared to phytoplankton markers. This argues strongly against an origin specifically from methanotrophic bacteria, since the latter would have lipids highly depleted in ^{13}C (see above). The reasons for such heavy isotopic compositions in dinoflagellates are not clear, although high growth rates (e.g. Laws *et al.* 1995, Goericke *et al.* 1994) and differing seasonalities of blooming might well result in an enrichment in ^{13}C of the lipids which they synthesise. The presence of 4-methylsteroids in OM indicates either the presence of dinoflagellate algae or methanotrophic bacteria (methane cycle). The sources of these components (whether dinoflagllates or methanotrophic bacteria) can then be unravelled using CSIA, thereby providing valuable information on the nature of the palaeoenvironment.

3.1.6   Diatoms

Highly branched C_{25} and C_{30} isoprenoid (HBI) (e.g. **31**) alkanes encountered in OM (Robson and Rowland 1986, Sinninghe Damsté *et al.* 1989) are derived from several species of diatoms (Nichols *et al.* 1988, Volkman *et al.* 1992). $\delta^{13}C$ values of these components in OM are heavier than phytoplankton biomarkers (e.g. Freeman *et al.* 1994, Summons *et al.* 1993). This has been attributed to seasonal blooming of certain diatoms. Non-blooming diatoms are also reported to use a bicarbonate pumping mechanism (Summons *et al.* 1993). Presence of isotopically heavy HBI in OM indicates the presence of diatoms in palaeonvironment.

3.1.7 Haptophytes

Long-chain C_{37} and C_{38} alkenones (e.g. **32** de Leeuw *et al.* 1980, Volkman *et al.* 1980, Rechka and Maxwell 1988a, b, Thiel *et al.* 1997) are biosynthesised by certain haptophytes, e.g. *Emiliana huxleyi* and *Isochrysis galbana* (e.g. Marlowe *et al.* 1990) and their relative proportions have accordingly been used widely as a proxy for predicting sea surface temperatures (e.g. Prahl and Wakeham 1987, Eglinton *et al.* 1992, Prahl *et al.* 1995, Prahl *et al.* 1988). The individual stable carbon isotopic compositions of alkenones, in combination with other data may thus provide estimates of ancient pCO_2 levels (Jasper and Hayes 1990, Jasper *et al.* 1994). The ratio of long-chain alkenones and their stable carbon isotopic compositions is unaffected by zooplankton herbivory (Grice *et al.* 1998a).

$\delta^{13}C$ values (*ca.* -36‰) of some abundant C_{37} and C_{38} *n*-alkanes (e.g. **33**) in an Eocene-Paleocene Chinese hypersaline lacustrine sediment have been found to be significantly more negative than phytoplankton markers (such as cholestane with a $\delta^{13}C$ value of -28.7‰), inferring a specific origin from long chain alkenones in haptophytes (Grice *et al.* 1998d). These light values have been attributed either to different bloom periods of the haptophytes than of the other euphotic zone algae during the annual cycle, or to longer periods of varying environmental conditions. The presence of C_{37} and C_{38} alkenones (and sedimentary derivatives therefrom) in OM indicates the presence of haptophytes and $\delta^{13}C$ values of alkenones in combination with other data may therefore be used to estimate ancient marine pCO_2 levels.

3.1.8 Heterotrophic ciliates

The biomarker gammacerane (**34**), is thought to originate from tetrahymanol (**35**); gammaceran-3-β-ol; e.g. ten Haven *et al.* 1989). Tetrahymanol (**35**) has been found in photosynthetic bacteria (Kleeman *et al.* 1990), a fern (Zander *et al.* 1969), a fungus (Kemp *et al.* 1984) and in ciliates feeding only on bacteria (Harvey and McManus 1991). $\delta^{13}C$ values of gammacerane (**34** *ca.* -24‰) are consistent with an origin from heterotrophic ciliates partially feeding on a ^{13}C-rich source (Grice *et al.* 1998d). This might apply for example to chlorobiaceae living at a chemocline (cf. Sinninghe Damsté *et al.* 1995). In the above mentioned sample analysed by Grice *et al.* (1998d), isotopically heavy isorenieratane (**2**) and chlorobactane (**25**) were also present, indicating the presence of Chlorobiaceae at the chemocline and under periods of photic zone euxinia.

Also the isotopic composition of a C_{30} 17α, $21\beta(H)$ hopane (**36**, *ca.* -23‰) in the samples was also found to be similar to gammacerane (**34**). Since C_{30} hopane-3β-ol (**37**) occurs together with tetrahymanol (**35**) in certain bacterivous marine sauticociliates (Harvey and McManus 1991), it was suggested that the C_{30} 17α, $21\beta(H)$ hopane might have been derived from bacterivorous ciliates. However, a contribution from a heterotrophic bacterial source would still be possible. The presence of isotopically heavy gammacerane and C_{30} 17α, $21\beta(H)$ hopane in OM therefore emerges as a potentially valuable indicator of water column stratification (cf. Schoell *et al.* 1994b, Sinninghe Damsté *et al.* 1995).

Table 1. δ¹³C of biomarkers derived from organisms in palaeoevironments and what they can tell us about the biogeochemistry and carbon cycle.

Organism	Palaeoenvironmental information
Botryococcus braunii	Freshwater/brackish algae in temperate, tropical and Arctic latitudes. $p\mathrm{CO_2}$ indicator (last glacial maximum). Ancient hypersaline euxinic (H_2S) environments. Water column stratification.
Cyanobacteria	Blooming cyanobacteria (HCO_3^-). Cyanobacteria in euphotic zone (CO_2). Water column stratification.
Archeae Methanogens Methanotrophs Halophiles	 CH_4 cycle, anoxia. Full CH_4 cycle, anoxia. Hypersalinity.
Photosynthetic sulfur bacteria (Chlorobiaceae) Green and brown coloured	 Photic zone euxinia (H_2S, $-O_2$, microbial sulfur cycle).
Dinoflagellates	Dinoflagellate blooms.
Diatoms	Diatom blooms (HCO_3^-). Upwelling.
Haptophytes	Marine, haptophyte blooms. Sea surface temperature (SST). $p\mathrm{CO_2}$ in ancient marine environments.
Heterotrophic ciliates	Water column stratification (photic zone euxinia, H_2S, $-O_2$, microbial sulfur cycle).

4.   DISTRIBUTIONS OF STABLE CARBON
ISOTOPES IN NATURAL SYSTEMS

As illustrated above, recent advances in irm-GCMS have provided instructive $\delta^{13}C$ data on the sources of biomarkers preserved in sedimentary OM. However, an understanding of the isotopic fractionations in modern biological systems is additionally required to interpret $\delta^{13}C$ of biomarkers in ancient sediments. This section outlines distributions of stable carbon isotopes in present day natural systems assuming them to be determined or influenced by (see Hayes 1993):

a) $\delta^{13}C$ of carbon source, and its availability.

b) Specific photosynthetic pathway (and complex associated fractionations) involved in uptake of CO_2 in photosynthesis.

c) Isotopic fractionations associated with biosynthesis in an organism.

d) Other factors including cell size/ cell geometry (Goericke *et al.* 1994, Popp *et al.* 1998) and growth rates of phytoplankton (Laws *et al.* 1995, Bidigare *et al.* 1997) and in respect of land plants plant water-use efficiency reported in C3, C4 and CAM plants (Ehleringer *et al.* 1993, see chapter by Pate in this volume) and pCO_2.

4.1 $\delta^{13}C$ of Carbon source

The carbon source utilised by photosynthetic organisms is CO_2 or in the case of aquatic organisms bicarbonate (Fig. 5). When CO_2 concentrations are too low to sustain photosynthesis, bicarbonate can be used only by certain phytoplankton. These organisms actively transport or pump bicarbonate into their cells, resulting in significant ^{13}C enrichment of their biomass (Sharkey and Berry 1985, Falkowski 1991). Photosynthesis in general leads to selective incorporation of ^{12}C into organic matter, which leads to ^{13}C enriched inorganic carbon, for example as carbonate, bicarbonate and gaseous CO_2 at 25°C. At isotopic equilibrium atmospheric CO_2 is enriched in ^{13}C by 1‰ relative to aqueous CO_2 (Wendt 1968), whereas bicarbonate is enriched in ^{13}C by 8‰ relative to dissolved CO_2 (Mook *et al.* 1974). But it must be remembered that these equilibria may in turn be perturbed by biological activity.

Methanotrophic bacteria or methane oxidisers use methane, methanol or formaldehyde as their carbon source which is generally provided by cohabiting methanogenic bacteria or methane producers (Fig. 5). Methanogens utilise hydrogen to reduce CO_2 to methane, or decarboxylate acetate to methane and CO_2. The isotopic fractionations occurring in the latter reactions lead to the production of isotopically light methane (Games and Hayes 1978) and large associated isotopic shifts have accordingly been reported in methane evolved by in both freshwater (δ^{13}C *ca.* -55‰ to *ca.* -90‰) and marine (δ^{13}C *ca.* -40‰ to *ca.* -55‰) environments (Whiticar *et al.* 1986). As a result of such fractionations methanotrophs feeding on isotopically light methane are typically highly depleted in ^{13}C (Summons *et al.* 1994).

For any heterotroph, one would expect its overall isotopic composition to be tightly determined by feeding patterns. However carbon retained as biomass is enriched in ^{13}C relative to that respired and/or egested and the expected isotopic difference concerned would be about $1 - 1.5$‰ per trophic level (DeNiro and Epstein 1978, Grice *et al.* 1998a and see also chapter by Smit in this volume). Hence, CO_2 produced from the bacterial oxidation of sinking organic matter is commonly depleted in ^{13}C relative to the inorganic carbon reservoir. Photosynthetic organisms (i.e. phytoplankton) that utilise this inorganic carbon are thus depleted in ^{13}C relative to the primary photosynthate. In the water column, the percentage of carbon which is re-fixed *via* the consumption of respired CO_2 and methane typically increases with stratification resulting in an overall depletion of ^{13}C down the water column. This and other effects are summarised in Figure 5.

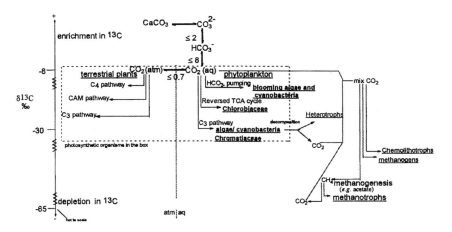

Figure 5. Schematic diagram showing approximate δ^{13}C values of organisms with different mechanisms of carbon dioxide fixation (after de Leeuw *et al.* 1994).

4.2 Specific photosynthetic pathway (and complex associated fractionations)

During photosynthesis, the magnitude of fractionation depends on various processes, including the amount of carbon available, the diffusion rate of CO_2 into the photoautotroph concerned and the specific photosynthetic pathway involved in its carbon fixation. There are 3 main possible pathways, C3, C4, CAM and the reversed TCA cycle (Fig. 5). The second and third of these pathways relate exclusively to higher plants (see chapter by Pate in this volume) so will not be considered here.

4.2.1 ^{13}C discrimination of the C3 pathway of photosynthesis

As a rough estimate 90 % of plants existing today operate using the C3 photosynthetic pathway. These include higher plants (ranging from oak trees to dandelions), phytoplankton including most cyanobacteria, algae, and some anoxygenic bacteria i.e. Chromatiaceae (purple sulfur bacteria). In C3 photosynthesis, CO_2 is fixed by the enzyme ribulose-1,5-biphosphate carboxylase-oxygenase (RuBisCO) into two molecules of the three carbon compound phosphoglyceric acid (PGA). The isotopic discrimination resulting from this enzyme is approximately 29‰ relative to the carbon source (Farquhar *et al.* 1982). It has been proposed that the limitation of carbon for RuBisCO is related to the ratio of internal and external pressures of CO_2 (Farquhar *et al.* 1982, 1989). In addition to the discrimination due to RuBisCO, a fractionation is also associated with the diffusion of CO_2 through the stomata in terrestrial and emergent plants, and is approximately 4.4‰ (Raven and Farquhar 1990). Thus, C3 plants have bulk δ^{13}C values ranging from -37‰ to -24‰ with a mean value *ca.* -27‰ (Deines 1980). For phytoplankton there are two isotope effects associated with the assimilation of CO_2, one causing a depletion of approximately 0.7‰ for diffusion of CO_2 through the aqueous boundary layer and the cell wall (O'Leary 1984), the other a depletion associated with RuBisCO-catalysed carbon fixation (Raven and Farquhar 1990). Microalgae generally have bulk δ^{13}C values in the range of -12‰ to -23‰. At reduced concentrations of CO_2, diffusion becomes rate limiting and the isotopic effect accompanying carbon fixation is accordingly not fully expressed (O'Leary 1984). As mentioned above, some types of phytoplankton utilise bicarbonate as their carbon source when CO_2 concentrations become very low due to high salinity, high temperature and high pH, or under conditions of high productivity such as during phytoplankton blooms.

4.2.2 ^{13}C discrimination associated with the reversed tricarboxylic acid cycle

The reversed tricarboxylic acid cycle involves two ferredoxin-dependent carboxylation reactions catalysed by the enzymes pyruvate synthase and α-oxoglutarate synthase, respectively (Evans and Buchanan 1965, Buchanan and Arnon 1965). This pathway generates one molecule of acetate from two molecules of CO_2 and is used for CO_2 fixation by anoxygenic bacteria of the family Chlorobiaceae and by the chemotrophic archaebacterium *Sulfolobus*. It is an energy-efficient process compared with the C3 pathway, making it a highly effective carbon fixation process in the Chlorobiaceae, living at very low light intensities (Omerod 1983). Indeed, it has been proposed, that *Chlorobium* bacteria lack the RuBisCO enzyme, and use only the reversed TCA cycle altogether (Buchanan *et al.* 1967, Sirevag 1974). This is supported by stable carbon isotope evidence (Sirevag *et al.* 1977) showing that *Chromatium strain D* (purple sulphur bacterium) and *Rhodospirillum rubrum* (non-purple sulphur bacterium) have isotopic discrimination characteristics similar to those of organisms known to use a C3 pathway. Conversely, *Chlorobium thiosulfatophilum* is significantly enriched in ^{13}C as to be expected from it operating the reversed tricarboxylic acid cycle.

4.3 Biosynthesis

As Hayes (1993) pointed out, "In all organisms, the synthesis of biological components is based on cellular pools of intermediates. Isotopic effects associated with metabolic reactions that produce or consume such intermediates affect, the ^{13}C contents of the biosynthetic products. At each point within the cellular reaction network, distribution of carbon among products will affect the isotopic compositions". For example, lipids that are biosynthesised from acetate are generally depleted in ^{13}C compared relative to other products of biosynthesis (i.e. proteins and carbohydrates) (Abelson and Hoering 1961, DeNiro and Epstein 1977). Thus, Abelson and Hoering (1961) demonstrated that plankton grown on glucose had lipid fractions depleted in ^{13}C by about 7‰ relative to the donor carbon of glucose. DeNiro and Epstein (1977) later confirmed this observation and showed that this depletion was related to a kinetic isotope effect, and involved a partitioning of pyruvate to other products of biosynthesis and acetyl-CoA *via* the pyruvate dehydrogenase complex. In keeping with this, compound specific isotope analysis (CSIA) of leaf *n*-alkane lipids isolated from waxes of plants utilising different photosynthetic pathways has shown them to be depleted in ^{13}C by up to 11‰ relative to the total carbon of whole leaf tissue (Rieley *et al.* 1993).

Monson and Hayes (1982) demonstrated that a secondary focus of isotopic fractionation exists during lipid biosynthesis. A kinetic isotope effect of approximately 1.02 was measured at the C-2 carbonyl carbon of acetyl-CoA (Melzer and Schmidt 1987). Furthermore, a few authors have observed a difference in isotopic compositions of *n*-alkyl fatty acids and isoprenoidal lipids (Hayes 1993 and references therein, Collister *et al.* 1994a). This differential is thought to relate to the ratio of carboxyl-derived carbon in fatty acids being 1:1, compared to 2:3 in polyisoprenoids. The building block for the latter would be isopentyl diphosphate (IPP). Condensation of three molecules of acetyl-CoA forms the C_5 isoprene precursor (mevalonic acid) with the loss of one molecule of CO_2 (mevalonate pathway, Fig. 6a). It has been predicted that the CO_2 emitted contains carbon from the carbonyl position of acetyl-CoA, which is in turn depleted in ^{13}C compared to the methyl group. Thus, it has been proposed that *n*-alkyl lipids (fatty acids) are generally depleted in ^{13}C by *ca.* 1.5‰ in comparison to polyisoprenoids (e.g. phytol, sterols) biosynthesised by the same organism (Hayes 1993). However, a later publication has shown that in methanotrophic bacteria, *n*-fatty acids are enriched in ^{13}C compared to polyisoprenoids (Summons *et al.* 1994). A possible explanation for this is that unrecognised biosynthetic fractionations may occur downstream from acetate during bacterial lipid biosynthesis. Recently an alternative means of biosynthesis of IPP has been discovered (pyruvate/glyceraldehyde-3-phosphate pathway (pyr/GAP pathway) Fig. 6b). This pathway occurs in prokaryotes (Rohmer *et al.* 1993, 1996), eukaryotes (Schwender *et al.* 1996) and higher plants (Lichtenthaler *et al.* 1997). If the pyr/GAP pathway is operational, then it would be expected that the polyisoprenoids would become enriched in ^{13}C compared to fatty acids since only one of the five carbons in IPP is derived from a carboxyl group likely to be depleted in ^{13}C. Another complication is that two pools of IPP can be present in the one organism. For example, it has been shown that in certain higher plants, isoprenoid phytol is biosynthesised in the chloroplast by the pyr/ GAP pathway, whereas sterols are biosynthesised in the cytosol by the mevalonate pathway (Lichtenthaler *et al.* 1997). Recent CSIA of lipids in thirteen different species of microalgae have shown that in general, the ^{13}C of phytol is consistently enriched by about 2 – 5‰ compared to that of the the C_{16} fatty acid (Schouten *et al.* 1998). However, sterols are variously found enriched in ^{13}C by 0‰ – 8‰ compared to the C_{16} fatty acid in all algae, a difference attributed to a different pool of IPP in the cytosol. Thus, the pathway used for IPP biosynthesis and the location of its biosynthesis is likely to have a profound effect on the carbon isotopic compositions of polyisoprenoids (sterols, phytol) biosynthesised by the same organism.

CSIA of lipids in the green sulfur bacterium, *Chlorobium limicola*, and in the purple sulfur bacterium, *Thiocapsca roseopersicina* have recently been investigated in order to examine the possible effects of the reversed TCA cycle on carbon isotopic compositions of lipids of these organisms (van der Meer *et al.* 1998). It was shown that lipids produced by the reversed TCA cycle were enriched in ^{13}C relative to biomass by 2 – 4‰, whereas isoprenoid lipids were depleted by 7 – 9‰ relative to straight chain lipids. This is opposite to the effect obtaining in organisms using RuBisCO to fix

CO_2 and comes as a consequence of the reversed sequence of fractionation effects mediated by the reversed TCA cycle. These trends were consistent with a partial use of this cycle in both green and purple sulfur bacteria.

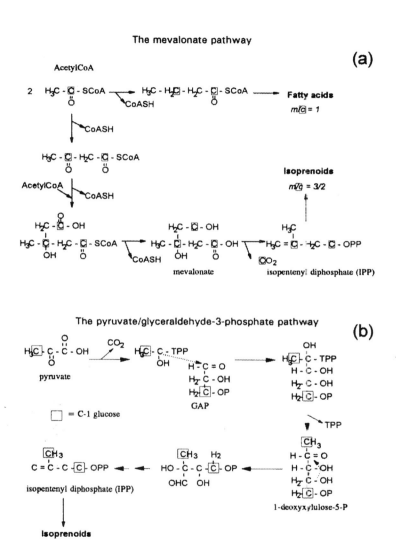

Figure 6. Biosynthesis of different fatty acids and isoprenoids in eukaryotes (Schouten *et al.* 1998) (a) biosynthesis of fatty acids from acetyl coenzyme A and biosynthesis of isoprenoids *via* IPP according to classical mevalonate pathway, and (b) biosynthesis of isoprenoids *via* IPP according to Pyr/GAP pathway (after Lichtenthaler *et al.* 1997), m/c = methyl to carboxyl ratio. Boxed areas in (a) indicate carboxyl carbon atoms. Boxed areas in (b) indicate carbon atoms derived from the C-1 carbon atom of glucose (from Schouten *et al.* 1998).

Two independent biosynthetic routes are supposedly involved in formation of the chlorophyll macrocycle and the lipid phytyl side chain (Galimov 1977, Madigan *et al.* 1989), although there is little fundamental data available on the primary relationship between carbon assimilation and chlorophyll biosynthesis in either eukaroytes or photosynthetic bacteria (Kennicutt *et al.* 1992). Individual amino acids do vary in their carbon and nitrogen isotopic contents, suggesting that isotopic compositions of chlorophylls might also vary appreciably between and within different organisms. The isotopic composition of amino acids has been investigated by Abelson and Hoering (1961), who suggested that amino acids isolated from total hydrolysates of protein fractions of an organism may vary widely in isotopic composition. However, certain consistent features were evident, for example the carboxyl carbon in amino acids proved to be significantly enriched to levels close to that of source CO_2.

Much further work on CSIA of lipids across a range of extant organisms, other than eukaryotes is obviously required to unravel isotopic fractionation effects occurring in lipid biosynthesis. Further complications would also need to be resolved when one considers the extent to which appreciable carbon isotopic fractionations are likely to occur during carbon acquisition or modified in response to factors such as growth rate (Laws *et al.* 1995, Bidigare *et al.* 1997), cell size/cell geometry (Goericke *et al.* 1994, Popp *et al.* 1998) and concentration and carbon isotopic composition of the carbon source (Goericke *et al.* 1994).

5.   CONCLUSIONS

It should be clear from this chapter that molecular carbon isotopic signatures of OM, determined using irm-GCMS, are likely to provide substantially more useful information on palaeoenvironments than one might gain simply from $\delta^{13}C$ analyses conducted on bulk biomass. Since the advent of irm-GCMS technology a wealth of information has been accumulated in the field of organic geochemistry, as illustrated by the numerous examples discussed in this chapter where CSIA of biomarkers derived from specific biochemicals has been employed as tracers for reconstructing the nature of ancient environments of deposition. A vast potential clearly now exists for application of irm-GCMS analysis to a multiple of environmental, geochemical, ecological or biochemical research areas. However, more CSIA data derived from contemporary organisms and recent sediments is still required to provide enough definitive information to provide more accurate interpretations of palaeoenvironments. Accordingly further application of irm-GCMS approaches to understanding the cycling of carbon (and even hydrogen), and to resolving metabolic relationships between compounds in living and extinct organisms will clearly be important to the success of future research. In particular a wealth of information has still to be gained at the molecular level towards a better understanding of the complex processes involved in relation to the intimate workings of the biogeochemical carbon cycles and the manner and extent of isotopic fractionations, discriminations and segregations which accompany the reactions apparently involved.

APPENDIX

1

2

3

4

5

6

R₁ = *i*-Bu
R₂ = Me

7

8

9

10

11

12

13

14

15

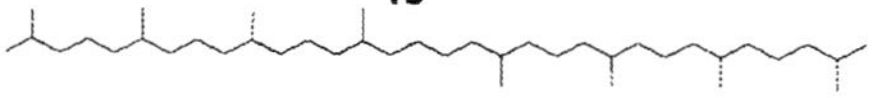

16

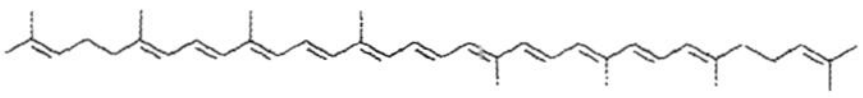

18

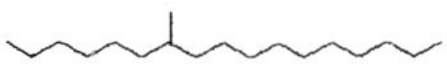

17

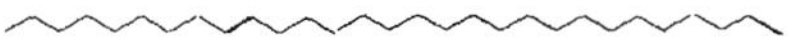

19

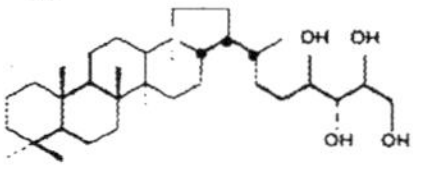

20

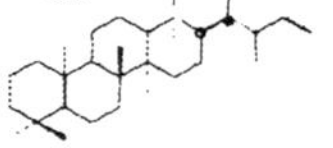

21

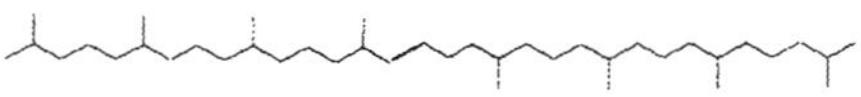

23

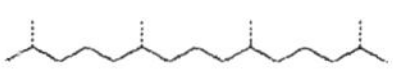

22

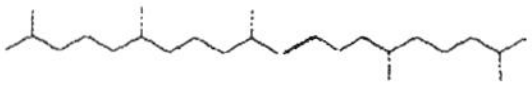

25

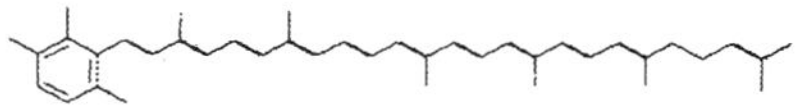

24

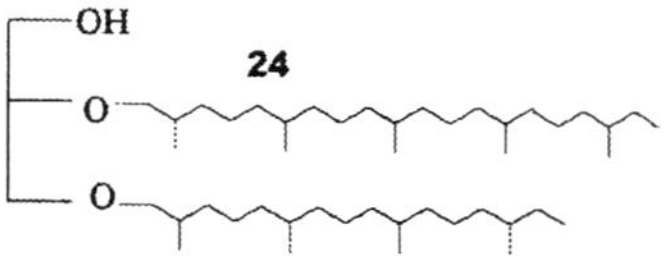

26

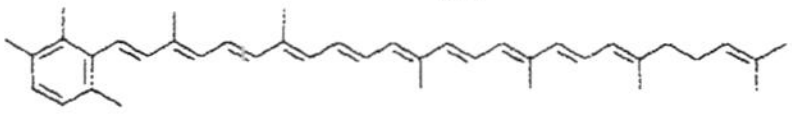

27

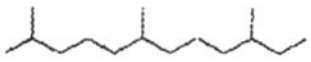

28

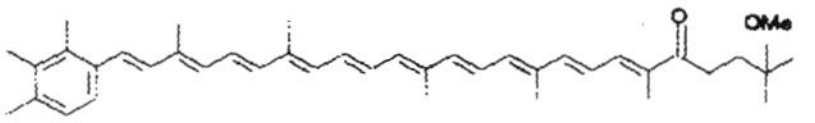

31

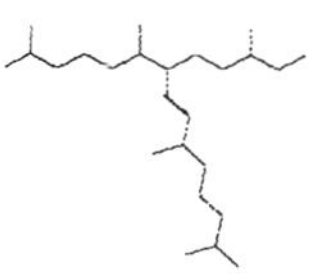

30

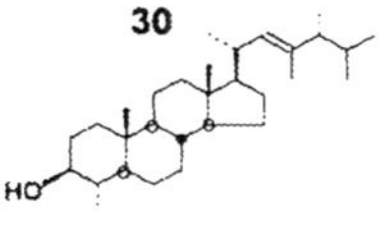

29

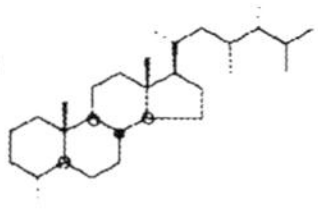

32

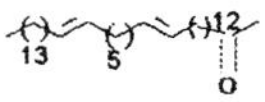

33

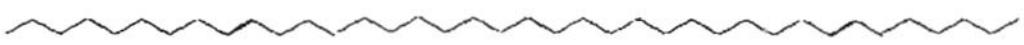

34

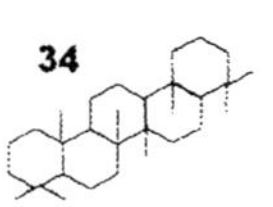

35

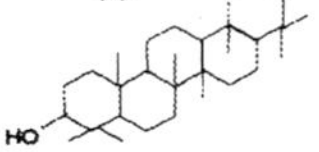

36

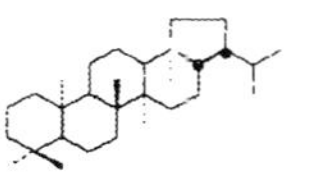

37

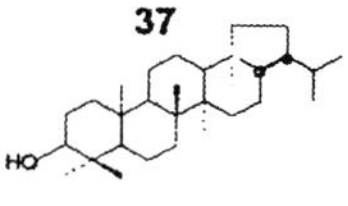

REFERENCES

Abelson, P.H., and Hoering, T.C. (1961) Carbon isotope fractionation in formation of amino acids by photosynthetic organisms. *Proceedings of National Academy of Science* 47, 623-632.

Albaiges, J. (1980) Identification and geochemical significance of long chain acyclic isoprenoid hydrocarbons in crude oils. In: *Advances of Organic Geochemistry* 1979, eds. A.G. Douglas and J.R. Maxwell, pp 19-28. Pergamon.

Armanios, C., Alexander, R., Kagi, R.I., and Sosrowidjojo, I. (1993) Molecular sieving of higher plant derived pentacyclics from crude oil. *In:* Oygard K (ed), Organic Geochemistry, Falch, Hurtigtrykk, Norway, 745-746.

Bidigare, R.R., Fluegge, A., Freeman, K.H., Hanson, K.L., Hayes, J.M., Hollander, D., Jasper, J.P., King, L.L., Laws, E.A., Milder, J., Millero, F.J., Pancost, R., Popp, B.N., Steinberg, P.A., and Wakeham, S.G. (1997) Consistent fractionation of ^{13}C in nature and in the laboratory: Growth-rate effects in some haptophyte algae, *Global Biogeochemical Cycles* 11, 279-292.

Bird, C.W., Lynch, J.M., Pirt, F.J., Reid, W.W., Brooks, C.J.W., and Middleditch, B.S. (1971) Steroids and squalene in *Methylococcus capsulatus* grown on methane. *Nature* 230, 473-474.

Boreham, C.J., Summons, R.E., Roksandic, Z., Dowling, L.M., and Hutton, A.C. (1994) Chemical, molecular and isotopic differentiation of organic facies in the Tertiary lacustrine Duaringa oil shale deposit. *Organic Geochemistry* 21, 685-712.

Brand, W.A., Tegtmeyer, A.R., and Hilkert, A. (1994) Compound specific isotope analysis: Extending toward $^{15}N/^{14}N$ and $^{18}O/^{16}O$. *Organic Geochemistry* 21, 585-594.

Brassell, S.C., Wardroper, A.M.K., Thompson, I.D., Maxwell, J.R., and Eglinton, G. (1981) Specific acyclic isoprenoids as biological markers of methanogenic bacteria in marine sediments. *Nature* 290, 693-696.

Brassell, S.C., Eglinton, G., Marlowe, I.T., Pflaumann, U., and Sarnthein, M. (1986) Molecular stratigraphy: a new tool for climatic assessment, *Nature* 320, 129-133.

Buchanan, B.B., and Arnon, D.I. (1965) Ferredoxin-dependent synthesis of labelled pyruvate from labelled acetyl-coenzyme-A and carbon dioxide. *Biochemistry and Biophysics Research Communications* 20, 163.

Buchanan, B.B., Evans, M.C.W., and Arnon D.I. (1967) Ferredoxin-dependent carbon assimilation in *Rhodospirillum rubrum*. *Archiv. Mikrobiology* 59, 32-40.

Burgoyne, T.W., and Hayes, J.M. (1998) Quantitative production of oxygen and hydrogen isotope data. *Chemical Geology* 72, 293-297.

Collister, J.W., Summons, R.E., Lichtfouse E., and Hayes, J.M. (1992) An isotopic biogeochemical study of the Green River oil shale. *Organic Geochemistry* 19, 265-276.

Collister, J. W., Rieley, G., Stern, B., Eglinton, G., and Fry, B. (1994a) Compound- specific $\delta^{13}C$ analyses of leaf lipids from plants with different carbon dioxide metabolisms. *Organic Geochemistry* 21, 619-627.

Collister, J.W., Lichtfouse, E., Hieshima, G., and Hayes, J.M. (1994b) Partial resolution of *n*-alkanes in the saline portion of the Parachute Creek Member, Green River Formation (Piceance Creek Basin, Colorado). *Organic Geochemistry* 21, 645-659.

David, M., Metzger, P., and Casadevall, E. (1988) Two cyclobotryococcenes from the B race of the green alga *Botryococcus braunii*. *Phytochemistry* 27, 2863-2867.

Deines, P. (1980) The terrestrial environment. *In:* Fritz P. and Fontes J.C. (*eds.*) Handbook of Environmental Geochemistry. 1, pp. 329-406. Elsevier, Amsterdam.

de Leeuw, J.W., van der Meer, F.W., Rijpstra, W.I.C., and Schenck, P.A. (1980) On the occurrence and structural identification of long chain unsaturated ketones and hydrocarbons in sediments. In *Advances in Organic Geochemistry* 1979 (edited by A.G. Douglas and J.R. Maxwell). pp. 211- 217. Pergamon Press, Oxford.

de Leeuw, J.W., Frewin, N.L., van Bergen, P.F., Sinninghe Damstè, J.S., and Collinson, M.E. (1994) Organic carbon as a palaeoenvironmental indicator in the marine realm. *In*: Bosence D. and Allison P. (*eds.*). Marine palaeoenvironmental analysis from fossils *Geol. Soc. Special Publ.* Blackwell Scientific Publications, Oxford.

DeNiro, M.J., and Epstein, S. (1977) Mechanism of carbon isotope fractionation associated with lipid synthesis. *Science* 197, 261-263.

DeNiro, M.J., and Epstein, S. (1978) Influence of diet on the distribution of nitrogen isotopes in animals. *Geochimica et Cosmochimica Acta* 42, 495-506.

De Rosa M., Gambacorta, A., Nicolaus, B., Ross, H.N.M., Grant, W.D., and Bu'Lock, J.D. (1982) An asymmetric archaebacterial diether lipid from alkaliphilic halophiles. *Journal of General Microbiology* 128, 343-348.

Dowling, L.M., Boreham, C.J., Hope, J.M., Murray, A.P., and Summons, R.E. (1995) Carbon isotopic composition of hydrocarbons in ocean-transported bitumens from the coastline of Australia. *Organic Geochemistry* 23, 729-737.

Eglinton, G., and Hamilton, R.J. (1963) The distribution of *n*-alkanes. In *Chemical Plant Taxonomy*. (edited by T. Swain) Academic Press, London. pp 187-217.

Eglinton, G., Bradshaw, S.A., Rosell, A., Sarnthein, M., Pflaumann, U., and Tiedemann, R, (1992) Molecular record of secular sea surface temperature changes on 100-year timescales for glacial terminations, I, II and IV, *Nature* 356, 423-426.

Ellis, L., Alexander, R.., and Kagi, R I. (1994) Separation of petroleum hydrocarbons using dealuminated mordenite molecular sieve-II. Alkylnaphthalenes and alkylphenanthrenes. *Organic Geochemistry* 21, 849-855.

Ehleringer, J.R., Hall, A.E., and Farquhar, G.D. (1993) Stable isotopes and plant carbon-water relations. Academic Press, San Diego, California

Evans, M.C.W., and Buchanan, B.B. (1965) Photoreduction of ferredoxin and its use in carbon dioxide fixation by a subcellular system from a photosynthetic bacterium. *Biochemistry* 53, 1420-1425.

Evans, M.C.W., Buchanan, B.B., and Arnon, D.I. (1966) A new ferredoxin-dependent carbon reduction cycle in a photosynthetic bacterium. *Biochemistry* 55, 928-934.

Falkowski, P.G. (1991) Species variability in the fractionation of ¹³C and ¹²C by marine phytoplankton. *Journal of Plankton Research Supplement.* 13, 21-28.

Farquhar, G.D., O'Leary, M.H., and Berry, J.A. (1982) On the relationship between carbon isotope discrimination and intercellular carbon dioxide concentration in leaves. *Australian Journal of Plant Physiology* 9, 121-137.

Farquhar, G.D., Ehleringer, J.R., and Hubick, K.T. (1989) Carbon isotope discrimination and photosynthesis. *Annual Review of Plant Physiology and Molecular Biology* 40, 503-537.

Freeman, K.H., Hayes, J.M., Trendel, J.M., and Albrecht, P. (1990) Evidence from carbon isotope measurements for diverse origins of sedimentary hydrocarbons. *Nature* 343, 254-256.

Freeman, K.,H., Wakeham, S.G., and Hayes, J.M. (1994) Predictive isotope biogeochemistry: hydrocarbons from anoxic marine basins. *Organic Geochemistry*, 21, 629-644.

Galimov, E.M. (1977) Investigation of the distribution of carbon isotopes in biogenic compounds. *In*: Proc. VIII International. Congress on Organic Geochemistry. Nauka. Moscow.

Games, L.M., and Hayes, J.M. (1978) Methane-producing bacteria: Natural fractionations of the stable carbon isotopes. *Geochimica et Cosmochimica Acta* 42, 1295-1297.

Goericke, R., Montoya, J.P., and Fry, B. (1994) Physiology of isotopic fractionation in algae and cyanobacteria. In *Stable Isotopes in Ecology and Environmental Science,* eds. K. Lajtha and R. H. Micener, pp 187-221. Blackwell Scientific Publications, Oxford.

Grice, K. (1995) Distributions and Stable Carbon Isotopic Compositions of Individual Biological Markers from the Permian Kupferschiefer (Lower Rhine Basin, N.W. Germany). PhD Thesis. University of Bristol, UK.

Grice, K., Gibbison, R., Atkinson, J.E., Eckardt, C.B., Schwark, L., and Maxwell, J.R. (1996a) 1*H*-Pyrrole-2,5-diones (maleimides) as indicators of anoxygenic photosynthesis in palaeowater columns. *Geochimica et Cosmochimica Acta.* 60, 3913-3924.

Grice, K., Schaeffer, P., Schwark, L., and Maxwell, J.R. (1996b) Molecular indicators of palaeoenvironmental conditions in an immature Permian shale (Kupferschiefer, Lower Rhine Basin, N. W. Germany) from free and sulfide-bound lipids. *Organic Geochemistry* 25, 131-147.

Grice, K., Schaeffer, P., Schwark, L., and Maxwell, J.R. (1997) Changes in palaeoenvironmental conditions during deposition of the Permian Kupferschiefer (Lower Rhine Basin, N.W. Germany) inferred from molecular and isotopic compositions of biomarker components. *Organic Geochemistry* 26, 677-690.

Grice, K., Klein Breteler, W.C.M., Schouten, S., Grossi, V., de Leeuw, J.W., and Sinninghe Damsté, J. S. (1998a) The effects of zooplankton herbivory on biomarker proxy records. *Paleoceanography* 13, 686-693.

Grice, K., Schouten, S., Nissenbaum, A., Charach, J., and Sinninghe-Damsté, J.S. (1998b) A remarkable paradox: freshwater algal (*Botryococcus braunii*) lipids in an ancient hypersaline euxinic ecosystem. *Organic Geochemistry* 28, 195-216.

Grice, K., Schouten, S., Nissenbaum, A., Charach, J., and Sinninghe-Damsté, J.S. (1998c) Isotopically heavy carbon in the C_{21} to C_{25} regular isoprenoids in halite-rich deposits from the Sdom Formation, Dead Sea Basin, Israel. *Organic Geochemistry* 28, 349-359.

Grice, K., Schouten, S., Peters, K.E., and Sinninghe Damsté, J.S. (1998d) Molecular isotopic characterisation of hydrocarbon biomarkers in Palaeocene-Eocene evaporitic, lacustrine source rocks from the Jianghan Basin, China. In *Advances of Organic Geochemistry. 18th International Meeting on Organic Geochemistry Organic Geochemistry* 29, 1745-1764.

Hartgers, W.A., Sinninghe Damsté, J.S., and de Leeuw, J.W. (1994a) A molecular and carbon isotopic study towards the origin and diagenetic fate of diaromatic carotenoids. *Organic Geochemistry* 22, 703-725.

Hartgers, W.A., Sinninghe Damsté, J.S., Requejo, A.G., Allan, J., Hayes, J.M., and de Leeuw, J.W. (1994b) Evidence for only minor contributions from bacteria to sedimentary organic carbon. *Nature* 369, 224-227.

Harvey, H.R., and McManus, G.B. (1991) Marine ciliates as a widespread source of tetrahymanol and hopan-3-β-ol in sediments. *Geochimica et Cosmochimica Acta* 55, 3387-3390.

ten Haven, H.L., de Leeuw, J.W., Sinninghe Damsté, J.S., Schenk, P.A., Palmer, S.E., and Zumberge, J.E. (1988) Application of biological markers in the recognition of palaeohypersaline environments. In: *Lacustrine Petroleum Source Rocks*, eds. A.J. Fleet, K. Kelts and M.R. Talbot, pp 123-130. 40, Geological Society special publication.

ten Haven, H.L., Rohmer, M., Rulkötter, J., and Bisseret, P. (1989) Tetrahymanol, the most likely precursor of gammacerane, occurs ubiquitously in marine sediments. *Geochimica et Cosmochimica Acta* 53, 3073-3079.

Hayes, J.M. (1983) *In*: Meinschein W.G. (*ed.*) Organic Geochemistry of Contemporaneous and Ancient Sediments, Great Lakes Section, Society of Economic Palaeontologists and Mineralogists, Bloomington, Indiana, pp. 5-31.

Hayes, J. M., Freeman, K.H., Popp, B.N., and Hoham, C.H. (1990) Compound-specific isotopic analyses: A novel tool for reconstruction of ancient biogeochemical processes. *Organic Geochemistry* 16, 1115-1128.

Hayes, J.M. (1993) Factors controlling ¹³C contents of sedimentary organic compounds: Principles and evidence. *Marine Geology* 113, 111-125.

Hoefs, M.E.L., Schouten, S., King, L.L., Wakeham, S.G., de Leeuw, J.W., and Sinninghe Damsté, J.S. (1997) Ether lipids of pelagic archeae in marine water columns. *Applied Environmental Microbiology* 63, 3090-3095.

Hollander, D.J., Sinninghe Damsté, J.S., Hayes, J.M., de Leeuw, J.W., and Huc A.Y. (1993) Molecular and bulk isotopic analyses of organic matter in marls of the Mulhouse Basin (Tertiary, Alsace, France). *Organic Geochemistry* 20, 1253-1262.

Huang, Z., Poulter, C.D., Wolf, F.R., Somers, T.C., and White, J.D. (1988) Braunicene, a novel cyclic C_{32} isoprenoid from *Botryococcus braunii. Journal of American Chemical Society* 110, 2959-3964.

Huang, Y., Murray, M., and Eglinton, G. (1995) Sacredicene, a novel monocyclic C_{33} hydrocarbon from sediment of Sacred Lake, a tropical freshwater lake, Mount Kenya. *Tetrahedron Letters* 36, 5973-5976.

Huang, Y., Street-Perrott, F.A., Perrott, R.A., Metzger P., and Eglinton, G. (1999) Glacial-interglacial environmental changes inferred from molecular and compound-specific δ¹³C analyses of sediments from Sacred Lake, Mt. Kenya. *Geochimica et Cosmochimica Acta* 63, 1383-1404.

Jasper, J.P., and Hayes, J.M. (1990) A carbon isotope record of CO_2 levels during the late Quaternary. *Nature* 347, 462-464.

Jasper, J.P., Hayes, J.M., Mix, A.C., and Prahl, F.G. (1994) Photosynthetic fractionation of ¹³C and concentrations of CO_2 in central equatorial Pacific during the last 225,000 years, *Paleoceanography* 9, 781-898..

Jiamo Fu., Sheng Guoying., and Dehan, L. (1988) Organic geochemical characteristics of major types of terrestrial petroleum source rocks in China. In *Lacustrine Petroleum Source rocks* (edited by A.J. Fleet, K. Kelts and M.R. Talbot) Geological Society Special Publication 40, pp 279-289.

Keely, B.J., Sinninghe Damsté, J.S., Betts, S.E., Yue, L., de Leeuw, J.W., and Maxwell, J.R. (1993) A molecular stratigraphic approach to palaeoenvironmental assessment and the recognition of changes in source inputs in marls of the Mulhouse Basin (Alsace, France) *Organic Geochemistry* 20, 1165-1187.

Kemp, P., Lander, D.J., and Orpin, C.G. (1984) The lipids of the rumen fungus *Piromonas communis. Journal of General Microbiology* 130, 27-37.

Kennicutt, M.C. (II), Bidigare, R.R., Macko, S.A., and Keeney-Kennicutt, W.L. (1992) The stable isotopic composition of photosynthetic pigments and related biomolecules. *Chemical Geology* 11, 235-245.

Kleeman, G., Poralla, K., Englert, G., Kjosen, H., Liaaen-Jensen, N., Neunlist, S., and Rohmer, M. (1990) Tetrahymanol from the phototrophic bacterium *Rhodopseudonomas palustris:* First report of a gammacerane triterpane from a prokaryote. *Journal of General Microbiology* 136, 2551-2553.

Kuypers, M.M.M., Pancost, R.D., and Sinninghe Damsté, J.S. (1999) A large abrupt fall in atmospheric CO_2 concentration during cretaceous times. *Nature* 399, 342-345.

Laws, E.A., Popp, B.N., Bigidare, R.R., Kennicutt, M.C., and Macko, S.A. (1995) Dependence of phytoplankton isotopic composition on growth rate and $[CO_2]_{aq}$: Theoretical considerations and experimental results. *Geochimica et Cosmochimica Acta* 59, 1131-1138.

Lichtenthaler, H.K., Schwender, J., Disch, A., and Rohmer, M. (1997) Biosynthesis of isoprenoids in higher plants chloroplasts proceeds *via* a mevalonate pathway. *FEBS Letters*. 400, 271-274.

Madigan, M.T., Takigiku, R., Lee, R.G., Gest, H., and Hayes, J.M. (1989) Carbon isotope fractionation by thermophilic phototrophic sulphur bacteria: Evidence for autotrophic growth in natural populations. *Applied and Environmental Microbiology* 55, 639-644.

Marlowe, I.T., Brassell S.C., Eglinton, G., and Green, J.C. (1990) Long-chain alkenones and alkyl alkenoates and the fossil coccolith record of marine sediments, *Chemical Geology* 88, 349-375.

Matthews, D.E., and Hayes, J.M. (1978) Isotope-ratio monitoring gas chromatography-mass spectrometry. *Analytical Chemistry* 50, 1465-1473.

Maxwell, J.R., Douglas, A.G., Eglinton, G., and McCormick, A. (1968) The botryococcenes-hydrocarbons of novel structure from the alga *Botryococcus braunii Kützing*. *Phytochemistry* 7, 2157-2171.

Melzer, E., and Schmidt, H.L. (1987). Carbon isotope effects on the pyruvate dehydrogenase reaction and their importance relative to the ^{13}C depletion in lipids. *Journal of Biological Chemistry* 262, 8159-8164.

Merrit, D.A. (1993) Continuous and highly precise analysis of stable isotopes of carbon and nitrogen in gas chromatographic effluents. PhD Thesis. Indiana University, Bloomington.

Metzger, P., Casadevall, E., Pouet, M.J., and Pouet, Y. (1985) Structures of some botryococcenes: branched hydrocarbons from the B-race of the green alga *Botryococcus braunii*. *Phytochemistry* 24, 2995-3002.

Metzger, P., Templier, J., Casadevall, E., and Couté, A. (1986) A *n*-alkatriene and some *n*-alkadienes from the A race of the green alga *Botryococcus braunii*. *Phytochemistry* 25, 1869.

Metzger, P., and Casadevall, E. (1987) Lycopadiene, a tetraterpenoid hydrocarbon from new strains of the green alga *Botryococcus braunii*. *Tetrahedron Letters* 28, 3931-3934.

Metzger, P., Largeau, C., and Casadevall, E. (1991) Lipids and macromolecular lipids of the hydrocarbon-rich microalga *Botryococcus braunii*. Chemical structure and biosynthesis. Geochemical and biotechnological importance. In: *Progress in the Chemistry of Organic Natural Products*, eds. W. Herz *et al.* pp. 1-70. 57, Springer Verlag.

Monson, K.D., and Hayes, J.M. (1982) Carbon isotopic fractionation in the biosynthesis of bacterial fatty acids. Ozonolysis of unsaturated fatty acids as a means of determining the intramolecular distribution of carbon isotopes. *Geochimica et Cosmochimica Acta* 46, 139-149.

Mook, W.G., Bommerson, J.C., and Staverman, W.H. (1974) Carbon isotope fractionation between dissolved bicarbonate and gaseous carbon dioxide. *Earth Planetary Science Letters* 22, 169-176.

Nichols, P.D., Volkman, J.K., Palmisano, A.C., Smith, G.A., and White D.C. (1988) Occurrence of an isoprenoid C_{25} di-unsaturated alkene and high neutral lipid content in Antarctic sea-ice diatom communities. *Journal of Phycology* 24, 90-96.

O'Leary, M.H. (1984) Measurement of the isotope fractionation associated with diffusion of carbon dioxide in aqueous solution. *Journal of Physical Chemistry* 88, 823-825.

Omerod, J.G. (1983) The phototrophic bacteria: Anaerobic life in the light. *In:* Omerod J.G. (*ed.*) Studies in Microbiology. 4. Blackwell Scientific publications.

Oren, A. (1993) Availability, uptake and turnover of glycerol in hypersaline environment. *FEMS Microbiology Ecology* 12, 15-23.

Popp, B.N., Laws, E. A., Bidigare, R.R., Dore, J.E., Hanson, K.L., and Wakeham, S.G. (1998) Effect of phytoplankton cell geometry on carbon isotopic fractionation. *Geochimica et Cosmochimica Acta* 62, 69-79.

Prahl, F.G., and Wakeham, S.G. (1987) Calibration of unsaturation patterns in long-chain ketone compositions for palaeotemperature assessment, *Nature* 330, 367-369.

Prahl, F.G., Muelhausen, L.A., and Zahnle, D.A. (1988) Further evaluation of long-chain alkenones as indicators of paleoceanographic conditions, *Geochimica et Cosmochimica Acta* 52, 2303-2310.

Prahl, F.G., Pisias, N., Sparrow, M.A., and Sabin, A. (1995) Assessment of sea-surface temperature at 42°N in the Californian Current over the last 30,000 years, *Paleoceanography* 10, 763-773.

Quandt, L., Gottschalk, G., Ziegler, H., and Stichler, W. (1977) Isotope discrimination by photosynthetic bacteria. *FEMS Microbiology Letters* 1, 125-128.

Raven, J.A., and Farquhar, G.D. (1990) The influence of N metabolism and organic acid synthesis on the natural abundance of isotopes of carbon plants. *New Phytology* 116, 505-529.

Rechka, J.A., and Maxwell, J.R. (1988a) Characterisation of alkenone temperature indicators in sediments and organisms. In *Advances in Organic Geochemistry* 1987 (edited by L. and L. Novelli). *Organic Geochemistry* 13, 727-734.

Rechka, J.A., and Maxwell, J.R. (1988b) Unusual long chain ketones of algal origin. *Tetrahedron Letters* 29, 2599-2600.

Rieley, G., Collister, J.W., Stern, B., and Eglinton, G. (1993) Gas chromatography/ isotope ratio mass spectrometry of leaf wax *n*-alkanes from plants of differing carbon dioxide metabolisms. *Rapid Communications in Mass Spectrometry* 7, 488-491.

Robson, J.N., and Rowland, S.J. (1986) Identification of novel widely distributed sedimentary acyclic sesterpenoids. *Nature*, 324, 561-563.

Rohmer, M., Knani, M., Simonin, P., Sutter, B., and Sahm, H. (1993) Isoprenoid biosynthesis in bacteria: A novel pathway for the early steps leading to isopentyl diphosphate. *Biochemistry* 295, 517-524.

Schaeffer, P., Adam, P., Werung, P., Bernasconi, S., and Albrecht, P. (1997) Molecular and isotopic investigation of free and S-bound lipids from an actual meromictic lake (Lake Cadagno, Switzerland). In *Advances of Organic Geochemistry. 18th International Meeting on Organic Geochemistry*. 57-58.

Schmidt K. (1978) Biosynthesis of carotenoids. *In*: Clayton R.K.. and Sistrom W.R. (*eds.*) Photosynthetic bacteria. Plenum Press, New York. 729-750.

Schoell, M., Schouten, S., Sinninghe-Damsté, J.S., de Leeuw, J.W., and Summons, R.E. (1994a) A molecular organic carbon isotope record of Miocene climate changes. *Science* 263, 1122-1125.

Schoell, M., Hwang, R.J., Carlson, R.M.K., and Whelan, J.E. (1994b) Carbon isotopic composition of individual biomarkers in gilsonites (Utah). In *Compound Specific Analysis in Biogeochemistry and Petroleum Research* (edited by M. Schoell and J.M. Hayes) *Organic Geochemistry* 21, 673-683.

Schouten, S., Schoell, M., Rijpstra, W.I.C., Sinninghe Damsté, J.S., and de Leeuw, J.W. (1997a) A molecular stable carbon isotope study of organic matter from the Miocene Monterey Formation (Pismo Basin). *Geochimica et Cosmochimica Acta* 61, 2065-2082.

Schouten, S., van der Maarel, M.J.E.C., Huber, R., and Sinninghe Damsté, J.S. (1997b) 2,6,10,15,19-Pentamethylicosenes in *Methanolobus bombayensis*, a marine methanogenic archaeon, and in *Methanosarcina mazei. Organic Geochemistry*, 26, 409-414.

Schouten, S., Klein Breteler, W.C.M., Blokker, P., Schogt, N., Rijpstra, W.I.C., Grice, K., Baas M., and Sinninghe Damsté, J.S. (1998) Biosynthetic effects on the stable carbon isotopic compositions of algal lipids: Implications for deciphering the carbon isotopic biomarker record. *Geochimica et Cosmochimica Acta* 62, 1397-1406.

Schwender, J., Seemann, M., Lichtenthaler, H.K., and Rohmer, M. (1996) Biosynthesis of isoprenoids (carotenoids, sterols, prenyl side-chains of chlorophylls and plastoquinone) *via* a novel pyruvate/glyceraldehyde 3-phosphate non-mevalonate pathway in the green algae Scenedesmus obliquus. *Biochemistry* 316, 73-80.

Sessions, A.L., Burgoyne, T.W., Schimmelmann, A., and Hayes, J.M. (1999) Organic Geochemistry, In Press.

Sharkey, T.D., and Berry, J.A. (1985) Carbon isotopic fractionation of algae as influenced by an inducible CO_2 concentrating mechanism. *In*: Lucas W.J. and Berry J.A. (*eds.*) Inorganic uptake by aquatic photosynthetic organism. American Society of Plant Physiology, Rockville, MD, 389-402.

Sinninghe Damsté, J.S., Rijpstra, W.I.C., Kock-Van Dalen A.C., de Leeuw, J.W., and Schenk P. (1989) Quenching of labile functionalised lipids by inorganic sulfur species. Evidence for the formation of sedimentary organic sulfur compounds at the early stage of diagenesis. *Geochimica et Cosmochimica Acta* 58, 1343-1355.

Sinninghe Damsté, J.S., Betts, S., Ling, Y., Hofmann, P.M., and de Leeuw, J.W. (1993a) Hydrocarbon biomarkers of different lithofacies of the Salt IV Formation of the Mulhouse Basin, France. *Organic Geochemistry* 20, 1187-1200.

Sinninghe Damsté, J.S., Wakeham, S.G., Kohnen, M.E.L., Hayes, J.M., and de Leeuw, J.W. (1993b). A 6,000-year sedimentary molecular record of chemocline excursions in the Black Sea. *Nature* 362, 827-829.

Sinninghe Damsté, J.S., Kenig, F., Koopmans, M., Köster, J., Schouten, S., Hayes, J.M., and de Leeuw, J.W. (1995) Evidence for gammacerane as an indicator of water column stratification. *Geochimica et Cosmochimica Acta*. 59, 1895-1900.

Sirevag, R. (1974) Further studies on carbon dioxide fractionation in *Chlorobium*. *Archives of Microbiology* 98, 3-18.

Sirevag, R., Buchanan, B.B., Berry, J.A., and Troughton, J.H. (1977) Mechanisms of CO_2 fixation in bacterial photosynthesis studied by the carbon isotope fractionation technique. *Archives of Microbiology* 112, 35-38.

Spooner, N., Rieley, G., Collister, J.W., Lander, M., Cranwell, P.A., and Maxwell J.R. (1994) Stable carbon isotopic correlation of individual biolipids in aquatic organisms and a lake bottom sediment. *Organic Geochemistry* 21, 823-827.

Summons, R.E., and Powell, T.G. (1986) *Chlorobiaceae* in Palaeozoic seas revealed by biological markers, isotopes and geology. *Nature* 319, 763-765.

Summons, R.E., Volkman, J.K., and Boreham, C.J. (1987a) Dinosterane and other steroidal hydrocarbons of dinoflagellate origin in sediments and petroleum. *Geochimica et Cosmochimica Acta* 51, 557-566.

Summons, R.E., and Powell, T.G. (1987b) Identification of aryl-isoprenoids in source rocks and crude oils: Biological markers for the green sulfur bacteria. *Geochimica et Cosmochimica Acta* 51, 557-566.

Summons, R.E., Thomas, J., Maxwell, J.R., and Boreham, C.J. (1992) Secular and environmental constraints on the occurrence of dinosterane in sediments. *Geochimica et Cosmochimica Acta* 56, 2437-2444.

Summons, R.E., Barrow, R.A., Capon, R.J., Hope, J.M., and Stranger, C. (1993) The structure of a new C_{25} isoprenoid alkane biomarker from diatomaceous microbial communities. *Australian Journal of Chemistry* 46, 907-915.

Summons, R.E., Jahnke, L.L., and Roksandic, Z. (1994). Carbon isotopic fractionation in lipids from methanotrophic bacteria: Relevance for interpretation of the geochemical record of biomarkers. *Geochimica et Cosmochimica Acta* 58, 2853-2863.

Summons, R.E., Hope, J.M., Dowling, L.M., Jahnke, L.L., Largeau, C., and Metzger, P. (1996) Carbon isotope fractionation in lipid biosynthesis by algae and cyanobacteria. In:

Organic Geochemistry: Developments and Applications to Energy, Climate, Environment and Human History, eds. J.O. Grimalt and C. Dorronsoro, pp.6-7. A.I.G.O.A. San Sebastian, Spain.

Summons, R.E., Jahnke, L.L., Hope, J.M., and Logan, G.A. (1998) Geoplipids from cyanobacteria: New studies of old rocks and hot spring microbial mats from Yellowstone National Park. Australian Organic Geochemistry Conference, Canberra: 28-30 September. 10-11.

Teixidor, P., Grimalt, J.O., Pueyo, J.J., and Rodriguez-Valera, F. (1993) Isopranylglycerol diethers in non-alkaline evaporitic environments. *Geochimica et Cosmochimica Acta* 57, 4479-4489.

Thiel, V., Jenisch, A., Landmann, G., Reimer, A., and Michaelis, W. (1997) Unusual distributions of long-chain alkenones and tetrahymanol from the highly alkaline Lake Van, Turkey, *Geochimica et Cosmochimica Acta* 61, 2053-2064.

Tindall, B.J. (1992) The family *Halobacteriaceae*. In: *The Prokaryotes*. A handbook on the biology of bacteria, eds. A. Balows, H.G. Trüper, M. Dworkin, W. Harder and K.H. Schleifer, pp 768-808. Chap 34. Springer Verlag.

Tyson, R.V. (1995) *Sedimentary Organic Matter. Organic Facies and Palynofacies*, pp. 309-315. Chapman and Hall. New York.

van der Meer, M.T.J., Schouten, S., and Sinninghe Damsté, J.S. (1998) The effect of the reversed tricarboxylic acid cycle on the C-13 contents of bacterial lipids. *Organic Geochemistry* 28, 527-533.

Vinke, A., Schouten, S., Sephton, S., and Sinninghe Damsté, J.S. (1998) A newly discovered norisoprenoid, 2,6,15,19-tetramethylicosane, in Cretaceous black shales. *Geochimica et Cosmochimica Acta* 6, 965-970.

Volkman, J.K., Eglinton, G., Corner, E.D.S., and Forsberg, T.E.V. (1980) Long chain alkenes and alkenones in the marine coccolithophorid *Emiliania huxleyi*. *Phytochemistry* 19, 2619-2622.

Volkman, J.K., Barrett, S.M., Dunstan, G.A., and Jeffrey, S.W. (1992) Advances in the biomarkers chemistry of marine algae. Abstracts, Third Conference on Petroleum Geochemistry and Exploration in the Afro-Asian Region, Melbourne, February, 1992.

Wendt, I. (1968) Fractionation of carbon isotopes and its temperature dependence in the system CO_2 - gas - CO_2 in solution and bicarbonate - CO_2 solution. *Earth Planetary Science Letters* 4, 64-68.

Wakeham, S.G., Freeman, K.H., Pease, T.K., and Hayes, J.M. (1993) A photoautotrophic source for lycopane in marine water columns. *Geochimica et Cosmochimica Acta* 57, 159-166.

Whiticar, M.J., Faber, E., and Schoell, M. (1986) Biogenic methane formation in marine and freshwater environments: CO_2 reduction *vs.* acetate fermentation-Isotope evidence. *Geochimica et Cosmochimica Acta* 50, 693-709.

Zander, J.M., Caspi, E., Pandey, G.N., and Mitra, C.R. (1969) The presence of tetrahymanol in *Oleandra wallichii*. *Phytochemistry* 8, 2265-2267.

Index

Current Plant Science and Biotechnology in Agriculture

Current Plant Science and Biotechnology in Agriculture

Current Plant Science and Biotechnology in Agriculture

KLUWER ACADEMIC PUBLISHERS – DORDRECHT / BOSTON / LONDON